# Advances in Forestry Research

## Volume I

# Advances in Forestry Research Volume I

Edited by **Malcolm Fisher**

New York

Published by Callisto Reference,
106 Park Avenue, Suite 200,
New York, NY 10016, USA
www.callistoreference.com

**Advances in Forestry Research: Volume I**
Edited by Malcolm Fisher



International Standard Book Number: 978-1-63239-042-4 (Hardback)



Printed in the United States of America.

# Contents

# Preface

Forestry research is the application of scientific methods to investigate and understand the management and practice of forestry. It is used to effectively and sustainably use resources derived from forests. This topic studies major issues such as, products from forests, forest resource management, nutrient loss of soil, deforestation, environmental studies, forest and organisms' interaction, timber harvesting, economic importance of micro-organisms, ecological studies, biospheres of different regions, biofuel, etc. Forest research is an independent field of science, which discusses numerous concepts concerned with both human life and other organisms on planet earth such as trees, animals, micro-organisms, plants, aquatic creatures, and aquatic plants.

In this book, each chapter includes an extensive review of the literature as well as current theories and philosophies on forestry research. It covers an evaluation of various forests reserves and habitats across the globe. The book also analyses human impact on ecosystems. We have been fortunate to have an outstanding group of forest specialists from all over the world contributing to this publication. This book would be a significant source of information to a broad spectrum of readers interested in forestry, management and conservation.

At the end of the preface, I wish to thank the editorial team and all the contributors, for making this book a success. I also thank my family and friends for being a constant source of encouragement at every step of my research and for my future assignments.

**Editor**

# 1

# Clone-Specific Response in Leaf Nitrate Reductase Activity among Unrelated Hybrid Poplars in relation to Soil Nitrate Availability

**Julien Fortier,[1,2] Benoit Truax,[1] France Lambert,[1] Daniel Gagnon,[1,2,3] and Normand Chevrier[4]**

[1] *Fiducie de Recherche sur la Forêt des Cantons-de-l'Est, Eastern Townships Forest Research Trust, 1 Rue Principale, St-Benoît-du-Lac, QC, Canada J0B 2M0*
[2] *Centre d'Étude de la Forêt (CEF), Université du Québec à Montréal, C.P. 8888, Succursale Centre-Ville, Montréal, QC, Canada H3C 3P8*
[3] *Department of Biology, University of Regina, 3737 Wascana Parkway, Regina, SK, Canada S4S 0A2*
[4] *Département des Sciences Biologiques, Université du Québec à Montréal, C.P. 8888, Succursale Centre-Ville, Montréal, QC, Canada H3C 3P8*

Correspondence should be addressed to Julien Fortier, fortier.julien@courrier.uqam.ca

Academic Editor: Guy R. Larocque

In this field study, we used *in vivo* NRA activity in hybrid poplar leaves as an indicator of $NO_3^-$ assimilation for five unrelated hybrid poplar clones. We also examined if leaf NRA of these clones is influenced to the same extent by different levels of soil $NO_3^-$ availability in two riparian agroforestry systems located in pastures. Leaf NRA differences of more than one order of magnitude were observed between the clones, clearly showing their different abilities to reduce $NO_3^-$ in leaves. Clone DxN-3570, a *P. deltoides* x *P. nigra* hybrid (*Aigeiros* intrasectional hybrid), always had the highest leaf NRA during the field assays. This clone was also the only one to increase its leaf NRA with increasing $NO_3^-$ soil availability, which resulted in a significant Site x Clone interaction and a positive relationship between soil $NO_3^-$ concentration and NRA. All of the four other clones studied had one or both parental species from the *Tacamahaca* section. They had relatively low leaf NRA and they did not increase their leaf NRA when grown on the $NO_3^-$ rich site. These results provide evidence that $NO_3^-$ assimilation in leaves varies widely among hybrid poplars of different parentages, suggesting potential preferences for N forms.

## 1. Introduction

Poplars (*Populus* spp.) are commonly planted for production and restoration purposes in different plantation systems, environments (agricultural land, abandoned farmland, clearcut forest, contaminated sites, riparian buffers, intercropping systems, etc.), and climates [1–7]. A large number of poplar hybrids exist throughout the world and they exhibit wide variations in functional traits (i.e., morphological, physiological, and phenological characteristics) [8, 9].

Because they can be clonally propagated with ease, poplars from the *Tacamahaca* and the *Aigeiros* sections and their hybrids form the basis of most poplar breeding programs worldwide [10]. Although they are all pioneer species that mainly occur in coastal, riparian, alluvial, and bottomland environments, poplars from these two sections differ widely in their natural distributions, as they are adapted to different soil and climatic conditions [11]. Poplars from the *Tacamahaca* section are mostly associated with riparian or wetland habitats [11]. They are widely distributed in the northern latitudes, with some species growing to the latitudinal or altitudinal limits of trees [11]. Poplars from the *Aigeiros* section are better adapted to riparian soils in the bottomlands of temperate and arid regions [11, 12].

Poplars from the *Tacamahaca* and the *Aigeiros* sections also show marked differences in their N-form preferences

[13, 14]. This is because organic matter mineralization processes are strongly influenced by habitat characteristics such as climate (temperature and precipitation), soil pH, and water table depth [15–17]. For example, balsam poplar (*Populus balsamifera*), growing as far north as 69° latitude in Alaska [11], is able to assimilate directly both $NH_4^+$ and amino acids, with no apparent preference [18], and its physiological capacity to assimilate $NH_4^+$ is greater than for $NO_3^-$ [13]. Fertilisation trials also show that *P. balsamifera* and poplar hybrids with one parental species related to the *Tacamahaca* section, tended to grow best with $NH_4^+$ fertilisation over $NO_3^-$ fertilisation [19]. Conversely, a high $NO_3^-:NH_4^+$ ratio in the soil solution strongly stimulates root development in Eastern cottonwood (*P. deltoides*) [20].

Nitrogen form preferences have been shown to be closely linked to the ability of tree species to reduce $NO_3^-$ from the soil solution [21–23]. Because nitrate reductase (NR) is a substrate-induced enzyme [24], *in vivo* nitrate reductase activity (NRA) in the leaves and in roots has proven to be a useful indicator of $NO_3^-$ assimilation in a wide range of species, management practices, and environmental conditions [23, 25–31].

In poplars, Dykstra [32] detected important differences in leaf NRA between hybrid poplar clones *P. tristis* x *balsamifera* and *P. deltoides* x *nigra*. In free-growing populations, similar observations were reported by Al Gharbi andHipkin [25] when *P. alba*, *P. deltoides* x *nigra*, and *P. tremula* were compared. More recently, Black et al. [33] reported that NRA in *P. tremula* x *alba* was at least 10-fold greater in leaves than in stems or in roots at all nitrate availabilities. These authors concluded that most nitrate assimilation occurs in poplar leaves, a finding that was later corroborated by Rosenstiel et al. [34] in Eastern cottonwood (*P. deltoides*). Balsam poplar also showed slightly higher NRA in leaves than in roots when substrate was not limiting [29].

Relationships between soil N availability and NRA in poplars have also been documented. Min et al. [21] showed that NRA in roots and in leaves of trembling aspen (*P. tremuloides*) is rapidly induced following $NO_3^-$ exposure, with leaves having the greatest activity. Higher NRA was also observed in the N fertilisation treatment for *P. deltoides* x *P. nigra* hybrid [35]. Nitrate reductase activity in leaves and roots was also greater when *P. tremula* x *alba* was grown in the highest external $NO_3^-$ concentration [33].

In this field study, we use *in vivo* NRA activity in hybrid poplar leaves as an indicator of $NO_3^-$ assimilation [29] in five unrelated hybrid poplar clones: (1) *Populus trichocarpa* x *P. deltoides* (TxD-3230), (2) *P. deltoides* x *P. nigra* (DxN-3570), (3) *P.* x *canadensis* x *P. maximowiczii* (DNxM-915508), (4) *P. nigra* x *P. maximowiczii* (NxM-3729), and (5) *P. maximowiczii* x *P. balsamifera* (MxB-915311). We hypothesize that the intrasectional hybrid from the *Aigeiros* section (DxN hybrid) should exhibit a higher NRA than the intersectional hybrids (TxD, NxM, and DNxM hybrids) and the intrasectional hybrid from the *Tacamahaca* section (MxB hybrid). This hypothesis should be supported because parental species of the DxN hybrid are adapted to temperate floodplain environments, where $NO_3^-$ is the dominant N-form [36], while parental species of the MxB hybrid are adapted to riparian habitats and wetland ecotones of colder climates (Alaska, northern Canada, Siberia) [11], where $NH_4^+$ is the dominant N-form in soils [37], although hydrologically driven $NO_3^-$ input occurs periodically [29]. In this study, we also examine if leaf NRA of these unrelated hybrid poplar clones is influenced to the same extent by different levels of soil $NO_3^-$ availability within two riparian agroforestry systems located in pastures.

## 2. Materials and Methods

*2.1. Study Sites.* In May 2003, two multiclonal hybrid poplar riparian buffers were planted along headwater streams in the Eastern Townships region of southern Quebec, Canada. The buffer had cumulated five years of growth in 2007, the year of the study. The two riparian buffer study sites (Bromptonville: 45°29 N; 71°59 W; Magog: 45°14 N; 72°07 W) are located in pastures of the regional landscape unit of Sherbrooke [38]. This landscape unit is characterised by gentle slopes, a continental subhumid moderate climate, a growing season of 180–190 days, and a precipitation regime of 1000–1100 mm/year [39]. Cattle densities at the two pasture sites are 0.6 cow $ha^{-1}$ at Bromptonville and 0.2 cow $ha^{-1}$ at Magog. The Bromptonville site is fertilized each year with cow manure, while the Magog site receives no fertilisation. The soil of the Bromptonville site developed on glacial outwash, deposited over lacustrine clay [40]. It is well-drained and named "Sheldon sandy loam" [40]. The soil of the Magog site developed on glacial till. It is imperfectly drained and named "Magog stony loam" [40]. Much higher aboveground biomass (including leaves) have also been measured at the fertile site of Bromptonville, one year following this study (6th growing season) [1]. Site and soil characteristics are presented in Table 1. Additional site and soil characteristics are available in other related studies [1, 7, 41].

Five unrelated hybrid poplar clones were used in this study: (1) *Populus trichocarpa* x *P. deltoides* (TxD-3230), (2) *P. deltoides* x *P. nigra* (DxN-3570), (3) *P.* x *canadensis* x *P. maximowiczii* (DNxM-915508), (4) *P. nigra* x *P. maximowiczii* (NxM-3729), and (5) *P. maximowiczii* x *P. balsamifera* (MxB-915311) (Table 2). The five poplar clones were chosen because they have different growth patterns, physiological characteristics, and because they had been selected for superior disease resistance/tolerance and growth characteristics in trials in southern Quebec [42].

*2.2. Experimental Design.* A randomized block design was used at each of the two sites, with 4 blocks (replicates) and 5 hybrid poplar clones (treatments) for a total of 40 experimental plots. Each block contains 5 experimental plots (one clone per plot). Plots are 4.5 m wide and 9 m long (40.5 $m^2$). Each plot contains 9 trees from a single clone (3 rows, 3 trees/row). Each tree is spaced 3 m on the row and the rows are 1.5 m apart. Tree rows were planted parallel to stream bank. A total of 180 hybrid poplars where planted at each site (36 trees of each clone) for a total of 360 hybrid poplars, two blocks on each side of the stream. This design

TABLE 1: Site and soil characteristics at the two riparian sites: elevation, mean aboveground poplar dry biomass (including leaves), soil drainage class, pH, and nutrient supply rate ($\mu g\ 10\ cm^{-2}\ 15\ d^{-1}$). The $NO_3^-:NH_4^+$ nutrient supply rate molar ratio is also indicated. For nutrient supply rate, PRS-Probes were buried during a 15 day period in late August 2007.

| Sites | Elev. (m) | Poplar biomass ($kg\ tree^{-1}$)[1] | Drainage[2] | pH[1] | $NO_3^-$ | $NH_4^+$ | $NO_3^-:NH_4^+$ | P | Ca | K | Mg | Mn |
|---|---|---|---|---|---|---|---|---|---|---|---|---|
| Bromptonville | 140 | 58.9 | Good | 6.36 | 69.2 | 11.4 | 1.8 | 8.37 | 1291 | 506 | 254 | 2.8 |
| Magog | 208 | 12.2 | Imperfect | 5.81 | 24.7 | 16.7 | 0.4 | 4.24 | 710 | 102 | 492 | 17.8 |
| SE | — | — | — | 0.03 | 9.7 | 1.0 | — | 1.16 | 57 | 19 | 23 | — |
| $P <$ | — | — | — | 0.001 | 0.01 | 0.01 | — | 0.05 | 0.001 | 0.001 | 0.001 | NS |

[1] Poplar aboveground biomass per tree data and pH data were taken from Fortier et al. [1].
[2] Soil drainage classes were obtained from Cann and Lajoie [40].

allowed us to test 5 poplar clones in two different riparian environments simultaneously, a common procedure in crop cultivar trials [43].

*2.3. Soil Nutrient Availability.* Nutrient availability in the hybrid poplar buffers was determined using Plant Root Simulator (PRS-Probes) technology from Western Ag Innovations Inc., Saskatoon, Canada. The PRS-probes consist of an ion exchange membrane encapsulated in a thin plastic probe, which is inserted into the ground with little disturbance of soil structure. Nutrient availability predicted with this method is generally significantly correlated with conventional soil extraction methods over a wide range of soil types [44].

Three pairs of probes (an anion and a cation probe in each pair) were buried along the middle row (of 3 rows parallel to stream bank) of poplars in each experimental plot (40 plots). At each site, burial length was 15 days, starting in mid-August 2007. After removal, probes were washed in the field with deionised water and returned to Western Ag Labs for analysis ($NO_3^-$, $NH_4^+$, P, K, Ca, Mg, and Mn). Composites were made by combining the three pairs of probes in each experimental plot. Probe supply rates are reported as $\mu$mol of nutrient $10\ cm^{-2}$ 15 $days^{-1}$ and are presented in Table 1.

*2.4. Nitrogen Mineralization Rate.* A sequential coring technique was used to measure nitrification and ammonification rates at the two riparian sites [45]. In each experimental plot, two pairs of hard PVC tubes (20 cm in length and 5.5 cm in diameter) were inserted 15 cm vertically into the soil. The first pair of tubes and their soil content were immediately removed and placed in a portable ice box and transported to the lab for extraction with 2 M KCl [46] within 24 h. The content of each pair of tubes was mixed thoroughly and duplicate extractions were made for each composite. During the following days, the concentration of $NO_3^-$ and $NH_4^+$ were determined using a Tecator FIAstar continuous flow analyzer.

The second pair of tubes remained in the soil for a 24 day period (from 17 July to 9 August 2007). They were capped with tape to prevent N loss from leaching by rain. A small hole was pierced laterally on each tube (1 cm from the top) to allow aeration. After 24 days, tubes were removed from the soil and the same procedure (as for initial tubes) was used in order to determine concentrations of N-forms. Nitrification was calculated as the $NO_3^-$ concentration of the soil at the end of the incubation period minus the $NO_3^-$ concentration at the beginning. Ammonification was calculated the same way.

*2.5. Nitrate Reductase Activity Assay*

*2.5.1. Enzymatic Kinetics.* In order to verify that the substrate ($KNO_3$) concentration that would be used in further NRA experiments was not limiting, we evaluated the effect of different substrate concentrations (0, 1, 5, 10, 15, 20, 40, 100, 150, and 200 mM) on NRA of two representative hybrid poplar clones (DxN-3570 and MxB-915311).

Measurement of the *in vivo* NRA in poplar leaves was done according to the method developed by Jaworski [47] and optimized for broadleaf tree species [30]. In early summer (June 29, 2007), leaves from the upper shoots were taken at noon at the Bromptonville site. For each of the two clones, a composite sample was made by combining two leaves from two different trees. The plastochron index was used to select fully expanded leaves at the same developmental stage [48]. Leaf plastochron index 7 (LPI 7) was selected for all samples. Samples were put in a plastic bag and immediately placed in a portable icebox (4°C). The sampling operations were always done within 1 h and plant material was brought back to the lab within 30 min after sampling.

At the lab, the leaves from each sample were cut into small pieces (2 × 2 mm) and 0.1 g of fresh tissue, in duplicate for each composite sample, was placed in a test tube containing 5 mL of incubating solution (pH 7.5), containing 100 mM phosphate buffer, 1.5% 1-propanol, and the different $KNO_3$ concentrations. Tissues samples and solutions were vortexed for 2 min to enhance infiltration of the assay medium. Each test tube was sealed and incubated in the dark for 1 h at 30°C. A blank was done for each composite sample. The enzymatic reaction was stopped by immersing the tubes in boiling water over five minutes. The colorimetric determination of the reaction was achieved by mixing 1 mL of incubation solution with 1 mL of 0.02% NED and 1 mL of sulphanilamide. After 30 min, the samples were centrifuged at 2000 ×g for 5 min and the supernatant was read in a spectrophotometer at 540 nm. Nitrate reductase activity is expressed as the amount

Table 2: Name, parentage, section, and origin of the five hybrid poplar clones.

| Clone number | Scientific name (common name) | Parentage | Section | Origin |
|---|---|---|---|---|
| 3230 | *P.* x *generosa* A. Henry (Boelare) | TxD | *Tacamahaca* x *Aigeiros* | Belgium |
| 3570 | *P.* x *canadensis* Moench | DxN | *Aigeiros* x *Aigeiros* | Belgium |
| 3729 | *P. nigra* x *P. maximowiczii* (NM6) | NxM | *Aigeiros* x *Tacamahaca* | Germany |
| 915311 | *P. maximowiczii* x *P. balsamifera* | MxB | *Tacamahaca* x *Tacamahaca* | Québec |
| 915508 | *P.* x *canadensis* x *P. maximowiczii* | DNxM | (*Aigeiros* x *Aigeiros*) x *Tacamahaca* | Québec |

of $NO_2^-$ measured in the test tube after the 1 h incubation period, calculated for 1 g of dry leaf tissue.

*2.5.2. Effect of Leaf Age on Hybrid Poplar NRA.* The effect of leaf age on NRA was assessed for the five hybrid poplar clones. The same procedure as the one described above was employed, except that leaf samples from LPI 3 to LPI 9 were collected from the five clones at the Bromptonville site (July 2-3, 2007). Based on the enzymatic kinetics assay, we used a substrate ($KNO_3$) concentration of 40 mM.

*2.5.3. Determination of Hybrid Poplar NRA at the Two Riparian Buffer Sites.* In mid July (July 10 and 11, 2007) and late August (August 28 and 29, 2007), NRA assays were performed with plant material from the two riparian buffer sites (Bromptonville and Magog). In each experimental plot ($n$ of plots = 40), a composite sample was made by combining two leaves from two different trees of a single clone (one clone per plot). Leaf samples were collected in the same manner as in the previous assays and the same procedure was used for NRA determination. Based on the observations concerning the effect of leaf age on NRA, plant material corresponding to LPI 6 was used for all clones in this assay.

*2.6. Statistical Analysis.* ANOVA tables were constructed in accordance with Peterson [49], and degrees of freedom, sum of squares, mean squares, and $F$ values were computed. When a factor was declared statistically significant (Sites, Clones and Sites x Clones interaction), the standard error of the mean (SE) was used to evaluate differences between means for four levels of significance ($P < 0.1$, $P < 0.05$, $P < 0.01$, and $P < 0.001$). All of the ANOVAs were run with the complete set of data (2 sites, 5 clones, and 4 blocks = 40 experimental plots). Results of enzymatic kinetics and effect of leaf age on NRA are presented as means with standard deviations (SD).

## 3. Results

*3.1. Nitrogen Supply Rate, Soil N Concentration, and N Mineralization.* Nitrate supply rate was approximately three times higher at the Bromptonville site compared to the Magog site, while $NH_4^+$ supply rate was approximately 50% higher at the Magog site (Table 1). This resulted in very contrasting molar ratios of $NO_3^-:NH_4^+$ availability (1.8 at Bromptonville versus 0.4 at Magog) (Table 1).

Initial $NO_3^-$ concentration measured at the beginning of the mineralization study was approximately 60% higher at the Bromptonville site, while initial $NH_4^+$ concentration was more than twice higher at the Magog site (Table 3). After the incubation period (24 days), a three-fold increase in $NO_3^-$ concentration occurred at Bromptonville, while only a two-fold increase was observed at Magog. This resulted in a higher nitrification rate at Bromptonville. At both sites, $NH_4^+$ concentration in soil showed a decrease after the incubation period because of high nitrification rates. Because nitrification rate was higher at Bromtptonville, the decrease in $NH_4^+$ concentration was also higher, although ammonification rates were not statistically different between the two sites.

Strong positive linear relationships were obtained between $NO_3^-$ supply rate measured during a 15 day period with ion exchange membrane (PRS-probes) and initial $NO_3^-$ concentration in soil tube soils prior to incubation (Figure 1(a)), final $NO_3^-$ concentration in soil tube soils following a 24 day incubation period (Figure 1(b)) and nitrification rate in soil tube soils during a 24 day incubation period (Figure 1(c)).

*3.2. Nitrate Reductase Activity*. The enzymatic kinetics assay revealed that leaf NRA was much higher for clone DxN-3570 than for clone MxB-915311 at any substrate concentration (Figure 2). Results also show that both of these clones had their highest leaf NRA at a substrate concentration of 20 mM $NO_3^-$. In fact, for both of these clones a similar pattern of NR induction was observed. For substrate concentrations lower than 20 mM $NO_3^-$, leaf NRA increased very rapidly in relation to substrate availability and afterwards decreased slightly at higher $NO_3^-$ concentrations (Figure 2).

The effect of leaf age (developmental stage) on NRA was different from one clone to another (Figure 3). At all developmental stages, leaf NRA of clone DxN-3570 was higher than any other clones and showed a rapid decrease from LPI3 to LPI9. In younger leaves, clone NxM-3729 had a higher leaf NRA value than clones TxD-3230, MxB-915311, and DNxM-915508, at most developmental stages. For the three other clones (TxD-3230, MxB-915311, and DNxM-915508), leaf NRA was equal or below 1 $\mu$mol $NO_2^- g_{dw}^{-1} h^{-1}$, with a maximum value observed at LPI 3 (Figure 3). At almost all leaf developmental stages, clone DNxM-915508 had the lowest NRA.

For the two NRA assays at the two riparian sites (July 10-11 and August 28-29 2009), a significant Site x Clone interaction was detected by the ANOVA (Figure 4). During the July assay, NRA of clone DxN-3570 was statistically higher than any other clones at both sites. However, during

TABLE 3: Nitrification and ammonification rates measured during a 24 days period at the two riparian sites (July 17–August 9, 2007). Nitrate and ammonium concentrations at the beginning (initial) and at the end (final) of the *in situ* incubation period are also presented.

| Sites | $NO_3^-$ initial ($mg\,kg^{-1}$) | $NO_3^-$ final ($mg\,kg^{-1}$) | Nitrification rate ($mg\,kg^{-1}\,24\,d^{-1}$) | $NH_4^+$ initial ($mg\,kg^{-1}$) | $NH_4^+$ final ($mg\,kg^{-1}$) | Ammonification rate ($mg\,kg^{-1}\,24\,d^{-1}$) |
|---|---|---|---|---|---|---|
| Bromptonville | 5.07 | 16.95 | 11.87 | 1.23 | 0.29 | −0.93 |
| Magog | 3.13 | 6.29 | 3.15 | 2.92 | 2.71 | −0.21 |
| SE | 0.38 | 1.76 | 1.55 | 0.26 | 0.30 | — |
| $P <$ | 0.01 | 0.001 | 0.001 | 0.001 | 0.001 | NS |

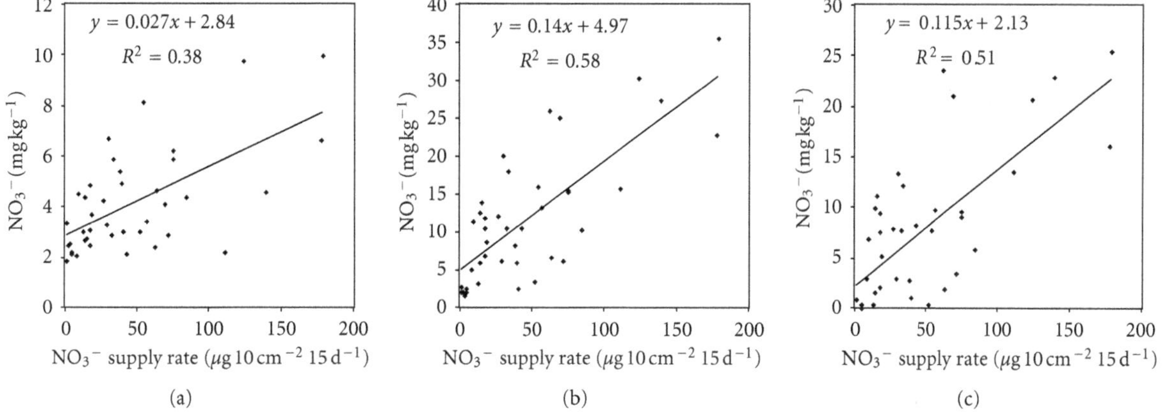


FIGURE 1: Linear relationships between soil $NO_3^-$ supply rate and soil tube *in situ* incubations in riparian soils for 3 variables: (a) initial $NO_3^-$ concentration in soil tube soils prior to incubation, (b) final $NO_3^-$ concentration in soil tube soils following a 24 day incubation period, and (c) nitrification rate in soil tube soils during a 24 day incubation period. All relationships are significant at $P < 0.001$.

the August assay, only NRA of clone DxN-3570 at the Bromptonville site was statistically higher than any other clones at both sites. During both assays, clone DxN-3570 had a significantly higher NRA at the Bromptonville site than at the Magog site, while NRA of the four other clones was generally higher at the Magog site or equivalent to what was observed at the Bromptonville site. The NRA of clone DxN-3570 at Bromptonville was almost twice as high in the August assay as in the July assay, while little difference in NRA was observed at the Magog site between the two assay dates. In fact, at the Bromptonville site, NRA of all clones was generally higher in the August assay compared to the July assay. This trend was not observed at the Magog site.

*3.3. Nitrate Reductase Activity in Relation to Soil Nitrate Concentration.* Significant relationships between $NO_3^-$ concentration observed following a 24 day incubation period in riparian soils and nitrate reductase activity (NRA) were only observed for two of the studied clones: DxN-3570 ($R^2 = 0.39$, $P < 0.1$) and MxB-915311 ($R^2 = 0.63$, $P < 0.05$) (Figure 5). A positive relationship (power function) between $NO_3^-$ and NRA was observed for clone DxN-3570, while a negative relationship (power function) between those two variables was observed for clone MxB-915311.

## 4. Discussion

In this study, leaf NRA differences of more than one order of magnitude were observed between the hybrid poplar clones, a clear indication of their different abilities to reduce $NO_3^-$ in leaves. Clone DxN-3570 had the greatest leaf NRA (1) during the enzyme kinetic assays at different substrate concentrations (Figure 2), (2) in leaves of different developmental stages (Figure 3), and (3) during both assays under field conditions in two contrasted sites in terms of soil $NO_3$ availability (Figure 3). Large differences in leaf NRA have also been reported in other species of *Populus* [25, 32].

In addition, large differences in nitrification rate and soil $NO_3^-$ availability across the two riparian sites (Tables 1 and 3) induced a clone-specific response in leaf NRA. This resulted in a significant Site x Clone interaction during the mid July and late August assays, with clone DxN-3570 having a significantly higher leaf NRA at the Bromptonville site, where $NO_3^-$ availability is high, compared to the Magog site (Figure 4). Conversely, leaf NRA of the four other clones (TxD-3230, MxN-3729, MxB-915311, and DNxM-915508) was generally much lower and similar between clones and between sites (Figure 2). This suggests that clone DxN-3570 has a particular ability to activate the $NO_3^-$ reduction process in its leaves in response to a higher $NO_3^-$ availability in the soil.

Based on previous findings, higher leaf NRA from all clones could be expected at Bromptonville given the higher soil $NO_3^-$ availability at this site. As reported by Black et al. [33] and Rosenstiel et al. [34], higher leaf NRA values were found in poplars grown on soils with higher external $NO_3^-$ concentrations. Our field-based study showed that this relationship was only true for clone DxN-3570. This resulted in a positive relationship between soil $NO_3^-$ availability

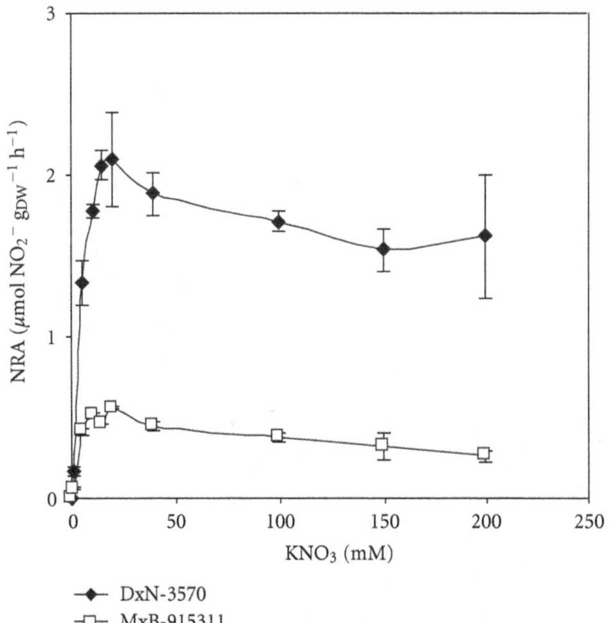


FIGURE 2: Effect of substrate concentration on nitrate reductase activity (NRA) of clone DxN-3570 and clone MxB-915311. All data points are means of duplicate samples and vertical bars represent SD of the means.

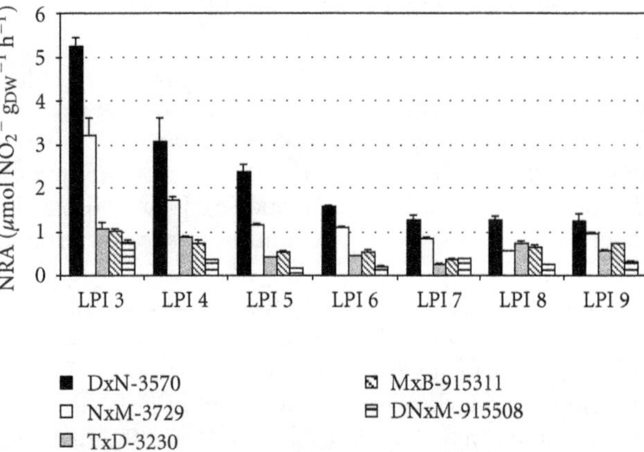


FIGURE 3: Effect of leaf age (developmental stage) on nitrate reductase activity (NRA) of five hybrid poplar clones. Leaf plastochron index (LPI) is used as an indicator of leaf age. All data points are means of duplicate sample analyses and vertical bars represent SD of the means.

measured across the two sites and leaf NRA for this clone ($R^2 = 0.39$, $P < 0.1$) (Figure 5), a positive relationship that was not observed for the other clones.

It could also be argued that the $NO_3^-$ reduction process in hybrid poplars grown at the Bromptonville site was diluted in a much larger biomass compared to the Magog site (Table 1). An inverse relationship between dry biomass and NRA was also reported in *Robinia pseudoacacia* [50]. However, this dilution effect was not observed for clone DxN-3570 in this study.

As proposed earlier by Dykstra [32], we suggest that important leaf NRA differences among unrelated hybrid poplars reflect the genetic assemblage, or parentage, of the five clones studied. The parental species of clone DxN-3570 (*P. deltoides* and *P. nigra*) are typical colonisers of floodplain habitats of temperate and arid climates, where best growth occurs on sandy loam soils [11]. In those habitats, the predominant N-form in soils is generally $NO_3^-$ because of the warm climate and good soil drainage (water table below 30 cm of soil surface) [17, 36].

Studies on Eastern cottonwood (*P. deltoides*) also suggests that a 60–80% $NO_3^-$ (balanced with $NH_4^+$) solution optimizes whole-plant growth [14]. Moreover, Woolfolk and Friend [20] found that greatest total root length, specific root length, and N concentration of roots in enriched patches occurred at the 80 : 20 $NO_3^- : NH_4^+$ ratio. These observations are consistent with the particular ability of the clone DxN-3570 to increase its NRA in riparian environments with high $NO_3^-$ availability, as observed at Bromptonville (Tables 1 and 3, Figure 4). High leaf NRA in response to a high $NO_3^- : NH_4^+$ ratio in soils has also been reported in red ash (*Fraxinus pennsylvanica*), a common early succession species of rich alluvial bottomlands [23].

The relatively low leaf NRA in hybrid poplar clones that have a *Tacamahaca* section genetic contribution may be related to the soil N-form preferences of these clones. It is well known that boreal species, such as balsam poplar, show a marked preference for the uptake of $NH_4^+$ over $NO_3^-$, as observed for other species growing in cold climates [13]. In addition, south of the boreal forest, balsam poplar typically colonises wet soils found at the edges of streams and lakes, in swamps and depressions [11]. In those habitats, where the water table is near the soil surface, ammonification is generally the dominant N-mineralization process and $NH_4^+$ generally accumulates in the top soil [17]. During fertilisation trials, DesRochers et al. [19] also pointed out that *P. balsamifera* and two other clones related to the *Tacamahaca* section were better adapted for $NH_4^+$ uptake rather than that of $NO_3^-$.

Furthermore, balsam poplar and other poplars related to the *Tacamahaca* section are known to have high concentrations of tannins and other phenolic compounds in their leaves and fine roots [51, 52]. Once entered into the soil system through litter fall and fine root decomposition, these secondary compounds have a major effect on the N-cycle, including a decrease in N-fixation from competitive species such as alder (*Alnus*), an increase in N immobilisation and a decrease in nitrification [51, 52]. Consequently, the physiological traits of low leaf NRA in clones related to the *Tacamahaca* section (Figures 2, 3 and 4) is consistent with the ability of poplars from this section to modify the environment, in order to reduce $NO_3$ availability, at their own advantage.

Our observations also suggest a negative relationship between leaf NRA and soil $NO_3^-$ concentration ($R^2 = 0.63$, $P < 0.05$) for the *Tacamahaca* intrasectional hybrid (MxB) (Figure 5). This is potentially the sign of a negative feedback of prolonged exposure to high soil $NO_3^-$ concentrations on leaf NRA.

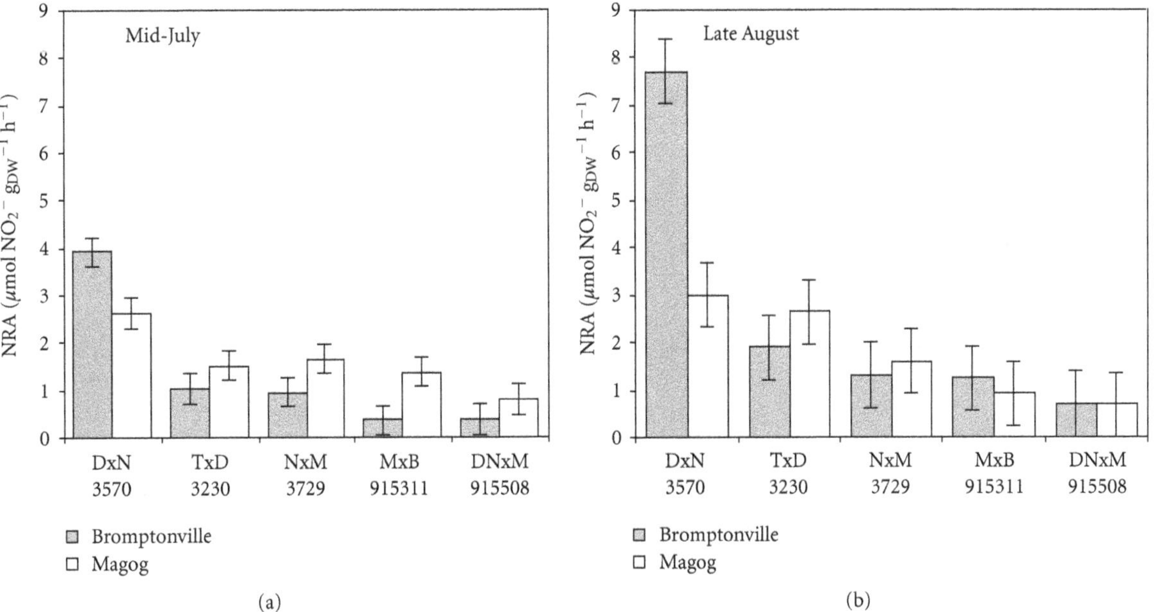


FIGURE 4: Nitrate reductase activity (NRA) of the five hybrid poplar clones at the two riparian buffer sites. On the left (a), the NRA assay was done in mid-July (July 10 and 11, 2007) and Site x Clone interaction is significant at $P < 0.05$. On the right (b), NRA assay was done in late August (August 28 and 29, 2007) and Site x Clone interaction is significant at $P < 0.01$. Vertical bars represent SE of the means.

Nitrate reductase activity observed in this study was neither a good indicator of mean aboveground biomass accumulation nor of N accumulation because the clone with the highest NRA (DxN-3570) (Figures 2, 3, and 4) had the lowest biomass growth and N accumulation among the five clones studied [1, 41]. In a greenhouse study with 12 week-old trees originating from cuttings, Dykstra [32] also reported that leaf NRA was not an index of N assimilation in hybrid poplars. But, this author observed that the poplar clone with the highest NRA had a significantly higher dry stem weight, dry leaf weight, and height growth. This was not the case in our field study during the fifth growing season. However, further research is needed to determine if NRA might be a good indicator of growth among hybrid poplar clones of the same parentage.

Other components of productivity that have important genetic variation among poplar species may have affected clone growth, whether they are structural or physiological [8]. This includes leaf morphology and leaf growth [53], leaf photosynthetic capacity [54], nutrient requirements [55], nutrient-use efficiency [56], water-use efficiency [57], light-use efficiency [58], phenotypic plasticity in response to differential N availability [59], early-rooting ability and rooting patterns [60–62], size, distribution and orientation of leaves and branches [63–65], wood density [66], and so forth. A favourable combination of many of these morphological, physiological and phenological characteristics explains the superior growth of selected hybrid poplar clones [8].

Finally, a clearer portrait of NRA in the set of clones studied here may have been obtained if NRA in roots had been tested in parallel to leaf NRA. Given that very low NRA was found in the leaves of some clones (DNxM-915508 and MxB-915311), it could also be suggested that $NO_3^-$ assimilation is relatively important in the roots of some clones and hybrid types, as shown in trembling aspen and balsam poplar [21, 29]. Given the high genetic variability in physiological traits of poplar species, generalities such as that nitrate assimilation is almost entirely restricted to leaves in poplars [33], should be made with caution.

Although different poplar species and hybrids may have soil N-form preferences, trees from the genus *Populus* are considered generalist pioneer species, capable of thriving on low and high $NO_3^-$ or $NH_4^+$ sites [67]. Still, future research is needed to clearly understand soil N-form preferences among unrelated hybrid poplar clones and its potential relationship with NRA. If this relationship is proven, NRA could provide interesting information on the suitability and adaptability of different hybrid poplar clones to different fertilisation treatments [36] or to various plantation environments (clearcut forests, cultivated fields, abandoned farmland, riparian buffers, contaminated sites, etc.). In that context, it may be important to understand which N-form is preferred by different parental species from both *Tacamahaca* and *Aigeiros* sections, and how hybridization between and within those two sections influences N-form preferences and mechanisms of N assimilation.

## 5. Conclusion

This study has shown large variation in leaf NRA among hybrid poplars, with NRA differences of more than one order of magnitude for the hybrids studied (DxN, TxD, NxM, DNxM, and MxB). Clone DxN-3570, an *Aigeiros*

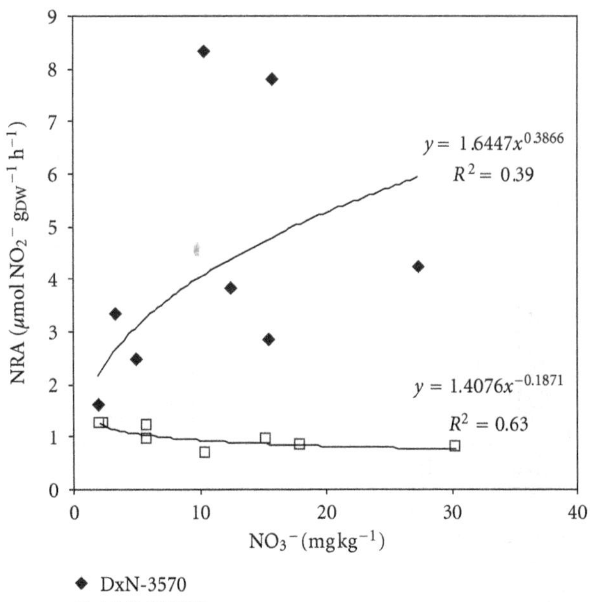


FIGURE 5: Relationship between $NO_3^-$ concentration observed following a 24 day incubation period in riparian soils and nitrate reductase activity (NRA) for clone DxN-3570 ($P < 0.1$) and MxB-915311 ($P < 0.05$). NRA values in this figure were obtained by averaging NRA values of the mid-July and late August assays.

intrasectional hybrid, always had the highest leaf NRA during the assays under field conditions. Across the two riparian sites, this clone was also the only one to increase its leaf NRA with increasing $NO_3^-$ soil availability, which resulted in a significant Site x Clone interaction, but also a positive relationship between soil $NO_3^-$ concentration and NRA. All four other clones studied had one or both parental species from the *Tacamahaca* section. They also had relatively low leaf NRA and they did not increase their leaf NRA when grown on the $NO_3^-$ rich site, which sharply contrasts with the physiological response observed for the DxN hybrid in terms of leaf NRA. These results suggest that $NO_3^-$ assimilation in leaves varies widely among hybrid poplars of different parentages suggesting potential soil N-forms preferences under genetic control.

## Acknowledgments

The authors gratefully acknowledge funding received from the Ministère des Ressources naturelles et de la Faune du Québec (MRNF), the Ministère de l'Agriculture, des Pêcheries et de l'Alimentation du Québec (MAPAQ), Agriculture et Agroalimentaire Canada (AAC), and the Conférence régionale des élus de l'Estrie. The authors are very grateful to the landowners, M. Beauregard and J. Lamontagne, who allowed the planting of the buffers on their farms. The authors would also like to thank Nathalie Boulanger, Pierre-Olivier Émond, Guillaume Fleury, and Marie-Claude Giroux who assisted with field work. Thanks are also due to Claire Vasseur of the Biodôme de Montréal for facilitating soil analyses. Finally, a scholarship from the Fiducie de recherche sur la forêt des Cantons-de-l'Est to J. Fortier is gratefully acknowledged.

## References


[1] J. Fortier, D. Gagnon, B. Truax, and F. Lambert, "Biomass and volume yield after 6 years in multiclonal hybrid poplar riparian buffer strips," *Biomass and Bioenergy*, vol. 34, no. 7, pp. 1028–1040, 2010.

[2] I. Laureysens, J. Bogaert, R. Blust, and R. Ceulemans, "Biomass production of 17 poplar clones in a short-rotation coppice culture on a waste disposal site and its relation to soil characteristics," *Forest Ecology and Management*, vol. 187, no. 2-3, pp. 295–309, 2004.

[3] L. A. Licht and J. G. Isebrands, "Linking phytoremediated pollutant removal to biomass economic opportunities," *Biomass and Bioenergy*, vol. 28, no. 2, pp. 203–218, 2005.

[4] J. A. Stanturf, C. van Oosten, M. D. Coleman, and C. J. Portwood, "Ecology and silviculture of poplar plantations," in *Poplar Culture in North America*, D. I. Dickmann, J. G. Isebrands, J. E. Eckenwalder, and J. Richardson, Eds., pp. 153–206, NRC Research Press, National Research Council of Canada, Ottawa, ON, Canada, 2001.

[5] M. Oelbermann, R. Paul Voroney, and A. M. Gordon, "Carbon sequestration in tropical and temperate agroforestry systems: a review with examples from Costa Rica and southern Canada," *Agriculture, Ecosystems and Environment*, vol. 104, no. 3, pp. 359–377, 2004.

[6] B. Truax, D. Gagnon, J. Fortier, and F. Lambert, "Yield in 8 year-old hybrid poplar plantations on abandoned farmland along climatic and soil fertility gradients," *Forest Ecology and Management*, vol. 267, pp. 228–239, 2012.

[7] J. Fortier, D. Gagnon, B. Truax, and F. Lambert, "Understory plant diversity and biomass in hybrid poplar riparian buffer strips in pastures," *New Forests*, vol. 42, no. 2, pp. 241–265, 2011.

[8] H. D. Bradshaw, R. Ceulemans, J. Davis, and R. Stettler, "Emerging model systems in plant biology: poplar (*Populus*) as a model forest tree," *Journal of Plant Growth Regulation*, vol. 19, no. 3, pp. 306–313, 2000.

[9] J. E. Eckenwalder, "Descriptions of clonal characteristics," in *Poplar Culture in North America. Part B*, D. I. Dickmann, J. G. Isenbrands, J. E. Eckenwalder, and J. Richardson, Eds., chapter 13, pp. 331–382, NRC Research Press, National Research Council of Canada, Ottawa, ON, Canada, 2001.

[10] D. E. Riemenschneider, B. J. Stanton, G. Vallée, and P. P. Périnet, "Poplar breeding strategies. Part A," in *Poplar Culture in North America*, D. I. Dickmann, J. G. Isebrands, J. E. Eckenwalder, and J. Richardson, Eds., chapter 2, pp. 43–76, NRC Research Press, National Research Council of Canada, Ottawa, ON, Canada, 2001.

[11] D. I. Dickmann and Y. A. Kuzovkina, "Poplars and willows of the world, with emphasis on silviculturally important species," Working Paper IPC/9-2, FAO Forest Management Division, Rome, Italy, 2008.

[12] J. E. K. Cooke and S. B. Rood, "Trees of the people: the growing science of poplars in Canada and worldwide," *Canadian Journal of Botany*, vol. 85, no. 12, pp. 1103–1110, 2007.

[13] F. S. Chapin, K. Van Cleve, and P. R. Tryon, "Relationship of ion absorption to growth rate in taiga trees," *Oecologia*, vol. 69, no. 2, pp. 238–242, 1986.

[14] W. T. M. Woolfolk, *Influence of ammonium:nitrate ratio on growth and N accumulation of Populus deltoides [M.S. thesis]*, Mississippi State University, Starkville, Miss, USA, 2000.

[15] M. C. Leirós, C. Trasar-Cepeda, S. Seoane, and F. Gil-Sotres, "Dependence of mineralization of soil organic matter on temperature and moisture," *Soil Biology and Biochemistry*, vol. 31, no. 3, pp. 327–335, 1999.

[16] K. G. Cassman and D. N. Munns, "Nitrogen mineralization as affected by soil moisture, temperature, and depth," *Soil Science Society of America Journal*, vol. 44, no. 6, pp. 1233–1237, 1980.

[17] M. Hefting, J. C. Clément, D. Dowrick et al., "Water table elevation controls on soil nitrogen cycling in riparian wetlands along a European climatic gradient," *Biogeochemistry*, vol. 67, no. 1, pp. 113–134, 2004.

[18] K. Kielland, J. McFarland, and K. Olson, "Amino acid uptake in deciduous and coniferous taiga ecosystems," *Plant and Soil*, vol. 288, no. 1-2, pp. 297–307, 2006.

[19] A. DesRochers, R. Van Den Driessche, and B. R. Thomas, "The interaction between nitrogen source, soil pH, and drought in the growth and physiology of three poplar clones," *Canadian Journal of Botany*, vol. 85, no. 11, pp. 1046–1057, 2007.

[20] W. T. M. Woolfolk and A. L. Friend, "Growth response of cottonwood roots to varied NH4:$NO_3$ ratios in enriched patches," *Tree Physiology*, vol. 23, no. 6, pp. 427–432, 2003.

[21] X. Min, M. Y. Siddiqi, R. D. Guy, A. D. M. Glass, and H. J. Kronzucker, "Induction of nitrate uptake and nitrate reductase activity in trembling aspen and lodgepole pine," *Plant, Cell and Environment*, vol. 21, no. 10, pp. 1039–1046, 1998.

[22] X. Min, M. Y. Siddiqi, R. D. Guy, A. D. M. Glass, and H. J. Kronzucker, "A comparative study of fluxes and compartmentation of nitrate and ammonium in early-successional tree species," *Plant, Cell and Environment*, vol. 22, no. 7, pp. 821–830, 1999.

[23] B. Truax, F. Lambert, D. Gagnon, and N. Chevrier, "Nitrate reductase and glutamine synthetase activities in relation to growth and nitrogen assimilation in red oak and red ash seedlings: effects of N-forms, N concentration and light intensity," *Trees*, vol. 9, no. 1, pp. 12–18, 1994.

[24] W. H. Campbell, "Nitrate reductase and its role in nitrate assimilation in plants," *Physiologia Plantarum*, vol. 74, no. 1, pp. 214–219, 1988.

[25] A. Al Gharbi and C. R. Hipkin, "Studies on nitrate reductase in british angiosperms. I. A comparison of nitrate reductase activity in ruderal, woodland-edge and woody species," *New Phytologist*, vol. 97, no. 4, pp. 629–639, 1984.

[26] F. Lambert, B. Truax, D. Gagnon, and N. Chevrier, "Growth and N nutrition, monitored by enzyme assays, in a hardwood plantation: effects of mulching materials and glyphosate application," *Forest Ecology and Management*, vol. 70, no. 1–3, pp. 231–244, 1994.

[27] M. A. Nicodemus, K. F. Salifu, and D. F. Jacobs, "Nitrate reductase activity and nitrogen compounds in xylem exudate of *Juglans nigra* seedlings: relation to nitrogen source and supply," *Trees*, vol. 22, no. 5, pp. 685–695, 2008.

[28] N. Smirnoff, P. Todd, and G. R. Stewart, "The occurrence of nitrate reduction in the leaves of woody plants," *Annals of Botany*, vol. 54, no. 3, pp. 363–374, 1984.

[29] L. Koyama and K. Kielland, "Plant physiological responses to hydrologically mediated changes in nitrogen supply on a boreal forest floodplain: a mechanism explaining the discrepancy in nitrogen demand and supply," *Plant and Soil*, vol. 342, no. 1-2, pp. 129–139, 2011.

[30] B. Truax, D. Gagnon, and N. Chevrier, "Nitrate reductase activity in relation to growth and soil N forms in red oak and red ash planted in three different environments: forest, clear-cut and field," *Forest Ecology and Management*, vol. 64, no. 1, pp. 71–82, 1994.

[31] B. Truax, D. Gagnon, F. Lambert, and N. Chevrier, "Nitrate assimilation of raspberry and pin cherry in a recent clearcut," *Canadian Journal of Botany*, vol. 72, no. 9, pp. 1343–1348, 1994.

[32] G. F. Dykstra, "Nitrate reductase activity and protein concentration of two Populas clones," *Plant Physiology*, vol. 53, pp. 632–634, 1974.

[33] B. L. Black, L. H. Fuchigami, and G. D. Coleman, "Partitioning of nitrate assimilation among leaves, stems and roots of poplar," *Tree Physiology*, vol. 22, no. 10, pp. 717–724, 2002.

[34] T. N. Rosenstiel, A. L. Ebbets, W. C. Khatri, R. Fall, and R. K. Monson, "Induction of poplar leaf nitrate reductase: a test of extrachloroplastic control of isoprene emission rate," *Plant Biology*, vol. 6, no. 1, pp. 12–21, 2004.

[35] R. T. W. Siegwolf, R. Matyssek, M. Saurer et al., "Stable isotope analysis reveals differential effects of soil nitrogen and nitrogen dioxide on the water use efficiency in hybrid poplar leaves," *New Phytologist*, vol. 149, no. 2, pp. 233–246, 2001.

[36] H. Rennenberg, H. Wildhagen, and B. Ehlting, "Nitrogen nutrition of poplar trees," *Plant Biology*, vol. 12, no. 2, pp. 275–291, 2010.

[37] K. M. Klingensmith and K. V. Cleve, "Patterns of nitrogen mineralization and nitrification in floodplain successional soils along the Tanana River, interior Alaska," *Canadian Journal of Forest Research*, vol. 23, no. 5, pp. 964–969, 1993.

[38] A. Robitaille and J. P. Saucier, *Paysages Régionaux Du Québec Méridional*,, Les publications du Québec, Ste-Foy, QC, Canada, 1998.

[39] J.-P. Saucier, J.-F. Bergeron, P. Grondin, and P. Robitaille, "Les régions écologiques du Québec méridional (3e version): un des éléments du système hiérarchique de classification écologique du territoire mis au point par le ministère des Ressources naturelles du Québec," *L'Aubelle*, vol. 124, pp. 1–12, 1998.

[40] D. B. Cann and P. Lajoie, *Etudes des Sols des Comtés de Stanstead, Richmond, Sherbrooke et Compton dans la Province de Québec*, Ministère de l'Agriculture, Ottawa, Canada, 1943.

[41] J. Fortier, D. Gagnon, B. Truax, and F. Lambert, "Nutrient accumulation and carbon sequestration in 6-year-old hybrid poplars in multiclonal agricultural riparian buffer strips," *Agriculture, Ecosystems and Environment*, vol. 137, no. 3-4, pp. 276–287, 2010.

[42] P. Périnet, H. Gagnon, and S. Morin, *Liste des Clones Recommandés de Peuplier Hybride par Sous-Région écologique au Québec (Révision Février 2001)*, Direction de la Recherche Forestière, MRN, Québec, Canada, 2001.

[43] R. G. D. Steel and J. H. Torrie, *Principles and Procedures of Statistics*, McGraw-Hill, New York, NY, USA, 1980.

[44] P. Qian, J. J. Schoenau, and W. Z. Huang, "Use of ion exchange membranes in routine soil testing," *Communications in Soil Science & Plant Analysis*, vol. 23, no. 15-16, pp. 1791–1804, 1992.

[45] R. J. Raison, M. J. Connell, and P. K. Khanna, "Methodology for studying fluxes of soil mineral-N *in situ*," *Soil Biology and Biochemistry*, vol. 19, no. 5, pp. 521–530, 1987.

[46] D. R. Keeney and D. W. Nelson, "Nitrogen: inorganic forms," in *Methods of Soil Analysis: Part 2*, A. L. Page, R. H. Miller,

and D. R. Keeney, Eds., pp. 643–698, American Society of Agronomy, Madison, Wis, USA, 1982.

[47] E. G. Jaworski, "Nitrate reductase assay in intact plant tissues," *Biochemical and Biophysical Research Communications*, vol. 43, no. 6, pp. 1274–1279, 1971.

[48] P. R. Larson and J. G. Isebrands, "The plastochron index as applied to developmental studies of cottonwood," *Canadian Journal of Forest Research*, vol. 1, no. 1, pp. 1–11, 1971.

[49] R. G. Peterson, *Design and Analysis of Experiments*, Marcel-Dekker, New York, NY, USA, 1985.

[50] K. H. Johnson, B. C. Bongarten, and L. R. Boring, "Effects of nitrate on in viva nitrate reductase activity of seedlings from three open-pollinated families of Robinia pseudoacacia," *Tree Physiology*, vol. 8, pp. 381–389, 1991.

[51] J. P. Schimel, R. G. Cates, and R. Ruess, "The role of balsam poplar secondary chemicals in controlling soil nutrient dynamics through succession in the Alaskan taiga," *Biogeochemistry*, vol. 42, no. 1-2, pp. 221–234, 1998.

[52] J. A. Schweitzer, M. D. Madritch, J. K. Bailey et al., "From genes to ecosystems: the genetic basis of condensed tannins and their role in nutrient regulation in a *Populus* model system," *Ecosystems*, vol. 11, no. 6, pp. 1005–1020, 2008.

[53] C. R. Ridge, T. M. Hinckley, R. F. Stettler, and E. Van Volkenburgh, "Leaf growth characteristics of fast-growing poplar hybrids *Populus trichocarpa x P. deltoides*," *Tree Physiology*, vol. 1, pp. 209–216, 1986.

[54] T. S. Barigah, B. Saugier, M. Mousseau, J. Guittet, and R. Ceulemans, "Photosynthesis, leaf area and productivity of 5 poplar clones during their establishment year," *Annales des Sciences Forestieres*, vol. 51, no. 6, pp. 613–625, 1994.

[55] P. Heilman and R. J. Norby, "Nutrient cycling and fertility management in temperate short rotation forest systems," *Biomass and Bioenergy*, vol. 14, no. 4, pp. 361–370, 1998.

[56] L. S. Lodhiyal and N. Lodhiyal, "Nutrient cycling and nutrient use efficiency in short rotation, high density central Himalayan Tarai poplar plantations," *Annals of Botany*, vol. 79, no. 5, pp. 517–527, 1997.

[57] T. J. Blake, T. J. Tschaplinski, and A. Eastham, "Stomatal control of water use efficiency in poplar clones and hybrids," *Canadian Journal of Botany*, vol. 62, no. 7, pp. 1344–1351, 1984.

[58] M. G. R. Cannell, L. J. Sheppard, and R. Milne, "Light use efficiency and woody biomass production of poplar and willow," *Forestry*, vol. 61, no. 2, pp. 125–136, 1988.

[59] J. E. K. Cooke, T. A. Martin, and J. M. Davis, "Short-term physiological and developmental responses to nitrogen availability in hybrid poplar," *New Phytologist*, vol. 167, no. 1, pp. 41–52, 2005.

[60] D. S. Green, E. L. Kruger, and G. R. Stanosz, "Effects of polyethylene mulch in a short-rotation, poplar plantation vary with weed-control strategies, site quality and clone," *Forest Ecology and Management*, vol. 173, no. 1–3, pp. 251–260, 2003.

[61] J. A. Zalesny, R. S. Zalesny, D. R. Coyle, R. B. Hall, and E. O. Bauer, "Clonal variation in morphology of *Populus* root systems following irrigation with landfill leachate or water during 2 years of establishment," *Bioenergy Research*, vol. 2, no. 3, pp. 134–143, 2009.

[62] R. M. A. Block, K. C. J. Van Rees, and J. D. Knight, "A review of fine root dynamics in *Populus* plantations," *Agroforestry Systems*, vol. 67, no. 1, pp. 73–84, 2006.

[63] R. Ceulemans, R. F. Stettler, T. M. Hinckley, J. G. Isebrands, and P. E. Heilman, "Crown architecture of *Populus* clones as determined by branch orientation and branch characteristics," *Tree Physiology*, vol. 7, pp. 157–167, 1990.

[64] J. G. Isebrands and D. A. Michael, "Effects of leaf morphology and orientation on solar radiation interception and photosynthesis in *Populus*," in *Crown and Canopy Structure in Relation To Productivity*, T. Fujimori and D. Whitehead, Eds., pp. 359–381, Forestry and Forest Products Research Institute, Ibaraki, Japan, 1986.

[65] J. M. Dunlap and R. F. Stettler, "Genetic variation and productivity of *Populus trichocarpa* and its hybrids. X. Trait correlations in young black cottonwood from four river valleys in Washington," *Trees*, vol. 13, no. 1, pp. 28–39, 1998.

[66] A. Pliura, S. Y. Zhang, J. MacKay, and J. Bousquet, "Genotypic variation in wood density and growth traits of poplar hybrids at four clonal trials," *Forest Ecology and Management*, vol. 238, no. 1–3, pp. 92–106, 2007.

[67] X. Min, M. Y. Siddiqi, R. D. Guy, A. D. M. Glass, and H. J. Kronzucker, "A comparative kinetic analysis of nitrate and ammonium influx in two early-successional tree species of temperate and boreal forest ecosystems," *Plant, Cell and Environment*, vol. 23, no. 3, pp. 321–328, 2000.

2

# A Spatial Index for Identifying Opportunity Zones for Woody Cellulosic Conversion Facilities

**Xia Huang,[1] James H. Perdue,[2] and Timothy M. Young[1]**

[1] *Center for Renewable Carbon, University of Tennessee, Knoxville, TN 37996-4570, USA*
[2] *USDA Forest Service, Southern Research Station, 2506 Jacob Drive, Knoxville, TN 37996-4570, USA*

Correspondence should be addressed to Timothy M. Young, tmyoung1@utk.edu

Academic Editor: John Stanturf

A challenge in the development of renewable energy is the ability to spatially assess the risk of feedstock supply to conversion facilities. Policy makers and investors need improved methods to identify the interactions associated with landscape features, socioeconomic conditions, and ownership patterns, and the influence these variables have on the geographic location of potential conversion facilities. This study estimated opportunity zones for woody cellulosic feedstocks based on landscape suitability and market competition for the resource. The study covered 13 Southern States which was a segment of a broader study that covered 33 Eastern United States which also included agricultural biomass. All spatial data were organized at the 5-digit zip code tabulation area (ZCTA). A landscape index was developed using factors such as forest land cover area, net forest growth, ownership type, population density, median family income, and farm income. A competition index was developed based on the annual growth-to-removal ratio and capacities of existing woody cellulosic conversion facilities. Combining the indices resulted in the identification of 592 ZCTAs that were considered highly desirable zones for woody cellulosic conversion facilities. These highly desirable zones were located in Central Mississippi, Northern Arkansas, South central Alabama, Southwest Georgia, Southeast Oklahoma, Southwest Kentucky, and Northwest Tennessee.

## 1. Introduction

Energy, its availability and use, is fundamental to a sustainable economy. The 20th century was marked by rapid growth and increased prosperity in the world. By 2020, the world's energy consumption is predicted to be 40% higher than it is today [1]. Key sources of oil are located in complex geopolitical environments that increase economic risk. Since the 1970s, macroeconomists have viewed changes in the price of oil as an important source of economic fluctuations, as well as a paradigm for global shock, likely to affect many economies simultaneously [2].

Renewable energy is projected to be one of the fastest growing industries in the US agricultural and forest sectors. As Elbehri [3] noted replacing petroleum products with bio-based fuels and energy presents several technical, economic, and research challenges, one of which is the availability of biomass feedstock. Elbehri [3] also noted that lack of biomass production capacity, high relative costs of production, logistics, and transportation of feedstocks are all potential constraints that need to be better understood. This study directly addresses Elbehri's [3] thesis by developing physical landscape and socioeconomic data for use by decision-makers interested in identifying opportunity zones for biomass-using facilities.

A plethora of literature exists on the economic availability of biomass [4–18]. A recent report by the US Department of Agriculture and Department of Energy concluded that 1.3 billion tons of biomass are available annually for energy production [18, 19].

A major difficulty addressed in this study that biomass production and access to this biomass in the field are not always directly related in a spatial context to decision-makers interested in mill siting. Improved information and methods for biomass markets that display and visualize the costs of supply and logistics from farm to forest gate to collection or conversion facilities may improve knowledge essential for market formation. The supply of biomass may be more constrained when relying on a supply network that is independent of the production facility for the raw material, for example, *facility relies on gate prices and does not have company or contract-engaged suppliers*. Decisions made in response to societal objectives frequently result in (more or less) permanent physical occupation of areas of land, for example, *buildings, roadways, preserves*, and so forth. Therefore, emerging opportunities that compete against existing uses of property or raw resources are often socially constrained or permanently denied regardless of economic viability.

This study identifies opportunity zones in a spatial context for woody cellulosic feedstocks available to potential conversion industries, for example, *biorefineries, wood pellet mills, biopower*, and so forth. The opportunity zones are derived from the use of landscape suitability and competition indices. Landscape features (*measure to which a competing land use is physically restricted by current land use*) may adversely impact economically viable competing uses of property and thereby restrict biomass access and positive location decisions. Spatial competition is particularly important for access to biomass resources. Existence of competing biomass using facilities reduces the probability of making a positive location decision and this impact decreases with distance from competition.

Specific objectives of the study were (1) compile data on the physical and socioeconomic characteristics of the landscape and display this data in a spatial context at the 5-digit zip code tabulation resolution for 33 eastern United States; (2) develop an index from the spatial data that would discriminate the landscape to identify opportunity zones for biomass-using facilities; and (3) integrate objectives (1) and (2) with the Biomass Site Assessment Tool (BioSAT), http://www.biosat.net/, as an example of application of the spatial data for practitioners.

## 2. Methods

*2.1. Datasets.* This study involved organizing large volumes of data collected from various sources, including the US Census Bureau [20], US Forest Service [21], US National Land Cover Database [22], US National Elevation Dataset [23], US Department of Agriculture National Agricultural Statistic Service [24], US Environmental Protection Agency [25], and state mill directories.

Another resource that was used to illustrate how this data could be helpful to possible users was the integration of the BioSAT model with the study [26]. The BioSAT model was used in this study to estimate the availability of woody cellulose for procurement zones within a 128.8 km (80 mile) one way travel distance which may not be concentric, that is, the shape of such zones rely on the available transportation network and biomass supply. National forests, parks, urban areas, and other restricted areas were not considered in BioSAT when estimating availability. Travel times and distances were estimated from Microsoft MapPoint 2006. Road networks in MapPoint were a combination of the Geographic Data Technology, Inc. (GDT) and Navteq data. GDT data were used for rural areas and small to medium size cities (e.g., *rural paved two-lane roads, privately owned driveway, pedestrian walkway*). Navteq data were used for major metropolitan areas (e.g., *roads with turn restrictions, physical barriers and gates, one-way streets, restricted access* and *relative road heights*). In the BioSAT model, estimates of all-live total biomass, as well as average annual growth, removals, and mortality were obtained from the Forest Inventory and Analysis Database (FIADB) version 3.0.

All records were organized at the US Census Bureau 5-digit ZIP Code Tabulation Area (ZCTA) level [20]. There were 10,016 ZCTAs in the 13-state (Alabama, Arkansas, Florida, Georgia, Kentucky, Louisiana, Mississippi, North Carolina, Oklahoma, South Carolina, Tennessee, Texas, Virginia) study region which corresponded to 10,016 potential analytical polygons or opportunity zones for biorefineries using woody cellulose. The average area size for 5-digit ZCTAs in the 13-state study regions was 209.84 $km^2$. Twelve variables (Table 1) were used in a spatial context as geographical landscape and socioeconomic factors with the BioSAT model in determining opportunity zones.

The research methodology used in this study has four main components: (1) estimation of forest biomass availability; (2) measurement of landscape suitability of forest biomass access; (3) analysis of a spatial market competition for forest biomass resources; and (4) visualization of biomass opportunity zones. Each of these components is described in the following section.

*2.2. Estimation of Biomass Availability.* Forest biomass annual growth and removal quantity data were collected at the county level from Forest Inventory and Analysis Database (FIADB) version 3.0 (Figure 1(a)), and reallocation was done for each of the 10,016, 5-digit ZCTAs using a geographic information system (GIS) technology. National land cover data [22] and digital raster map were used to identify forestland. In the digital raster map, each pixel represents one particular land cover class, that is, *water, urban, forest, or cropland*, and so forth (Figure 1(b)). Forest biomass annual growth and removal quantities were proportionally allocated to each 5-digit ZCTA using the county boundary, 5-digit ZCTA, and the land cover image data with GIS spatial overlay techniques.

Due to the mismatch of county boundary and 5-digit ZCTA (i.e., *some 5-digit ZCTAs cross county*), each forest biomass county was split into multiple area parts via the 5-digit ZCTA area shape and assigned a unique 5-digit ZCTA identifier. By overlaying each area part with the land cover image layer, the numbers of pixels in all land cover classes within each area were estimated (Figure 1(c)). By summing up the pixels of deciduous forest, evergreen forest, and mixed forest, which together represented forestland, in the unit of county, a forestland pixel ratio for each area part to its belonging county was calculated and the forest biomass quantity in every area part was derived for this pixel ratio (Figure 1(d)). A summed quantity value for all area

TABLE 1: Geographical landscape and socio-economic factors used in study.

| Variable | Original data resolution | Unit | Data sources |
|---|---|---|---|
| Population density | 5-digit ZCTA | People/mile$^2$ | U.S. Census Bureau (2010) population density in each 5-digit ZCTA. |
| Farm net income | County | Dollar | USDA NASS Census Agriculture (2007) farm net income in each county. |
| Road density | 5-digit ZCTA | km/km$^2$ | U.S. Census Bureau (2010) road length |
| Crop cultivated land area ratio | | | |
| Forest land area ratio | 5-digit ZCTA | Percent | U.S. National land Cover Database (2006) |
| Urban Land area ratio | | | |
| Water area ratio | | | |
| Slope | 5-digit ZCTA | Percent | U.S. National Elevation Dataset (1999) NED 1 arc second |
| Ecoregions Level III | Ecoregions | — | U.S. EPA (2011) |
| Timberland annual growth-to-removal ratio | County | — | Forest Inventory and Analysis—The Timber Products Tools (TPO) (2009) |
| Lands in public preserves | 5-digit ZCTA | — | U.S. Forest Service (2009) |
| Primary wood-using mill locations | 5-digit ZCTA | — | U.S. Forest Service (2009) and state mill directories |

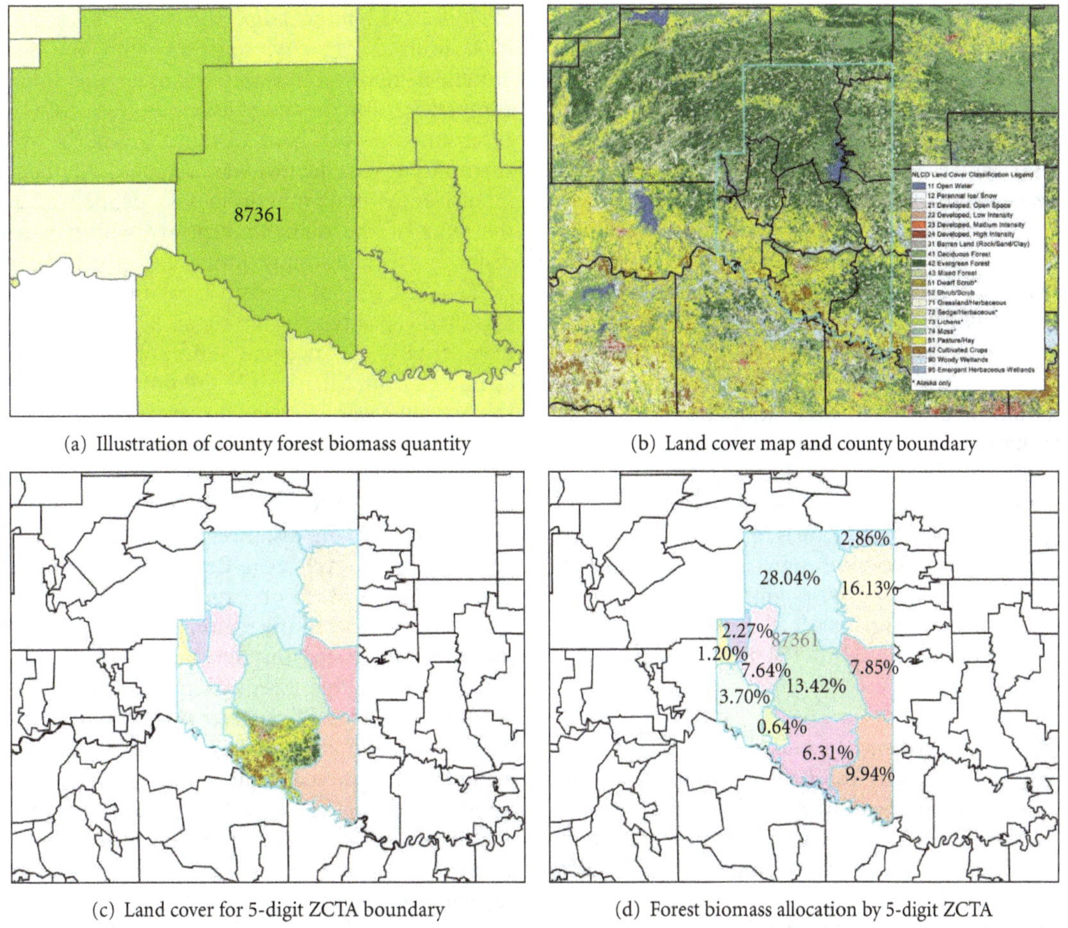


(a) Illustration of county forest biomass quantity

(b) Land cover map and county boundary

(c) Land cover for 5-digit ZCTA boundary

(d) Forest biomass allocation by 5-digit ZCTA

FIGURE 1: Illustration of forest biomass allocation at the level of 5-digit ZCTA.

parts belonging to the same 5-digit ZCTA was then calculated as the forest biomass quantity in this 5-digit ZCTA.

*2.3. Landscape Suitability Index.* The availability of forest biomass, as well as other forest resources, is physically constrained by a set of factors from the natural and socialeconomic environment [27]. The landscape suitability index explores the biophysical environment and its impacts on forest biomass access in a spatial context. Several criteria and their combinations including current land use characteristics, land ownership, and socialeconomic and/or legal constraints were used to preselect opportunity zones for forest biomass using facilities. The suitability index assumes that the presence of harvestable forests, access to abundant forest resource supply, and minimal socialeconomic restrictions from human activity (e.g., *urban development, suburban sprawl, national parks*, etc.) provides optimal conditions for woody cellulosic conversion facilities. Attributes of "forest land area ratio," "slope," as well as "suitable ecoregions for forests" determined the spatial degree of the presence of harvestable forests. The attribute "timberland annual growth-to-removal ratio" was an indicator of forest net growth. The variables "population density," "farm net income," "median family income," and "road density" were used to estimate socialeconomic indicators.

The final suitability index value was organized into ordinal levels based on expert judgment to estimate the amount and accessibility of forest biomass given the aforementioned possible constraints (Table 2). "High" suitability was considered to be a suitable opportunity zone for woody cellulosic conversion facilities relative to "moderate" or "low" suitability which would be less desirable as potential opportunity zones for woody cellulosic conversion facilities. "Unsuitable" land areas as defined by EPA ecoregions classification (The US EPA ecological regions (ecoregions) comprehensively categorize and outline main features of each unit across North America according to a variety of biological, physical, and human factors. Each ecoregion marks a geographic area with a shared climate, terrain and similar vegetation, hydrology, wildlife and land use/human activities throughout.) [25] depicted areas that are not ecologically suitable for forest production, for example, *desert in western Texas*, or *mountain tops of Smoky Mountains*. "Exclusion" zones referred to land areas that will not support forest production given socioeconomic and/or ownership type, for example, *national parks, military bases*, or *urban areas with population density over 58 people per square kilometer* (58 people/km$^2$ equals to 150 people/mile$^2$) [28], see Figure 2.

Table 2: Definitions of landscape suitability index.

| Level | Description |
|---|---|
| High suitability | Lands suitable for forest production only, for example, forests of northern Minnesota |
| Moderate suitability | Lands that have moderate capability for being only in forest production |
| Low suitability | Lands may be easily converted to agricultural production from forestland |
| Unsuitable | Land areas as defined by ecoregion classification that are not suitable for forest or agricultural production, for example, desert in western Texas, mountain tops of Smoky mountains |
| Exclusion | Land areas that will not support forest or agricultural production given socioeconomic and/or legal constraints, for example, national parks, military bases, urban areas with population density >58 people/km$^2$ [28], and so forth |

*2.4. Competition Index.* Potential forest biomass availability is also strongly influenced by the level of competition for the resource. Resource competition is usually negatively correlated with forest biomass availability unless the potential supply is a byproduct of existing harvesting operations such as forest residues [29].

A "zone-of-influence" model was developed in this study. The zone-of-influence model assumes the procurement zones associated with existing demand points or mills may not be concentric and that neighboring mills have procurement zones that occupy the same space and overlap [29–31]. The zone of influence model developed in this study used existing primary wood-using operating capacities (e.g., *sawmills, OSB mills, pulp*, and *paper mills*, etc.) assuming 80% utilized capacity, together with the forest annual growth-to-removal ratio to estimate the intensiveness of competition for the forest biomass resource (Figure 3). (A 128.8 km one-way haul distance given the road network surrounding each facility was assumed). Mathematically, the intensiveness of competition was defined as a percent of the sum of the demand capacity within a fixed driving distance over the supply annual net growth for each 5-digit ZCTA. (The sum of demand capacities was the total value of allocated capacities, assuming an 80% utilization rate). The fixed driving distance was 128.8 km one-way haul distance estimated from Microsoft MapPoint 2006 road network. The initial value was adjusted by the annual net growth and growth-to-removal ratio. Each facility's operating capacity, based on the forest area coverage characteristics, was proportionally allocated to the neighboring supply 5-digit ZCTAs for the fixed driving distance.

Six ordinal levels were developed defining the intensity of competition based on expert judgment (Table 3). Codes 1 to 5 are considered the regions where a positive annual net forest growth existed, but the intensity of competition for the resource was different. For example, the "highest" intensity of competition for the resource was considered regions where the adjusted annual operating capacities of all primary wood-using facilities exceeded the annual net forest growth of the forest resources. The "least" intensive competitive regions were those where only 10% or less of the annual net forest growth was consumed by the adjusted annual operating capacities of primary wood-using facilities. Code 6, "No or negative supply net growth", considered the regions where either no forest resources existed or forest annual mortality rate exceeded the annual growth rate.

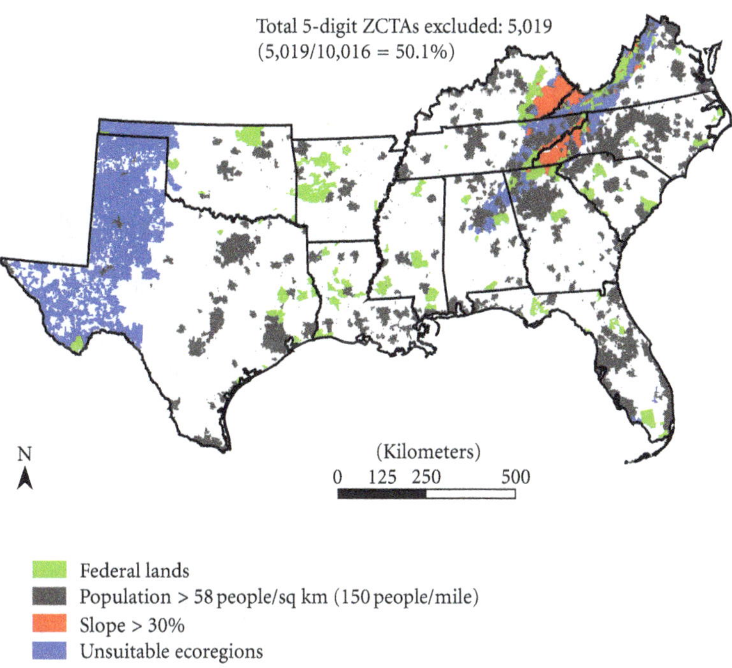


FIGURE 2: 5-digit ZCTAs excluded in the 13-state study region.

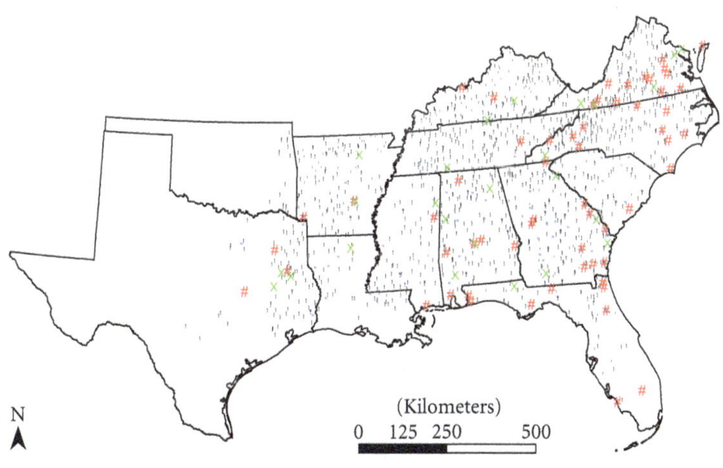


FIGURE 3: Primary wood-using facilities in the 13-state study region.

*2.5. Visualization of Opportunity Zones.* GIS methods are highly effective for visualization mapping of spatial opportunity zones for biomass availability and accessibility when a study area consists of more than 10,000 location units (5-digit ZCTAs), as was the case in this study. Two sets of maps were produced in this study. The first set was spatial opportunity zones for woody cellulosic conversion facilities using the aforementioned landscape suitability and competition indices. The second set of maps illustrates the spatial pattern of the competition intensity for the resource

Table 3: Definitions of competition index on resource utilization.

| Code | Description |
|---|---|
| 1 | *Least competitive*—less than 10% supply net growth consumed by existing capacity, and annual supply net growth over 10,000 dry ton/year, and timberland annual growth-to-removal ratio greater than 1.5 |
| 2 | *Less moderate competitive*—less than 50% supply net growth consumed by existing capacity, and positive annual supply net growth, and timberland annual growth-to-removal ratio greater than 1.0 |
| 3 | *Moderate competitive*—less than 100% supply net growth consumed by existing capacity, and positive annual supply net growth, and timberland annual growth-to-removal ratio greater than 1.0 |
| 4 | *More moderate competitive*—less than 100% supply net growth consumed by existing capacity, and positive annual supply net growth, and timberland annual growth-to-removal ratio less or equal than 1.0 |
| 5 | *Most competitive*—equal or more than 100% supply net growth consumed by existing capacity |
| 6 | *No or negative supply net growth* (or timberland annual growth-to-removal ratio equals to 0) |

Table 4: Criteria for four levels of landscape suitability.

| Level | Criteria |
|---|---|
| High suitability | Forest area ratio greater than 30%; and timberland annual growth-to-removal ratio greater than 1.5; and ecoregions defined as mostly forestland; and slope lower than 30%; and population density less than 39 people/km$^2$ |
| Moderate suitability | Forest area ratio greater than 10%; and timberland annual growth-to-removal ratio greater than 1; and ecoregions defined suitable for forestland; and slope equal or lower than 30% and population density equal or less than 58 people/km$^2$ |
| Low suitability | Forest area ratio greater than 10%; and timberland annual growth-to-removal ratio equal or less than 1; and ecoregions defined suitable for forestland or cropland; and slope equal or lower than 30% and population density equal or less than 58 people/km$^2$ |
| Unsuitable for forests | Forest area ratio equal or less than 10%; or timberland annual growth-to-removal ratio less than 0; or ecoregions defined as mostly cropland; or negative farm net income but median family income greater than $49,445 or road density higher than 5 km/km$^2$ |

assuming a fixed haul distance from each existing wood-using facility.

## 3. Results and Discussion

*3.1. Opportunity Zones Using the Landscape Suitability Index.* The four indicators of "federal lands," "population density," "slope," and "unsuitable ecoregions" were used to "exclude" ZCTAS from the study region. "Federal lands" included lands in ownership by the Bureau of Indian Affairs, Department of Defense, Fish and Wildlife Service, Forest Service, National Aeronautics and Space Administration, National Park Service, Tennessee Valley Authority, and US Department of Agriculture Research Center. ZCTAs with "population density" (>58 people/km$^2$) were excluded given the results of previous research [28]. ZCTAs with slopes greater than 30% were excluded given the limitations of ground-based harvest capabilities in the Eastern United States and also given the results of previous research related to soil disturbance [32]. Seven "unsuitable" Level III ecoregions excluded areas from mountain tops in Smoky Mountains, grassland in West Oklahoma, deserts in West Texas, and marshland and swampland in Southern Florida [25]. These four criteria together resulted in an initial exclusion of 5,019, 5-digit ZCTAs or approximately 50.1% of the total 5-digit ZCTAs in the 13-state study region (Figure 2).

The criteria to assess the other four levels of landscape suitability are given in Table 4. Two threshold values for "forest area ratio" (10% and 30%) were selected based on the definition of Food and Agriculture Organization of the United Nations [33] and United Nations Framework for Climate Change Convention [34]. (Food Agriculture Organization of the United Nations [33] defines forests as "lands with a tree crown cover equal or more than 10% of the area". United Nations Framework for Climate Change Convention [34] defines natural forests should be with greater than 30% of tree canopy cover for deciduous forests, evergreen forests, and mixed forests). Three threshold values for "timberland annual growth-to-removal ratio" (0, 1.0, and 1.5) were used to measure existing forest annual net growth and the potential for further harvesting. An annual net growth-to-removal ratio of 1.0 indicates forest net growth equals removals. A ratio of 1.5 indicates that 50% of the forest annual net growth exceeds removals which in this study was assumed a desirable metric. "Slope less than 30%" was considered slopes where timber harvesting activities were barely impacted [35]. US Census Bureau [36] found the transition between rural and urban land uses occurred when the population density was about 39 people/km$^2$ (equivalent to 100 people/mile$^2$). "Farm net income" and "median family income" together were used as a proxy measure to separate farm and nonfarm population. Median family income exceeding $49,445 (i.e., *the median value of median family income in 2010* [20]) and income from nonfarm sources greater than farming (i.e., *negative farm net income*) indicated most family dependent on nonfarm income. "Road density", which is highly correlated with population density, was also considered as a proxy or indirect measure of forest parcel size [37]. A threshold value of exceeding 5 km/km$^2$ was considered as very high road density and that when this level of road density occurs the probability of forest harvesting dramatically declines [38, 39].

Regions that had forest area ratio greater than 30%, timberland annual growth-to-removal ratios greater than 1.5, ecoregions defined as mostly forestland, slopes less than 30%, and less than 39 people/km$^2$ were considered areas that were highly suitable for forest productions. Based on these criteria, high suitable opportunity zones for facilities

Table 5: Criteria for four levels of the combined landscape suitability and competition indices.

| Level | Landscape suitability criteria | Competition criteria |
|---|---|---|
| High suitability | Forest area ratio greater than 30%; and timberland annual growth-to-removal ratio greater than 1.5; and ecoregions defined as mostly forestland; and slope lower than 30%; and population density less than 39 people/km$^2$ | Competition index ≤ 4 |
| Moderate suitability | Forest area ratio greater than 10%; and timberland annual growth-to-removal ratio greater than 1; and ecoregions defined suitable for forestland; and slope equal or lower than 30% and population density equal or less than 58 people/km$^2$ | Competition index ≤ 4 |
| Low suitability | Forest area ratio greater than 10%; and timberland annual growth-to-removal ratio equal or less than 1; and ecoregions defined suitable for forestland or cropland; and slope equal or lower than 30% and population density equal or less than 58 people/km$^2$ | Competition index ≤ 5 |
| Unsuitable for forests | Forest area ratio equal or less than 10%; or timberland annual growth-to-removal ratio less than 0; or ecoregions defined as mostly cropland; or negative farm net income but median family income greater than $49,445 or road density higher than 5 km/km$^2$ | Competition index ≤ 6 |

relying on woody cellulosic feedstocks were located along the Central Mississippi, northwest and southeast Alabama, north Arkansas, west Georgia, east Oklahoma, and areas in Kentucky, Tennessee and Virginia close to Smokey Mountains (Figure 4).

A strength of the data analyses was that socioeconomic data which is collected at the 5-digit ZCTA resolution was incorporated in the data overlays and therefore were not aggregated which maintained data integrity for these variables. A potential weakness of the study was the deaggregation of forest inventory data which implies that the opportunity zones have improved validity as the procurement area for a potential site location increase in area.

*3.2. Opportunity Zones Combining the Landscape Suitability and Competition Indices.* The spatial pattern of competition intensity within a 128.8 km one-way haul distance zone is displayed in Figure 5. Regions that had less than 10% supply net growth consumed by existing capacity, annual supply net growth greater than 10,000 dry tons/year, and timberland annual growth-to-removal ratio greater than 1.5 were considered least competitive. Regions that had greater than 100% supply net growth consumed by existing adjusted capacity were considered highly competitive. The criteria to assess the opportunity zones by the combined landscape suitability and competition indices are given in Table 5.

"High" suitability areas from the landscape suitability index when combined with high competition intensity (>5) resulted in a reduction of 395 ZCTAs, most of which were located in northwest Alabama. Given the aforementioned criteria, a total of 592 ZCTAs were considered highly desirable opportunity zones for forest biomass availability. These preferred zones were located in Central Mississippi, Northern Arkansas, South central Alabama, Southwest Georgia, Southeast Oklahoma, Southwest Kentucky, and Northwest Tennessee (Figure 6).

*3.3. Opportunity Zones Combined with BioSAT Model.* One potential value to the practitioner from the aforementioned analyses is in the siting of biomass-using facilities. An example of use for practitioners involved in plant siting would be to combine the analyses with the BioSAT model. As cited earlier [26], the BioSAT model (http://www.biosat.net/) can be used to assess more detailed economic information for any particular opportunity zone such as harvesting costs, transportation costs, stumpage costs, marginal cost curves, and so forth. Information related to the producers' marginal costs can be used to derive the important supply curve information necessary for potential users of wood cellulosic feedstocks. This may be important in developing markets such as woody cellulosic renewable energy where detailed assessment of the economic viability of mill location is essential.

In this study the BioSAT model was used to derive more detailed economic information for one of the high suitability opportunity zones located in central Mississippi (Figure 7). For the sake of illustration, the BioSAT model was run using ZCTA 39090 (Kosciusko MS) as the demand location for woody cellulosic feedstocks, specifically southern pine pulpwood (*pinus* spp.). The associated biobasin for ZCTA 39090 (Kosciusko MS) assuming a 120 mile haul distance is displayed in Figure 8. The associated marginal cost curve for demand ZCTA 39090 (Kosciusko MS) is also displayed in Figure 8. Marginal costs increase from approximately $48 to $66/dry ton over a maximum supply of southern pine pulpwood of 773,096 dry tons.

## 4. Conclusions

Renewable energy is projected to be one of the fastest growing industries in the US agricultural and forest sectors. However, replacing petroleum products with renewable energy presents technical, economic, and research challenges, one of which is the availability of biomass feedstock. This study directly addresses this problem by developing spatial geographic information for potential users of the woody cellulosic feedstocks for a 13-state study region in the Southern United States. The spatial geographic data accounts for landscape features, socioeconomic factors, and competition

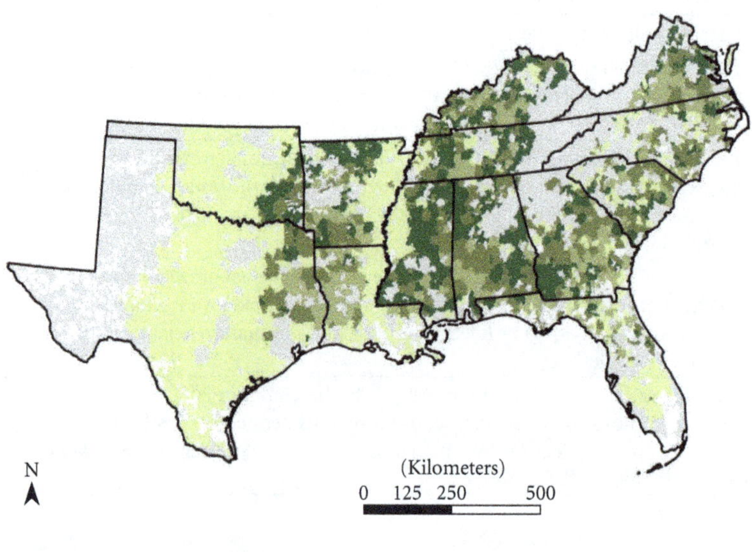


Landscape suitability index of forest biomass

- High (987, 5-digit ZCTAs)
- Moderate (1,046, 5-digit ZCTAs)
- Low (713, 5-digit ZCTAs)
- Unsuitable (2,251, 5-digit ZCTAs)
- Exclusion (5,019, 5-digit ZCTAs)

FIGURE 4: Opportunity zones for woody biomass-using facilities identified by landscape suitability index in the 13-state study region.

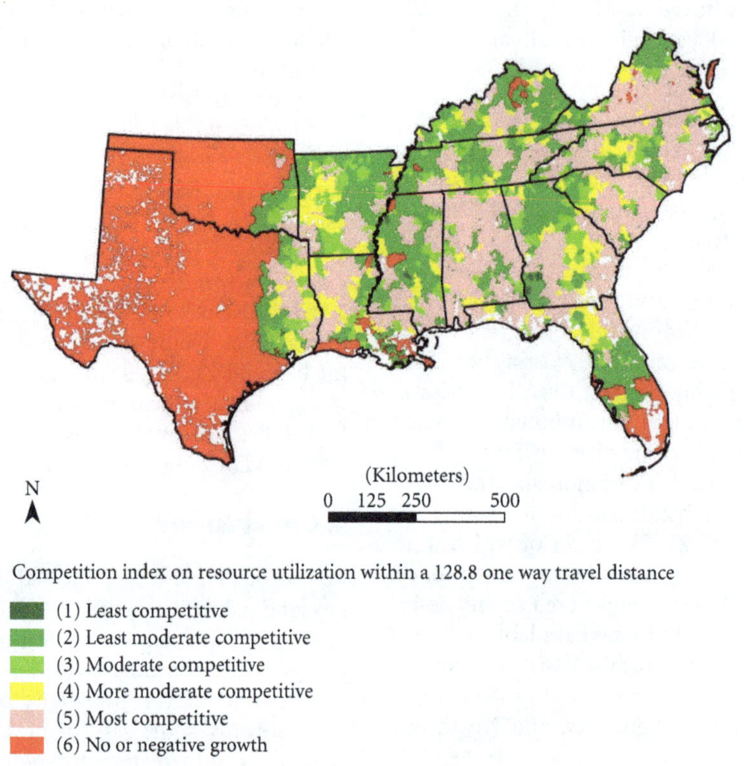


FIGURE 5: Competition index on resource utilization within a 128.8 km one way travel distance.

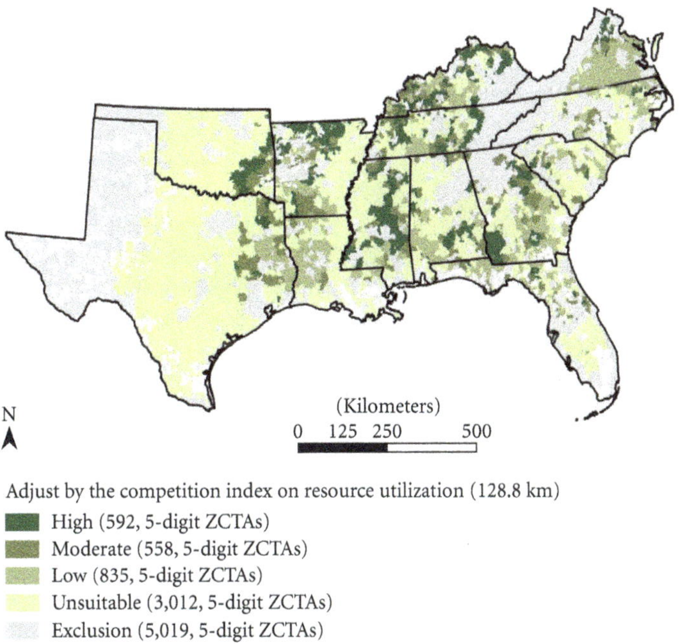


FIGURE 6: Opportunity zones identified by the combined landscape suitability and competition indices in the 13-state study region.

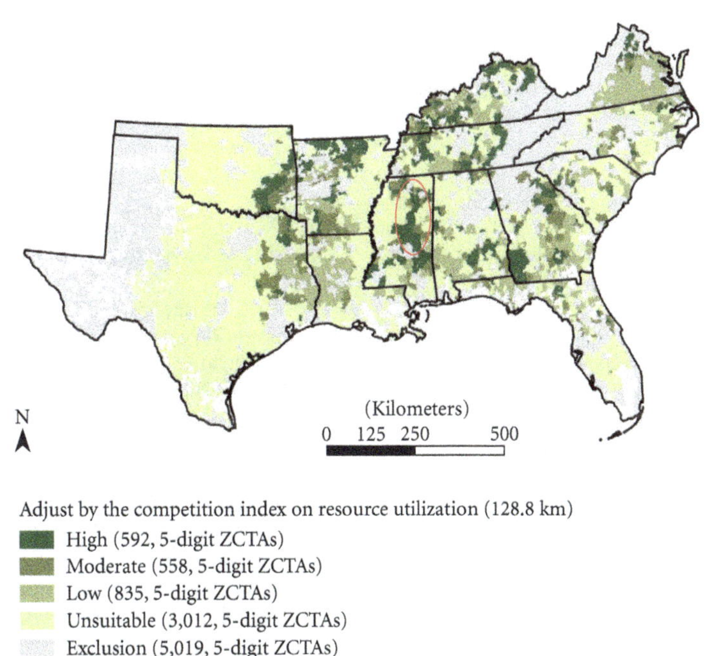


FIGURE 7: Opportunity zone in central Mississippi (highlighted in red) for mill location at ZCTA 39090 (Kosciusko MS).

for the resource, organized at the 5-digit zip code resolution. Landscape and competition indices were developed in the study and combining these indices in a spatial geographic context derives a classification of "opportunity zones" for potential users of woody cellulosic feedstocks. A total of 592, 5-digit ZCTAs were considered highly desirable opportunity zones for woody cellulosic feedstocks. These preferred zones were located in Central Mississippi, Northern Arkansas, South central Alabama, Southwest Georgia, Southeast Oklahoma, Southwest Kentucky, and Northwest Tennessee.

Project work is ongoing in developing short rotation woody crop (SRWC) data layers. The SRWC data layers will incorporate soils data, climatology data, growth modeling, and economic cost analyses. The SRWC data layers will provide dedicated energy crop analyses as a feedstock source for practitioners interested in siting scenarios using SRWC.

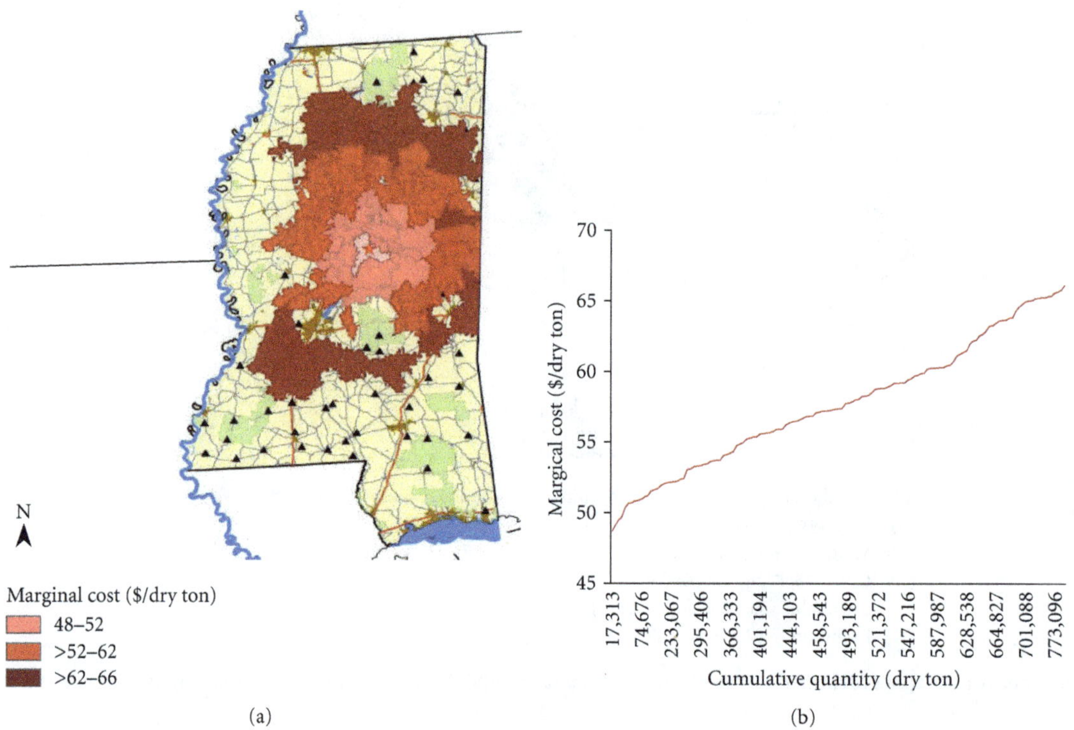


FIGURE 8: Spatial representation of biobasin for ZCTA 39090 (Kosciusko MS) and associated marginal cost curve for pine pulpwood (*pinus* spp.) from the BioSAT model.


## Acknowledgments

This research was funded by US Forest Service Southern Research Station under contract agreement 07-CR-11330115-087, Southeastern Sun Grant Center, and University of Tennessee Agricultural Experiment Station.


## References


[1] Energy Information Administration, "International energy outlook, 2008," Report DOE/EIA-0484(2008), http://www.tulane.edu/~bfleury/envirobio/readings/International%20Energy%20Outlook%2008.pdf.

[2] O. J. Blanchard and J. Gali, "The macroeconomic effects of oil price shocks: why are the 2000s so different from the 1970s?" in *Proceedings of NBER ME Conference on International Dimensions of Monetary Policy*, Catalonia, Spain, 2007, http://www.crei.cat/people/gali/pdf_files/bgoil07wp.pdf.

[3] A. Elbehri, *The Changing Face of the U.S. Grain System*, Economic Research Service. U.S., Department of Agriculture, Washington, DC, USA, 2007.

[4] T. M. Young and D. M. Ostermeier, "IFCHIPSS—the industrial fuel chip supply simulation model," Final Report for Contract with Southeastern Regional Biomass Energy Program as administered by the Tennessee Valley Authority, 1989.

[5] T. M. Young and D. M. Ostermeier, "The economic availability of woody biomass for the southeastern United States," *Bioresource Technology*, vol. 37, no. 1, pp. 7–15, 1991.

[6] T. M. Young, D. M. Ostermeier, J. D. Thomas, and R. T. Brooks Jr., "Computer model simulates supply, cost of chips," *Forest Industries*, vol. 118, no. 8, pp. 20–21, 1991.

[7] A. Lunnan, "Agriculture-based biomass energy supply—a survey of economic issues," *Energy Policy*, vol. 25, no. 6, pp. 573–582, 1997.

[8] M. E. Walsh, "U.S. Bioenergy crop economic analyses: status and needs," *Biomass and Bioenergy*, vol. 14, no. 4, pp. 341–350, 1998.

[9] M. E. Walsh, "Method to estimate bioenergy crop feedstock supply curves," *Biomass and Bioenergy*, vol. 18, no. 4, pp. 283–289, 2000.

[10] J. DiPardo, "Outlook for biomass ethanol production and demand," U.S. Energy Information Administration, 2000, http://www.eia.doe.gov/oiaf/analysispaper/pdf/biomass.pdf.

[11] D. G. de la Torre Ugarte and D. E. Ray, "Biomass and bioenergy applications of the POLYSYS modeling framework," *Biomass and Bioenergy*, vol. 18, no. 4, pp. 291–308, 2000.

[12] D. G. de la Torre Ugarte, B. C. English, R. J. Menard, and M. Walsh, "Conditions that influence the economic viability of ethanol from corn stover in the Midwest of the USA," *International Sugar Journal*, vol. 108, no. 1287, pp. 152–156, 2006.

[13] D. G. de la Torre Ugarte, B. C. English, and K. Jensen, "Sixty billion gallons by 2030: economic and agricultural impacts of ethanol and biodiesel expansion," *American Journal of Agricultural Economics*, vol. 89, no. 5, pp. 1290–1295, 2007.

[14] Biomass Research and Development Board, "Increasing feedstock production for biofuels economic drivers, environmental implications, and the role of research," 2008, http://www.usbiomassboard.gov/pdfs/increasing_feedstock_revised.pdf.

[15] Western Governor's Association, "Strategic assessment of bioenergy development in the West—spatial analysis and supply curve development," University of California, Davis, Calif, USA, 2008, http://www.treesearch.fs.fed.us/pubs/34631.

[16] G. Perez-Verdin, D. L. Grebner, C. Sun, I. A. Munn, E. B. Schultz, and T. G. Matney, "Woody biomass availability for

bioethanol conversion in Mississippi," *Biomass and Bioenergy*, vol. 33, no. 3, pp. 492–503, 2009.

[17] C. S. Galik, R. Abt, and Y. Wu, "Forest biomass supply in the southeastern United States—implications for industrial roundwood and bioenergy production," *Journal of Forestry*, vol. 107, no. 2, pp. 69–77, 2009.

[18] U.S. Department of Energy, R. D. Perlack, and B. J. Stokes, "U.S. billion-ton update, biomass supply for a bioenergy and bioproducts industry," ORNL/TM-2011/224, Oak Ridge National Laboratory, Oak Ridge, Tenn, USA, 2011.

[19] R. D. Perlack, L. L. Wright, A. F. Turhollow, R. L. Graham, B. J. Stokes, and D. C. Erbach, "2005Biomass as feedstock for a bioenergy and bioproducts industry: the technical feasibility of a billion-ton annual supply," Publication DOE/GO-102995-2135/ORNL TM-2005/66, OAR Ridge National Laboratory, OAR Ridge, Tenn, USA, http://feedstockreview.ornl.gov/pdf/billion_ton_vision.pdf.

[20] U.S. Census Bureau, "2010 Census ZIP code tabulation areas [data file]," 2010, http://www.census.gov/geo/ZCTA/zcta.html.

[21] U.S. Forest Service, "Lands in public preserves [data file]," Personal contact on 08/2011, 2009.

[22] U.S. National Land Cover Database, "Multi-resolution land characteristics consortium [Data file]," 2006, http://www.mrlc.gov/nlcd2006.php.

[23] U.S. National Elevation Dataset, "National elevation dataset 1 arc second [data file]," 2010, http://seamless.usgs.gov/ned1.php.

[24] U.S. Department of Agriculture National Agricultural Statistics Service, "Census of agriculture: farm net income [data file]," 2007, http://quickstats.nass.usda.gov.

[25] U.S. Environmental Protection Agency, "Ecoregions of the United States [data file]," 2011, http://www.epa.gov/wed/pages/ecoregions/level_iii_iv.htm.

[26] J. H. Perdue, T. M. Young, and T. G. Rials, "The biomass site assessment model—BioSAT," Final Report For U.S. Forest Service, Southern Research Station , The University of Tennessee, Knoxville, Tenn, USA, 2011.

[27] B. J. Butler, Z. Ma, D. B. Kittredge, and P. Catanzaro, "Social versus biophysical availability of wood in the Northern United States," *Northern Journal of Applied Forestry*, vol. 27, no. 4, pp. 151–159, 2010.

[28] D. N. Wear, R. Liu, J. M. Foreman, and R. M. Sheffield, "The effects of population growth on timber management and inventories in Virginia," *Forest Ecology and Management*, vol. 118, no. 1–3, pp. 107–115, 1999.

[29] C. J. Chu, F. T. Maestre, S. Xiao et al., "Balance between facilitation and resource competition determines biomass-density relationships in plant populations," *Ecology Letters*, vol. 11, no. 11, pp. 1189–1197, 2008.

[30] J. Weiner, P. Stoll, H. Muller-Landau, and A. Jasentuliyana, "The effects of density, spatial pattern, and competitive symmetry on size variation in simulated plant populations," *American Naturalist*, vol. 158, no. 4, pp. 438–450, 2001.

[31] J. Weiner and C. Damgaard, "Size-asymmetric competition and size-asymmetric growth in a spatially explicit zone-of-influence model of plant competition," *Ecological Research*, vol. 21, no. 5, pp. 707–712, 2006.

[32] M. Kimsey, D. Page-Dumroese, and M. Coleman, "Assessing bioenergy harvesting risks: geospatially explicit tools for maintaining soil productivity in Western US forests," *Forests*, vol. 2, no. 3, pp. 797–813, 2011.

[33] Food and Agriculture Organization of the United Nations (FAO), "Definitions of forests [definition]," 2000, http://www.fao.org/DOCREP/005/Y4171E/Y4171E10.htm.

[34] United Nations Framework for Climate Change Convention (UNFCCC), "Definitions of forests [definition]," 2001, http://www.fao.org/DOCREP/005/Y4171E/Y4171E10.htm.

[35] F. R. Greulich, D. P. Hanley, J. F. McNeel, and D. Baumgartner, *A Primer for Timber Harvesting*, Washington State University Extension, Pullman, Wash, USA, 1999, http://faculty.washington.edu/greulich/Documents/eb1316.pdf.

[36] U.S. Census Bureau, "The urban and rural classifications," Chapter 12, 2010, http://www.census.gov/geo/www/GARM/Ch12GARM.pdf.

[37] R. I. McDonald, G. Motzkin, M. S. Bank, D. B. Kittredge, J. Burk, and D. R. Foster, "Forest harvesting and land-use conversion over two decades in Massachusetts," *Forest Ecology and Management*, vol. 227, no. 1-2, pp. 31–41, 2006.

[38] The World Bank, "Transports: roads & highways [definition]," 2011, http://www.worldbank.org/html/extdr/thematic.htm.

[39] B. J. Butler and S. L. King, "Assessment and mapping of forest parcel sizes," in *Proceedings of the 5th Annual Forest Inventory and Analysis Symposium*, New Orleans, La, USA, 2003, http://www.treesearch.fs.fed.us/pubs/14259.

# 3

# Participatory Forest Carbon Assessment and REDD+: Learning from Tanzania

**Kusaga Mukama,[1] Irmeli Mustalahti,[2] and Eliakimu Zahabu[3]**

[1] *District Natural Resources Office, Liwale District Council, P.O. Box 23, Liwale, Tanzania*
[2] *Department of Political and Economic Studies, University of Helsinki, P.O. Box 59, 00014 Helsinki, Finland*
[3] *Department of Forest Mensuration and Management, Sokoine University of Agriculture, P.O. Box 3011, Morogoro, Tanzania*

Correspondence should be addressed to Irmeli Mustalahti, irmeli.mustalahti@helsinki.fi

Academic Editor: Piermaria Corona

Research initiatives and practical experiences have demonstrated that forest-related data collected by local communities can play an essential role in the development of national REDD+ programs and its' measurement, reporting, verification (MRV) systems. In Tanzania, the national REDD+ Strategy aims to reward local communities participating in forest management under Participatory Forest Management (PFM). Accessing carbon finances requires among other things, accurate measurements of carbon stock changes through conventional forest inventories, something which is rarely done in PFM forests due to its high cost and limited resources. The main objective of this paper is to discuss experiences of Participatory Forest Carbon Assessment (PFCA) in Tanzania. The study revealed that villagers who participated in PFCA were able to perform most steps for carbon assessment in the field. A key challenge in future is how to finance PFCA and ensure the technical capacity at local level.

## 1. Introduction

Reducing Emissions from Deforestation and Forest Degradation (REDD) is a financial mechanism of the United Nations Framework Convention on Climate Change (UNFCCC) which would provide developing countries with incentives to reduce carbon emissions from forests. The scope of the likely REDD+ mechanism has broadened to REDD+ to accommodate different country interests such as natural forests, protected areas and forests under community-based management. Under REDD+, developing countries that are effectively protecting their forests through conservation, sustainable forest management, and enhancement of forests carbon stock will be recognized and eligible for carbon payments.

The development of national forest monitoring and carbon accounting systems forms the basis for a future system of measurement, reporting, verification (MRV) for forest-related emissions reductions. At the national level advanced satellite sensor technology and mapping systems will be used for the assessment of forest carbon resources and to provide a level of detail for large geographic areas in a cost-effective manner. More bottom-up community-based forest monitoring methods are developed in order to support operational forest management and monitoring. Research initiatives and practical experiences have demonstrated that forest-related data collected by local communities can play an essential role in the development of national REDD+ programs and its MRV systems [1].

At the 16th Conference of Parties (CoP 16) of the UNFCCC, held in Cancun in 2010, REDD+ mechanism and MRV activities were agreed to be financed and implemented after 2012. In this regard, Tanzania is being supported by various development partners including the government of Norway to undertake demonstration activities so that it can report on best practices. The resulting best practices and experiences will form a basis for negotiation on the future global climate change regime and provide methods that can be scaled up by other Parties under REDD+. The national REDD+ framework and the draft national REDD+ strategy for Tanzania recognize that the REDD+ initiative will provide incentives for local communities participating in forest management [2, 3]. Accessing carbon finances through

REDD+ requires, among other things, measurements of carbon stock changes in forests.

It has been found that carbon assessment by professionals is associated with high costs while the same work can be carried out by communities managing their forests by the use of Participatory Forest Carbon Assessment (PFCA) methods. In Tanzania, the PFCA methods were previously developed and tested under the Kyoto: Think Global Act Local project (K: TGAL). However, this testing was limited to only few localities and involved forests of small size (28–600 ha). For the wider application of the technique, some more testing was required. This study therefore tested the applicability of the developed PFCA technique in Angai Villages Land Forest Reserves (AVLFR). The AVLFR has large forest area (139,420 ha), therefore expected to have high carbon stock potential among the participatory forest management (PFM) forests in Tanzania.

The overall objective of this paper is to discuss the PFCA technique for assessing carbon stock. Empirically, the study is based on the three selected villages around AVLFR. The research had specific objectives: (a) to assess local communities' perception and willingness to be involved in the REDD+ initiative; (b) to assess local communities' capability and the costs to carry out PFCA in three villages surrounding AVLFR; (c) to determine forest carbon stock in three villages' forest area constituting AVLFR. The key questions in this paper are how local communities perceive the PFCA and what assistance, technical and financial resources, do they need to carry out the PFCA.

## 2. Case Study Area

Participatory Forest Management (PFM) in Tanzania was introduced in 1990s following the failure of centralised forest management system. PFM is stipulated in the Forest Policy of 1998 and operationalised by the Forest Act No. 14 of 2002 [4]. The law recognizes two main types of PFM, namely, Joint Forest Management (JFM) and Community Based Forest Management (CBFM). JFM is based on a management agreement between local communities and government authorities regarding the management of central or local government forest reserves. Within JFM, forest ownership remains with the government while the local communities are duty bearers and in turn get user rights and access to some forest products and services [4, 5]. CBFM takes place in forests on village lands that have been surveyed and registered under the provisions of the Village Land Act No. 5 of 1999 and the Forest Act No. 14 of 2002. Under CBFM, villagers take full ownership of village land forest reserves.

In Tanzania, PFM is so far implemented in 2,323 villages with coverage of 4.12 million hectares out of 35.3 million hectares of forestland in the country [6]. These statistics include Village Land Forest Reserves (VLFR) under CBFM with a total area of 2.35 million hectares in 1,460 villages. However, in spite of these achievements there are little incentives accrued by local communities compared to the costs incurred in PFM [7, 8]. These costs include damage to their crops by animals, difficulties in patrolling vast forest boundaries without transport, and attending risks that are not compensated or rewarded. On the other hand, PFM in Tanzania has been observed to have high potential for achieving the REDD+ objectives which could provide the financial incentives required for sustainable PFM implementation [9].

This study was conducted in three selected villages of Mihumo, Ngongowele, and Ngunja that are among 13 villages surrounding the AVLFR which occupies a total area of 139,420 hectares in Liwale district, Southeast Tanzania. During recent years, the villages have been growing and District Council has advised most of the villages to be divided which means that in future there will be 24 villages instead of 13. The case study villages cover 26,703 hectares of forest, which is equivalent to about 19% of the total AVLFR. The forest is characterized by dry miombo, closed dense forest, riverine and wet miombo with some high valuable timber species such as *Brachystegia sp.*, *Julbernardia sp.*, and *Pterocarpus angolensis*, locally known as mninga. The forest has many different tree species where 133 tree species were identified in 2004 [10]. The forest is also an important catchment area because it is the source of many rivers and streams which support not only the human population but also wildlife. Identified rivers for Mihumo village are Angai, Mihumo, and Nangula, while Nakawale and Nchonda rivers originate from forest belonging to Ngunja. Angai river is the only main river which crosses through Ngongowele VLFR. However, these rivers are seasonal and during prolonged drought most of them dry up. Angai and Mihumo rivers are the major source of water for people living in Mihumo, Kiangara, Kibutuka, Liwale B, and Mikunya villages. Villagers in Ngunja, Ngongowele, and Mihumo have access to water supply from deep and shallow water wells. However, during the dry season most of shallow wells dry up, at which point the villagers fetch water from rivers. During the dry season, some villagers in Ngunja and Ngongowele get water from Ruhuhu River which is about five kilometres from the villages centre and it originates in unreserved catchments area in Ngongowele village.

## 3. Methodology

This study is part of an action research project on the role of participatory forest management in the mitigation of and adaptation to climate change (https://blogs.helsinki.fi/tzredd-actionresearch/). Participatory Action Research (PAR) methods were used to empower villagers to participate in forest carbon assessment as well as to study possibilities to implement REDD+ intervention in AVLFR. According to Chambers [11], PAR is an integrated approach involving the participation of community members to investigate social reality, build local skills and capacity for the purpose of increasing community independence and self sufficiency. Action research originally emerged from the conviction that science should benefit human society and not separate its research from issues of relevance and context [12].

For this study both primary and secondary methods of data collection were used based on their applicability and usefulness towards achieving the research objectives in the case study villages. In this study, Participatory Rural Appraisal (PRA) methods were used to introduce the

Table 1: List of villagers selected to form VFAT for the three studied villages.

| Village name | No | Trainee gender | Position at the village | Age | Education |
|---|---|---|---|---|---|
| MIhumo | (1) | Male | VNRC chairman | 50 | Standard VII |
| | (2) | Male | VNRC secretary | 21 | Standard VII |
| | (3) | Female | VNRC members | 35 | Standard VII |
| | (4) | Male | Community RP | 64 | Standard IV |
| | (5) | Male | Community RP | 50 | Standard VII |
| | (6) | Male | VG member | 45 | Standard VII |
| | (7) | Female | Community RP | 30 | Standard VII |
| | (8) | Female | VNRC member | 42 | Standard VII |
| Ngongowele | (1) | Male | Community RP | 42 | Standard VII |
| | (2) | Male | Community RP | 70 | Informal Ed. |
| | (3) | Male | VNRC secretary | 49 | Standard VII |
| | (4) | Male | Community RP | 42 | Standard VII |
| | (5) | Female | Community RP | 24 | Standard VII |
| | (6) | Male | VNRC member | 31 | Standard VII |
| | (7) | Male | Community RP | 51 | Adult Ed. |
| | (8) | Female | VG member | 31 | Standard VII |
| Ngunja | (1) | Male | Community RP | 40 | Standard VII |
| | (2) | Male | VNRC chairman | 29 | Standard VII |
| | (3) | Male | VNRC secretary | 26 | Standard VII |
| | (4) | Male | Community RP | 51 | Standard VII |
| | (5) | Female | Community RP | 37 | Standard VII |
| | (6) | Male | Community RP | 42 | Standard VII |
| | (7) | Male | VG chairman | 51 | Standard VII |
| | (8) | Female | Community RP | 49 | Standard VII |

Note: Ed.: Education, Community RP: Community representative.

research objective and REDD+ concept through PFCA in three of the 13 villages that own and manage AVLFR. PRA methods help to shift the positioning of scientific inquiry from the researcher, who acts as a facilitator, to the participants who guide the collection of data and their local analyses [13]. PRA methods allow local people to apply their indigenous knowledge, experience, and capacity to share information [14, 15]. Both men and women attended PRA exercises convened in each of the studied villages for two days. The group of participants included village government, Village Natural Resources Committee (VNRC) members, elders, and farmers. The PRA methods used included group discussion, pairwise ranking and scoring exercises, participatory forest mapping, a forest transect walk, participant observations, and focus group discussions.

*Group discussion* was used to introduce the research objectives and to collect the information on villagers' perception and their willingness to be involved in the REDD+ initiative. Participants were divided into two groups of men and women that constituted 15 participants each. The discussion of each group was guided by a checklist. The technique was used to select villagers to carry out forest carbon assessment and monitoring and to train them on the PFCA techniques. The PFCA was done using the field guide for assessing forest carbon by local communities developed by K: TGAL research project: the guide outlines steps that should be followed for assessing carbon of any forest. During the group discussions villagers were asked to select the members of the Village Forest Assessment Team (VFAT) who will participate in the PFCA. In each of the three case study villages, eight (8) villagers were selected from the village government, the VNRC, and the community at large to form the VFAT (Table 1). All members of the VFAT were permanent residents of their respective village, and most of them have standard seven educations except one elderly man from Ngongowele who had adult education, another elderly man from Mihumo who attained class four during the colonial era, and a third elderly man from Ngongowele who had informal education (Table 1). The age of the team members ranges from 21 to 70 years and 29% of them are women.

*Pairwise ranking and scoring* methods were employed to collect information on villagers' interest and perception regarding to the importance of the VLFR. First the researchers asked the participants to identify the importance of their VLFRs. Participants were divided into two groups of men and women. The grouping was meant to capture gender difference on the discussion issues. Thereafter matrix ranking was used to determine the villages' preference score for each perceived important factor of VLFR in each group.

*Focus group discussion* method was used to learn diversities among participants in participatory forest carbon assessment as well as evaluating the whole process. The focus group discussion is a qualitative research method which can be used, for example, in the programme development and evaluation context as well planning and needs assessment situations [16]. Focus group discussion was carried out in each village with the VFATs. These participants were trained and participated in PFCA in their VLFRs. The discussion with trained villagers was guided by a checklist which focused on which steps for carbon assessment they were able or unable to do including the use of forest inventory equipments, GPS, hypsometers, and establishment of permanent sample plots. This technique enabled the researchers to learn how different villagers understood the PFCA techniques. Moreover, the technique helped the consolidation and triangulation of information acquired through participant observation regarding the capacity of trained villagers to carry out PFCA.

Since the Angai VLFRs already have a georeferenced map in place, the participants were taught to use a GPS instead of a handheld computer. Nobody among the participants had seen a GPS before as was the case in the other studies under Kyoto: TGAL in Tanzania [9]. Thus the first task was to introduce the trainees to the basic parts of eTrax GPS and its operation such as the power key, installing batteries, the zoom in/out key, the menu/find key, and the quit/page key. Also they were taught how to take GPS coordinates for marking locations such as sample plots, strata, roads, rivers, and other features in the forest.

Participants were also exposed to different forest mensuration techniques and inventory equipment as outlined in the forest inventory guide, which is inline with IPCC Good Practice Guideline [17]. They were exposed to various forest inventory equipment and their uses including hypsometers, diameter tapes, tape measures, calipers, and relascopes. Similarly they were trained on taking tree measurements like diameter at breast height (DBH, measured at 1.3 m), total height of different tree shape, and form on flat/slope ground. At the end of the second day, participants carried out a practical exercise, using GPS and inventory equipment in the nearby village forest.

*Participatory Forest Mapping.* Participatory forest mapping is a mapmaking process that attempts to make the association between forest and local communities visible by using local knowledge [18]. In this study the technique was used to provide information about different vegetation type in the VLFR as perceived by villagers. Initially, the purpose of forest stratification to forest carbon assessment and monitoring was explained to participants. Thereafter the researcher facilitated the drawing of a sketch map by participants of their respective VLFR on large sheet of paper using marker pen. They were then asked to use different colours to show the boundary of each forest type/stratum on the sketch map based on their local knowledge and experience.

*Forest Transect Walk and Plot Measurements.* The transect walk is an information-gathering exercise in which results can be both qualitative and quantitative [19]. In this study, the tool was used to collect information on forest stratification after participatory forest mapping. Participants were requested to draw routes on sketch map for visiting each vegetation type in the field. Participants and researcher visited each vegetation type within respective VLFR to observe similarity and differences between vegetation types in the field and those earlier drawn on the map. Site visits can establish, if what is reported is correct, true and existing [20]. Participants and researcher went around the boundary of each identified stratum/vegetation type in the field where GPS coordinates were taken for producing the final map. Furthermore the same technique was employed for a pilot survey to determine variance. Participants and researcher walked within each stratum where basal area measurements were taken to determine standard deviations and means then calculate number of plots for each stratum.

Then the number of permanent sample plots ($n$) required in each stratum for the carbon inventory was determined using the following formula

$$n = \frac{t^2 \mathrm{CV}^2}{E^2}, \quad (1)$$

where; CV = Coefficient of variation = standard deviation/mean $t$ is the value of $t$ obtained from the student's distribution table at $n - 1$ degree of freedom of the pilot study plots at 5% probability.

According to the IPCC [17] a sampling error ($E$) of 5% is recommended for Land Use, Land Use Change, and Forestry (LULUCF) projects. However experience has shown that under certain circumstances, a 10% sampling error can be used to reduce costs while maintaining estimates within the precision of $\pm 10\%$ of the mean with a 95% confidence level [9]. In this study, number of sample plots and sampling errors adopted for the different vegetation types in the case study villages are shown in Table 2.

In this study, the permanent sample plots were laid systematically with a random start. Concentric circular plot with four circles of radius of 2 m, 5 m, 10 m, and 15 m was used. Major advantage of this type of plot hinges on reduced edge effects which may lead to possible counting error. In Tanzania, a similar plot size and shape is currently being used in National Forest Resources Monitoring and Assessment (NAFORMA).

After establishing permanent sample plots in the ground, the next step was to take tree variable measurements from the plots. From the different plot radii, the trees measurements taken are shown in Table 3.

Tree DBH was measured using tree caliper/DBH tape whereas the hypsometer was used for tree height measurements. The names of all measured trees and the counted regenerants were identified by local names by the participants. Data on each measurement were recorded along with the local tree species name on the inventory form. It was observed that most participants were able to identify the name of each tree species in their local name using their local knowledge and experience. Consequently the names of all measured trees were locally known and recorded.

*Participant observation* was employed in this study not only to collect information on the ability of trained villagers to carry out PFCA but also to observe the presence of disturbances in the VLFR. During the participant observations,

TABLE 2: Number of permanent sample plots for the studied Angai VLFRs.

| Forest name | Strata name | Strata area (ha) | Number of Plots | | |
|---|---|---|---|---|---|
| | | | $E = 10\%$ | $E = 15\%$ | Total |
| Ngunja | Lowland dry miombo | 2379 | 41 | | |
| | Upland dry miombo | 4247 | 39 | | |
| **sub total** | | **6626** | **80** | | **80** |
| Ngongowele | Dry miombo woodland | 8021 | 62 | | |
| | Closed dense forest | 181 | | 6 | |
| | Degraded riverline | 83 | | 11 | |
| **sub total** | | **8285** | **62** | **17** | **79** |
| Mihumo | Dry miombo woodland | 8169 | 78 | | |
| | Wet miombo | 1695 | 13 | | |
| | Closed dense forest | 1927 | | 11 | |
| **sub total** | | **11791** | **91** | **11** | **102** |
| **Grand Total** | | **26702** | **233** | **28** | **261** |

the researchers participated in day-to-day activities in the field in order to gain the confidence of the persons being studied, so that their presence did not interfere with their normal activities. The researchers were actively participating in PFCA while also documenting observations on how participants were using GPS to locate transects and mark permanent plots in the field, and how they were taking tree measurement using the hypsometer and tree caliper during carbon assessment in the field. In this way, the researcher had an opportunity to compare what the villagers learned through demonstration and practice and what they were really doing in the field. In the event a villager was taking a wrong measurement, the researchers interacted with trained villagers and ensured that correct measurements and proper recording were done. Therefore the role of the researchers in this study was not only to observe but also to facilitate the process in the field when participants failed to undertake some activities regarding the carbon assessment process.

## 4. Results and Discussion

Data collected through PFCA were analyzed so as to obtain forest carbon stock in terms of tons of biomass and carbon per hectare and other stand parameters including stand density in terms of number of stems per ha ($N$), basal area ($G$, $m^2$/ha), and volume ($V$, $m^3$/ha). Before analysis, a checklist of tree species was prepared for the three-village forest reserves studied. Botanical names were matched with the local names and each tree species was assigned a code number and arranged alphabetically for further analysis. As mentioned earlier, only sample trees were measured for total height, therefore a height-diameter equation was developed for each stratum for estimating the height of trees which were measured for DBH only.

For calculating tree volume and biomass, the use of local allometric equations for areas with similar geographical and vegetation types was recommended [17, 27, 28]. However in the absence of allometric equations, a common procedure is to compute stand volume and convert it to biomass. Since

TABLE 3: Plot size and tree variables measured.

| Plot radius | Tree variables measured |
|---|---|
| 2 m | Number of regenerants were counted |
| 5 m | 1 cm $\leq$ dbh $\leq$ 10 cm |
| 10 m | 10.1 cm $\leq$ dbh $\leq$ 20 cm |
| 15 m | Dbh $\geq$ 20.1 cm |

there is neither an equation for tree volume nor tree biomass developed for miombo woodland in Southern Tanzania, a general tree volume equation was used to calculate tree volume then converted to biomass by multiplying with a conversion factor of 0.5. Carbon was estimated from calculated biomass. According to MacDicken [29], it is assumed that 49% of biomass is carbon. The computed parameters were separated into eight diameter classes (DBH range (cm) from 0 to >70).

This paper intended to explore the potential of the AVLFR communities to benefit from the REDD+ initiative. It first assesses the perception and willingness of the local communities to participate in the REDD+ initiative. However, participation of the local communities in the REDD+ could be hindered by lack of information on their forest carbon stock changes. The study then assesses the capability and costs of the local communities to carry out PFCA. Lastly the forest carbon stocks in selected case study villages constituting the AVLFR are determined.

*4.1. Local Community Perception and Willingness to Participate in REDD+.* In Tanzania, it has been observed that households living in miombo woodlands derive more than 50% of their cash incomes from selling forest products [30]. For REDD+ intervention to be successful, it is necessary to create awareness for the purpose of introducing the new concept and assessing the communities' willingness to participate in activities. It is argued that forest communities must be informed about REDD+ initiative and have choices clearly presented to them, because changing livelihood strategies

Table 4: Scoring and ranking of the importance of VLFR in studied three villages.

| Importance of forest | Mihumo | | | | Ngongowele | | | | Ngunja | | | |
|---|---|---|---|---|---|---|---|---|---|---|---|---|
| | Score | | | Rank | Score | | | Rank | Score | | | Rank |
| | M | W | A | | M | W | A | | M | W | A | |
| Source of timber | 10 | 8 | 9 | 1 | 10 | 6 | 8 | 2 | 9 | 7 | 8 | 3 |
| Building materials | 8 | 4 | 6 | 4 | 8 | 6 | 7 | 3 | 7 | 5 | 6 | 4 |
| Beekeeping | 8 | 8 | 8 | 3 | 6 | 8 | 7 | 3 | 8 | 10 | 9 | 2 |
| Carbon sequestration | 10 | 8 | 9 | 1 | 12 | 10 | 11 | 1 | 12 | 12 | 12 | 1 |
| Source of fuel wood | 4 | 8 | 6 | 4 | 4 | 6 | 5 | 5 | 4 | 6 | 5 | 5 |
| Traditional medicine | 2 | 4 | 3 | 6 | 2 | 4 | 3 | 6 | 2 | 1 | 2 | 6 |
| Source of food (ming'oko and angadi) | 0 | 2 | 1 | 7 | 0 | 2 | 1 | 7 | 0 | 1 | 1 | 7 |

M: Men, W: women and A: Average.

or participating in a new programme may be too risky or unattractive to them [30].

During this research, the REDD+ concept was introduced in the case study community for two consecutive days using PRA exercises in each of the three studied villages. The first day was used for the introduction of the study objectives and REDD+ concept. The researcher explained to participants the role of forests in climate change mitigation through storage and sequestering carbon from the atmosphere. The possibilities of obtaining incentives through a REDD+ project in the future for villagers who are managing forests under PFM like AVLFR were also explained. Participants were also informed about the requirements needed for them to participate in the carbon trading. The first and foremost requirement is to prove that their forest management efforts result in halting deforestation and forest degradation and eventually lead to carbon enhancement or conservation. It was explained that data needed to determine the carbon benefits resulting from their actions could be obtained through PFCA. At this point the researcher requested participants to participate in the participatory forest carbon assessment and monitoring for their VLFR. In 2009, the information about REDD+ was still a totally new and abstract idea amongst the participants in the studied villages because none of them had heard about it before.

After the REDD+ introduction, a number of issues were raised by the participants. The main concern of the participants was when the carbon fund will be available for them if they decide to participate. While some participants wanted to know exactly when the carbon funds will be available, others asked whether the price of carbon per tree is competitive to valuable timber. They were told that the price of carbon is based on tonnage of $CO_2$ equivalent and is expected to be above US$ 5 per $tCO_2$; however, it is based on market price and can vary from time to time. Moreover participants from Mihumo asked if villagers would be able to harvest some valuable timber species although they would also like to participate in REDD+ initiative. They were informed that it is possible through sustainable forest management and harvesting. During the discussion, some people perceived that the REDD+ concept was designed to exclude them from accessing their VLFR, however they were assured that the design of REDD+ policy will guarantee provision of most of the benefits to the local communities.

Participants who attended PRA exercises for the first and second day were the same group of people. In each of the case study villages, participants were divided into two groups of men and women and asked to list and rank the importance of the VLFR through pairwise ranking. After completing pair wise ranking, the score for each item was determined by counting the number of times an item appeared on the table. The scores from the two groups were summed up and the average score obtained was ranked. The total score for all items in each group was 42 as illustrated in Table 4.

Both timber and carbon sequestration ranked the first priority regarding the importance of the VLFR in Mihumo while the low priority was food (ming'oko and angadi). The villagers perceived that they can reduce the harvestable wood volume to enhance carbon stock for carbon trading in future. The reason behind this was that they cannot rely only on carbon trading since they are not sure when the payment will be available. The participants in Mihumo village perceived that their VLFR has abundant valuable timber species that could be harvested for the generation of quick income from the logging when the VLFR management plan will be in place. They also expressed concern that they have been participating in forest management since 1994 without getting any financial benefit from their VLFR [31]. They pointed out that VLFR covers about 40% of their total village land, and forest under general land has been deteriorated heavily due to shifting cultivation, wildfire, and uncontrolled timber harvesting. Therefore they are looking forward to managing and harvesting their VLFR in a sustainable manner.

For Ngongowele and Ngunja, carbon sequestration was ranked high, as was the case for Mihumo, the low priority was wild food (Table 4). Participants in both villages perceived that carbon enhancement in their forests will not only enable them to gain income but will also reduce the impact of climate change in their village environment. They expressed the view that climate change had adverse impacts in their village including insufficient rainfall and prolonged drought which has led to low crop production, increased pests on cashew trees and some important rivers such as Angai and Ruhuhu drying up during the dry season. Also some short and deep water wells are said to be dry in the dry season, a situation which makes it necessary for the villagers to fetch water about five kilometres away. These climate change impacts are also reported elsewhere in the country [32, 33].

Further discussions with the participants in these villages revealed that unreserved forest under village general land in Ngongowele and Ngunja villages occupies about 73% and 45% of the village land, respectively. Both villages also noted that the village population is low because the average forest area per person in two villages is about 15 ha as compared to the average national figure of less than 1 ha of indigenous forest per inhabitant [34]. Villagers said that they meet most of their forest products needs from outside the AVLFR specifically in unreserved forests on general land. It was observed that the forests on general land have valuable resources which ensure the availability of forest products to the villagers. The interviewed villagers are therefore of the opinion that they could protect their already established VLFRs under the REDD+ project and could establish other buffer VLFRs in general land for logging and domestic uses. Results from PRA methods show that the food including angadi and ming'oko was ranked last followed by traditional medicines in the studied villages because these products are abundant in the general land that is near to the villages' centre while the VLFRs are far. The distance from the village centres to the forest boundaries ranges from 5 km to 20 km. Traditional medicine and wild food and fuelwood received high priority only from the women group whereas the rest received high priority from both men and women. However, the enhancement of the water source was included under carbon sequestration because participants perceived that carbon enhancement would ensure the availability of forest services including water.

Moreover, PRA results in the study area revealed that the major interest of both men and women in the group is to create economic benefits from the AVLFR. The villagers noticed the need to improve forest governance and sustainable management of forest resources as one way to achieve benefits from forest resources. Access to both timber and carbon funding or markets would help to diversify livelihoods and motivate people to manage and protect forest resources. Relying on REDD+ or, for example, voluntary carbon markets could create a dependence on carbon funds or markets and reward carbon sequestration over the provision of other forest-related benefits which could also generate income.

Generally, the interest of villagers to participate in the REDD+ is stimulated by the expected incentives and the observed impacts of climate change on their village environment. Also it is likely that REDD+ project could be accepted easily in the villages which are far from town with both low population and abundance forest resources, and these were clearly contributing factors to the local people's decision to participate in an REDD+ initiative.

This result is inline with other studies elsewhere in the country where villagers' willingness to participate in forest management was stimulated by the expected benefit and their perception of negative effect on environment [35, 36]. In the case of REDD+, the actual participation of villagers will depend on the value of carbon payment and immediate flow of carbon funds to the local communities. This is especially the case for local communities like those of AVLFRs who have been managing their valuable forest under PFM for years now without realizing any tangible benefit.

*4.2. Local Community Capability and the Current and Predicted Cost of Carrying out PFCA.* This section provides a summary of the ability of villagers to follow the carbon assessment procedure. It also presents the costs associated with the involvement of the villagers in the PFCA. Results from field observation and discussion with participants (Table 5) show that 90%, 75%, and 85% of the trainees managed to identify different vegetation types in Mihumo, Ngongowele, and Ngunja VLFR, respectively. It was observed that few participants (between 40% and 30%) were able to use GPS to mark vegetation/stratum boundary. During forest stratification, it was found that men, especially adults, are more knowledgeable on different vegetation types in the forest than youth and women. Women have limited knowledge on forest types because they are involved more in household and agricultural activities than men who are involved in hunting and logging, among other production activities. It was also revealed that neither district forest staff nor VFAT members were able to utilize the GPS coordinates for the production of stratified maps using GIS software. The reason behind this is the absence of GIS facilities at the district level. As a result, the maps were produced at the Sokoine University of Agriculture (SUA) GIS laboratory. This result is contrary to similar studies under Kyoto: TGAL elsewhere because in this study GPS was used for delineating stratum boundary instead of a handheld computer [9, 37].

On the pilot survey for determining sample size, results in Table 5 show that a high number of participants (85% in Mihumo, 75% in Ngongowele, and 80% in Ngunja) managed to use the relascope for measuring basal area in their VLFR. On the other hand all of them lacked knowledge on calculating the sample size from the collected data. Therefore data analysis for determining sample size was the responsibility of the researcher as was similarly reported in other studies under Kyoto: TGAL [9, 37, 38].

Prior to locating permanent sample plots on the ground, planning for fieldwork including transects and plot layout on the map was done by the researcher as was in the case of similar studies elsewhere [9, 38]. It was observed that all participants in each village were able to lay down 261 permanent sample plots in the studied VLFRs. However it was found that only 70%, 60%, and 65% of participants in Mihumo, Ngongowele, and Ngunja respectively, managed to take coordinates of the plot centre using GPS. Furthermore it was observed that on average only 40% of all participants were able to navigate through transect line from one plot to another.

Regarding measurements from the permanent plots, results in Table 5 show that 95% of participants in both Mihumo and Ngunja village succeeded in taking tree diameter at breast height (DBH) measurement using calipers as compared to 85% in Ngongowele. The variation in these villages is probably due to the fact that individuals from,

TABLE 5: Percent of participants who were able to do various steps for carbon assessment.

| No. | Steps for carbon assessment | Participants percent | | |
|---|---|---|---|---|
| | | Mihumo | Ngongowele | Ngunja |
| 1.0 | *Stratification* | | | |
| 1.1 | Identifying different vegetation type | 90 | 75 | 85 |
| 1.2 | Marking vegetation type/stratum boundary with GPS | 40 | 30 | 35 |
| 1.3 | GIS process to produce forest map and its strata. | None | None | None |
| 2 | *Piloting for determining sample size* | | | |
| 2.1 | Taking basal area measurements using relascope | 85 | 75 | 80 |
| 2.1 | Calculating sample size from the collected data | None | None | None |
| 3.0 | *Locating permanent sample plots on the map* | | | |
| 3.1 | Determining distance between transects and sample plots. | None | None | None |
| 3.2 | Locating transect line and sample plots on the map | None | None | None |
| 4.0 | *Locating permanent sample plots on the ground* | | | |
| 4.1 | Navigating through transect line with GPS in the forest | 45 | 35 | 40 |
| 4.2 | Marking permanent sample plots on the ground using GPS | 70 | 60 | 65 |
| 4.3 | Establishment of permanent plot on the ground | 100 | 100 | 100 |
| 5.0 | *Measurements taken from permanent sample plots* | | | |
| 5.1 | Measuring tree DBH using caliper | 95 | 85 | 95 |
| 5.2 | Measuring tree height using hypsometer | 38 | 25 | 38 |
| 5.3 | Tree identification in local name | 100 | 100 | 100 |
| 6 | *Calculating carbon stock from the collected data* | None | None | None |

for example, different genders and age groups have different knowledge and understanding. It was revealed that only 38% of the participants in both Mihumo and Ngunja managed to take tree height measurement using hypsometer while 62% of the remaining participants failed to use it. Even in the group discussions with key informants, participants expressed concern that some of them were not able to use GPS and hypsometer during the carbon assessment. This result is contrary to Zahabu [9] who reported that among other procedures villagers were able to use handheld computer and hypsometer. The difference might be attributed to limited training time adopted in the study; in the previous case study villages under Kyoto: TGAL project, the learning-by-doing process was ongoing for four consecutive years while in this study only one training and assessment period has thus far been done. The short research period was therefore not enough for participants to gain sufficient knowledge and experience in using the GPS and hypsometer. For this reason, they have appealed for more training and technical assistance from the district council and development partners to increase their knowledge and become competent on using the mentioned equipment. Besides the training, other assistance that is required includes field uniforms and gumboots, transport, inventory equipment, and a fund for allowances.

Results also show that all of participants in each village were able to identify all tree/shrub species encountered during carbon assessment because they have local knowledge regarding tree species identification. Zahabu [9] reported that local community knowledge is important in identifying trees species and various vegetation types in the forests. Regarding data analysis, the data were analyzed by the researcher, and the carbon assessment results were communicated to villagers through group discussion where village leaders and members of VFAT attended.

Table 6 indicates the main steps for PFCA with their respective costs for each VLFRs. The total cost spent to accomplish PFCA in Mihumo, Ngongowele, and Ngunja VLFRs was Tsh.8, 595,181/=, 7,303,481/= and 7,222,481/= equivalent to US$ 6,612, 5,618 and 5,556, respectively. This is equivalent to the cost per hectare for Mihumo, Ngongowele, and Ngunja of TShs.729/=, 882/=, and 1,090/= equivalent to US$ 0.56, $0.68 and $0.84, respectively. It was observed that although the total cost was increasing with increasing forest area, the trend is vice versa for the cost per hectare.

The cost per hectare in other studies under Kyoto:TGAL research in Tanzania for the first year of carbon assessment ranged between $5 and $53 for large forest (1020 ha) and small forest area (28.5 ha), respectively [9]. In this study, the observed cost per hectare is much lower compared to those reported by Zahabu [9]. Also the cost per hectare is much lower than other studies in India where the first year's work was estimated at $3 per hectare and the cost of the professional team was done at $5.50 per hectare [37]. The

Table 6: Transaction cost for PFCA.

| Components | Mihumo (11,792 ha) No. of days | Mihumo Cost Tsh. "000" | Mihumo Cost US$ | Ngongowele (8,285 ha) No. of days | Ngongowele Cost (Tsh.) Tsh. "000" | Ngongowele Cost US$ | Ngunja (6,626 ha) No. of days | Ngunja Tsh. "000" | Ngunja Cost US$ |
|---|---|---|---|---|---|---|---|---|---|
| Training, stratification, and piloting | 5 | 950 | 731 | 5 | 758 | 583 | 5 | 742 | 571 |
| Inventory equipments | | 1,394 | 1,072 | | 1,394 | 1,072 | | 1,394 | 1,072 |
| Forest carbon assessment | 17 | 5,052 | 3,886 | 10 | 4,252 | 3,271 | 10 | 4,187 | 3,221 |
| Data analysis and reporting | 20 | 1,200 | 923 | 15 | 900 | 692 | 15 | 900 | 692 |
| **Total** | **42** | **8,595** | **6,612** | **30** | **7,303** | **5,618** | **30** | **7,222** | **5,556** |
| **Cost/ha** | | **0.72** | **0.56** | | **0.88** | **0.68** | | **1.09** | **0.8** |

plausible explanation for this variation in cost per hectare is that the studied VLFRs are too large (>6000 ha) compared to forest area of the previous studies (<600 ha).

In all three villages, participants complained about the payments they received as compensation for their participation in the PFCA. Concerning the budget, villagers need to be involved in budgeting process for carbon monitoring in future. They also want that the grant for PFCA be disbursed to village government account as this will increase the level of people's participation; community members should participate in decision making, implementation, benefit sharing, and evaluation in order to own the process of development. For example, in order to learn from the process participants need to know and experience the costs of PFCA; otherwise they will not be able to weigh them against the potential benefits. Participants were only discontent with Tshs. 5,000/= per day for each person as incentive for their participation in forest carbon inventory in their village forests.

Forest carbon assessment is difficult and dangerous because the teams need to walk for a long time (at least 15 km per day) in the forest with a risk of being attacked by wild animals like elephants and lions. Also, when required to camp in the forests, the costs are higher compared to their earnings per day from the carbon assessment. Most participants are heads of families with farming and daily bread-earning responsibilities so their participation in carbon assessment prevents them from farming and other income generating activities compared to their fellow villagers. Based on the above reasons and their experience on carbon assessment, participants from all villages proposed to be paid Tsh. 15,000/= per day as an incentive for their participation in carbon assessment in the future. However, Zahabu [9] reported that the trainees were paid $5 which is equivalent to Tsh.6,550/= per day per trainee and this was appreciated in that study group. The difference is apparently because the payments in the previous studies were based on the government rate for hiring local labourers in the village, while the villagers' proposal in this study is based on opportunity costs and risks. Further, this study involved camping in the forests, a practice that was not done in the previous studies.

Table 7 shows that total cost for the PFCA is predicted to increase to 22%, 19%, and 12% for Mihumo, Ngongowele, and Ngunja VLFR, respectively, due to an increase in participants allowances from Tsh. 5000/= to 15,000/= and the need for more training. However this increase will not add much to the per hectare costs since these are much lower compared to other studies. The main challenge to this proposed rate is the availability of funds when the village governments will need to do carbon monitoring in the future. According to the village leaders, the village government lacks its own source of income that could be used for paying villagers to conduct carbon assessment.

*4.3. Forest Carbon Stock in Three Village Forest Reserves.* A total of 134 tree species were identified in the three studied village forest reserves including 93, 86, and 72 tree species for Mihumo, Ngongowele, and Ngunja village forest reserves. The number of tree species identified in this study is consistent with previous studies at AVLFR [10] and elsewhere in Tanzania [39–43]. Forest stand parameters such as number of stems, basal area, volume, biomass, and carbon per hectare in each stratum for the studied VLFRs are summarized in Table 8. The number of stems per hectare across the studied VLFRs in different strata ranged between 639 ± 215 and 903 ± 136 with the exception of the closed dense forest and encroached river basin (Table 8).

These results are comparable with other studies in miombo woodland and elsewhere in the country as shown in Table 10. However, the observed stems per hectare among the mentioned strata are lower than that reported in other studies in Duru-Haitemba, Kitulangaro GFR, and Kitulangalo Sokoine University of Agriculture Training Forest Reserve (KSUATFR) [9, 22, 24]. This could be attributed to the frequent occurrence of intense forest fires in the studied VLFR which affects the growth of young trees so the forests are moderately open. Moreover during the study, forest fires

Table 7: Predicted cost for PFCA.

| Components | Mihumo (11,792 ha) | | | Ngongowele (8,285 ha) | | | Ngunja (6,626 ha) | | |
|---|---|---|---|---|---|---|---|---|---|
| | No. of days | Cost Tsh. "000" | US$ | No. of days | Cost Tsh. "000" | US$ | No. of days | Cost Tsh. "000" | US$ |
| Training on uses of inventory equipments | 5 | 950 | 731 | 5 | 758 | 583 | 5 | 742 | 571 |
| Inventory equipments | | 1,394 | 1,072 | | 1,394 | 1,072 | | 1,394 | 1,072 |
| Forest carbon assessment | 17 | 6,959 | 5,353 | 13 | 5,651 | 4,347 | 10 | 5,070 | 3,900 |
| Data analysis and reporting | 20 | 1,200 | 923 | 15 | 900 | 692 | 15 | 900 | 692 |
| **Total** | **42** | **10,503** | **8,079** | **33** | **8,703** | **6,694** | **30** | **8,106** | **6,235** |
| **Cost/ha** | | **0.89** | **0.69** | | **1.05** | **0.81** | | **1.22** | **0.9** |

Village forest name spans all columns.

Table 8: Stand parameters for three studied VLFR.

| Forest name | Vegetation type | $N$ | $G$ (M$^2$/ha) | $V$ (M$^3$/ha) | Biomass (t/ha) | Carbon (t/ha) |
|---|---|---|---|---|---|---|
| Mihumo | Dry miombo | 870 ± 119 (14) | 9.8 ± 0.78 (7.99) | 68.95 ± 10.20 (14.79) | 34.48 ± 5.1 (14.79) | 17.24 ± 2.55 (14.79) |
| | Wet miombo | 639 ± 215 (34) | 9.21 ± 2.37 (25.77) | 67 ± 30.83 (46.02) | 33.5 ± 15.42 (46.02) | 16.75 ± 7.71 (46.02) |
| | Closed forest | 2824 ± 237 (8) | 29.23 ± 4.83 (16.5) | 339.59 ± 67.86 (19.98) | 169.79 ± 33.71 (19.98) | 84.89 ± 16.85 (19.98) |
| Ngongowele | Dry miombo | 731 ± 138 (19) | 11.37 ± 0.98 (8.58) | 77.46 ± 10.02 (12.93) | 38.73 ± 5.01 (12.93) | 19.36 ± 2.51 (12.93) |
| | Closed forest | 3305 ± 1402 (42) | 16.17 ± 5.24 (32.39) | 166.94 ± 88.69 (53.13) | 83.47 ± 44.35 (53.13) | 41.73 ± 22.17 (53.13) |
| | Encroached river basin | 253 ± 210 (83) | 5.85 ± 1.73 (29.59) | 28.82 ± 9.29 (32.25) | 14.41 ± 4.65 (32.25) | 7.2 ± 2.32 (32.25) |
| Ngunja | Lowland dry miombo | 903 ± 136 (15) | 9.82 ± 0.92 (9.42) | 73.49 ± 9.99 (13.59) | 36.75 ± 5 (13.59) | 18.37 ± 2.2.5 (13.59) |
| | Upland dry miombo | 795 ± 155 (19) | 9.98 ± 0.96 (9.67) | 72.05 ± 11.63 (16.13) | 36.03 ± 5.81 (16.13) | 18.02 ± 2.9 (16.13) |

The figures in brackets indicate precision level of estimates, that is, confidence intervals as percentage of mean value.

were observed in most strata. The number of stems per hectare in closed dense forest is 2824 and 3305 in Ngongowele and Mihumo VLFRs, respectively. The observed number of stems per hectare in these strata is not only relatively higher than that observed among strata in the studied VLFRs but also with other studies elsewhere in Tanzania (Table 9). The plausible reason for this would be attributed by absence of forest disturbance and the nature of the forests themselves.

The encroached river basin not only has a low stem density compared to the other strata in the three VLFRs but also compared with results reported in miombo elsewhere [9, 22, 24]. The plausible explanation for this is that farmers from Ngongowele have been clearing trees along Angai river basin (Nandete) for agriculture purposes.

Table 10 shows that on average the carbon stock in wet and dry miombo strata including lowland and upland in Ngunja VLFR ranged from 16.75 tC/ha to 19.36 tC/ha. There are numbers of reasons for low carbon stock in these strata including previous timber harvesting before it ceased in 2005 when the 13 Angai villages got legal land rights for the AVLFR, frequent high intensity fires, and the nature of the forests themselves. The closed forests strata were found to have high value of carbon stock, (84.89 tC/ha and 41.73 tC/ha in Mihumo and Ngongowele VLFR, resp.). The reason behind this high value is the presence of old growth *Brachystegia microphylla* trees and a lack of any human disturbance. The least amount of carbon stock found in the study was 7.2 tC/ha in the encroached river basin due to clearance of forest for rice farming.

Results in Table 10 also shows that the observed storage of carbon stock in Mihumo, Ngongowele, and Ngunja VLFR was 332.81–266.4 MgC, 163.42–139.01 MgC, and 120.23–102.56 MgC, respectively. The Mihumo VLFR stores considerably more carbon than the other two reserves. Although

Table 9: Stand parameters from various studies in miombo woodland.

| Author | Forest Name | $N$ | $G$ ($M^2$/ha) | $V$ ($M^3$/ha) | Biomass (t/ha) | Carbon (t/ha) |
|---|---|---|---|---|---|---|
| Zahabu (2008) [9] | KSUATFR | 628–694 | 7.9–9.9 | 55.3–74.8 | 35.2–45.9 | 17.6–22.9 |
| Zahabu (2008) [9] | Kimunyu | 701–845 | 7.9–8.8 | 72.4–88.2 | 39.7–45 | 19.86–22.5 |
| Njana (2008) [21] | Urumwa FR | 583 | 8.54 | 58.41 | — | — |
| Malimbwi (2003) [22] | Duru Haitemba | 1988 | 12.41 | 97.32 | — | — |
| Zahabu (2001) [23] | KSUATFR | 619 | 10.2 | 78 | — | — |
| Chamshama et al. (2004) [24] | Kitulangalo GFR | 1085 | 9 | 76 | 43.56 | |
| Malimbwi and Mugasha (2002) [25] | Handen Hill | 355 | 11.2 | 108.99 | — | — |
| Chamshama et al. (2004) [24] | KSUATFR | 1027 | 8.95 | 76.02 | 41.40 | |
| Nuru et al. (2009) [26] | Urumwa FR | 642 | 8.7 | 59.73 | | |

Table 10: Total carbon stock in each VLFR.

| Forest name | Stratum | Area (ha) | Carbon stock (tC/ha) | Total carbon stock (Mega tonnes of Carbon) | |
|---|---|---|---|---|---|
| | | | | Upper limit | Lower limit |
| Ngunja | Lowland dry miombo | 2379 | 18.37 ± 2.25 | 43.7 | 38.35 |
| | Upland dry miombo | 4247 | 18.02 ± 2.9 | 76.53 | 64.21 |
| **Total** | | | | **120.23** | **102.56** |
| Ngongowele | Dry miombo | 8021 | 19.36 ± 2.51 | 155.28 | 135.15 |
| | Closed forest | 181 | 41.73 ± 22.17 | 7.55 | 3.54 |
| | Encroached river basin | 83 | 7.2 ± 2.32 | 0.59 | 0.4 |
| **Total** | | | | **163.42** | **139.01** |
| Mihumo | Dry miombo | 8169 | 17.24 ± 2.55 | 140.83 | 120 |
| | Wet miombo | 1695 | 16.75 ± 7.71 | 28.40 | 15.32 |
| | Closed forest | 1927 | 84.89 ± 16.85 | 163.58 | 131.11 |
| **Total** | | | | **332.81** | **266.43** |

the VLFRs studied here have low carbon stock per hectare, the total amount of carbon they store is large because of their large areas (Table 10). It is clear that if the forest would be managed properly, the amount of carbon sequestration would be sufficient for carbon trading because of the total area under protection.

## 5. Conclusions

The experience of this study shows that the willingness of the villagers to participate in REDD+ initiative is motivated by the expected income they would like to receive from carbon enhancement activities and also by their perception of the negative effects that climate change might have on their environment. The study also revealed that in the long run, the actual commitment and participation of villagers in an REDD+ project will depend on a timely flow of carbon payments to the participating communities at the project level.

The present study demonstrates that PFCA is a useful technique at the project level because it ensures that the local communities' knowledge and experience can deliver carbon data at low cost through their participation. However, local communities' ability to carry out carbon assessments independently will require not less than two training periods for two consecutive years on PFCA techniques to allow them to properly understand the use of GIS and inventory equipment incorporated in the methodology. The experience gained in this study reveals that not all participants were able to neither develop a VLFR map from GPS coordinates nor analyze the collected data from the carbon inventory. Therefore the successful and sustainable implementation of PFCA at project level will depend on collaboration between local communities and a facilitating organization (NGO/District Council) for GIS and carbon data analysis. The technique tested here can be applied throughout PFM areas to generate carbon data, however, the training and methods need to be based on the local context and knowledge and conditions in local communities.

Findings from this study show that generally the case study VLFRs store carbon stock per hectare similarly to other miombo forests in Tanzania. On average the carbon stock in wet and dry miombo strata including lowland and upland in Ngunja VLFR ranged from 16.75 tC/ha to 19.36 tC/ha. Although the studied VLFRs have low carbon stock per hectare, they store a large amount of total carbon stock (ranging from 626.46 MgC to 508 MgC for upper and low limit, resp.) due to their vast area (26 703 ha). This ensures significant carbon stock and sequestration for carbon in future. Moreover medium tree diameters accounted for 60% of the average total carbon stock in all strata except for closed forest and encroached river basin. The possible reason for this distribution would be past harvesting of the large diameter trees in the past for timber businesses and the nature of the forests themselves. The trees of large diameters are responsible for 90% and 70% of the average total carbon stock in closed forest strata in Mihumo and Ngongowele respectively.

Under an REDD+ project, sustainable harvesting could be encouraged. However, proper management of carbon stock can be achieved only if the timber harvesting level is determined through the forest carbon assessment in permanent sample plots. This study demonstrates that timber harvesting could be conducted in all strata except for closed forest, wet miombo, and degraded river basin including a strip of twenty meters from the river bank of Angai, Nakawale, and Nangula. The reason is the unavailability of valuable timber species in these strata and water source protection for animal and human uses. The assessment demonstrated in this study will enable villagers to understand their forest growth potential and establish sustainable timber harvesting levels and at the same time allow carbon stock enhancement. For carbon management purposes in the case study, there is a need to conduct carbon monitoring consecutively before establishing timber harvesting level. Moreover local communities should harvest old grown trees and promote long-term regeneration of high carbon species.

It is apparent that the carbon stock stored in AVLFR faces threats from illegal timber harvesting, wildfires, and conversion of forest to agricultural use. If the observed carbon stocks are to be conserved and enhanced for the carbon payments, it is important that villagers be supported to take action to stop the prevailing disturbance inside AVLFR. A forest fire management programme is necessary because vast areas of AVLFR are torched every year during late dry season. In fact, management of these fires is an attractive option for enhancing carbon stock and reducing trace gas emission. Village governments and VNRCs should make a concerted effort to stop the prevailing forest disturbances in the AVLFR not only by increasing the number of patrols but also by training and educating community members on issues related to early burning and forest fire management. Moreover village governments could provide incentives/payments and food to patrolling teams.

There is a need for national level among the various development partners to recognize PFCA as an appropriate technique for carbon assessment in PFM forests and mainstream the technique into the PFM process to ensure sustainable availability of inventory data for both carbon credits and forest management planning. Moreover at national level, inventories and MRV systems should be reviewed to capture carbon data from REDD+ projects. A key challenge in future is how to finance PFCA in various REDD+ project areas and ensure the technical capacity at local level. Therefore central governments as well as development partners are requested to provide support to local communities to ensure sustainable management of carbon stock areas such as AVLFR. Both technical and financial support should continue in the following years until communities can independently carry out the PFCA and finance it for example through the REDD+ payments.

## Acknowledgments

This paper was presented at a conference called Reframing sustainability? Climate Change and North-South Dynamics, 10th-11th February, 2011, Helsinki, Finland. The authors want to thank Academy of Finland for funding this study. Grateful acknowledgement to the Liwale District Council and District Executive Director (DED), Mr. Moya, for permission to carry out this study. they also appreciate the assistance of Mr. Kilowoko, Mr. Mbujiro, Mr. Mzui, Mr. Namwewe, and Mr. Amanzi. they would like to express their appreciation for the cooperation and endurance shown by villagers who participated in carbon assessment in Mihumo, Ngongowele, and Ngunja and who worked tirelessly in harsh conditions and a threatening environment.

## References

[1] F. Danielsen, M. Skutsch, N. D. Burgess et al., "At the heart of REDD+: a role for local people in monitoring forests?" *Conservation Letters*, vol. 4, no. 2, pp. 158–167, 2011.

[2] United Republic of Tanzania (URT), "Draft: preparing for the REDD initiative in Tanzania. A synthesized consultative report for REDD task force in Tanzania," 2009, http://www.reddtz.org/.

[3] United Republic of Tanzania (URT), "Final draft: forest carbon partnership facility (FCPF), readiness preparation proposal (R-PP)," 2010, http://www.reddtz.org/.

[4] United Republic of Tanzania (URT), *The Forest Act no. 14*, Ministry of Natural Resources and Tourism. Government Printer, Dar es Salaam, Tanzania, 2002.

[5] L. A. Wily, *Villagers as Forest Managers and Government: "Learning to Let Go". The Case of Duru-Haitemba and Mgori Forests in Tanzania*, Forest participation Series 9, International Institute for Environment and Development (IIED), London, UK, 1997.

[6] United Republic of Tanzania (URT), *Participatory Forest Management in Tanzania. Facts and Figures*, Ministry of Natural Resources and Tourism, Forestry and Beekeeping Division, Dar es Salaam, Tanzania, 2008.

[7] R. E. Malimbwi and E. Zahabu, "REDD Experience in Tanzania," in *REDD, Forest Governance and Rural Livelihoods: The Emerging Agenda*, O. Springate-Baginski and E. Wollenberg, Eds., pp. 109–134, CIFOR, Bogor, Indonesia, 2010.

[8] I. Mustalahti and J. F. Lund, "Where and how can participatory forest management succeed? Learning from Tanzania, Mozambique, and Laos," *Society and Natural Resources*, vol. 23, no. 1, pp. 31–44, 2010.

[9] E. Zahabu, *Sinks and sources: a strategy to involve forest communities in Tanzania in global climate policy*, Ph.D. thesis, University of Twente, Enschede, The Netherlands, 2008.

[10] S. Dondeyne, A. Wijffels, L. B. Emmanuel, J. Deckers, and M. Hermy, "Soils and vegetation of Angai forest: ecological insights from a participatory survey in South Eastern Tanzania," *African Journal of Ecology*, vol. 42, no. 3, pp. 198–207, 2004.
[11] R. Chambers, "Beyond whose reality counts? New methods we now need. Studies in culture," *Journal of Organisations and Societies*, vol. 4, no. 2, pp. 279–301, 1998.
[12] H. Bradbury and P. Reason, "Action research: an opportunity for revitalising research purpose and practices," *Journal of Qualitative Social Work*, vol. 2, no. 2, pp. 156–175, 2003.
[13] J. A. Ericson, "A participatory approach to conservation in the Calakmul Biosphere Reserve, Campeche, Mexico," *Landscape and Urban Planning*, vol. 74, no. 3-4, pp. 242–266, 2006.
[14] G. C. Kajembe, *Indigenous Management Systems as a Basis for Community Forestry in Tanzania: A Case Study of Dodoma Urban and Lushoto Districts*, Tropical Resource Management Paper no. 6, Wageningen Agricultural University, Wageningen, The Netherlands, 1994.
[15] I. Mustalahti, "Sustaining participatory forest management: case study analyses of forestry assistance from Tanzania, Mozambique, Laos and Vietnam," *Small-Scale Forestry*, vol. 8, no. 1, pp. 109–129, 2009.
[16] R. Krueger and M. A. Casey, *Focus Groups: A Practical Guide for Applied Research*, Sage Publication, Thousand oaks, Calif, USA, 3rd edition, 2000.
[17] IPCC, *Good Practice Guidance for Land Use, Land-Use Changes and Forestry*, Institute of Global Environmental Strategies, Kanagawa, Japan, 2003.
[18] S. Di Gessa, "Participatory mapping as a tool for empowerment: experience and Lessons learned from the ILC network," 2008, http://www.landcoalition.org/pdf/08_ILC_Participatory_Mapping_Low.pdf.
[19] R. Kumar, *Research Methodology: A Step–by Step Guide for Beginners*, Sage Publication, London, UK, 2nd edition, 2005.
[20] D. Narayan, "What is participatory research?" in *Toward Participatory Research*, pp. 17–30, World Bank, Washington, DC, USA, 1996.
[21] M. A. Njana, *Arborescent species diversity and stocking in the Miombo woodland of Urumwa Forest Reserve and their contribution to livelihoods, Tabora, Tanzania*, M.S. thesis, Sokoine University of Agriculture, Morogoro, Tanzania, 2008.
[22] R. E. Malimbwi, *Inventory reports of Ayasanda, Bubu, Duru, Endagwe, Gidas Endanachan, Hoshan and Riroda Village Forest Reserves in Babati Manyara, Tanzania*, Land Management Programme, Babati District Council, Manyara, Tanzania, 2003.
[23] E. Zahabu, *Impact of charcoal extraction on the Miombo Woodlands: The case of Kitulangalo Area, Tanzania*, M.S. thesis, Sokoine University of Agriculture, Morogoro, Tanzania, 2001.
[24] S. A. O. Chamshama, A. G. Mugasha, and E. Zahabu, "Stand biomass and volume estimation for Miombo woodlands at Kitulangalo, Morogoro, Tanzania," *Southern African Forestry Journal*, no. 200, pp. 49–60, 2004.
[25] R. E. Malimbwi and A. G. Mugasha, *Reconnaissance Timber Inventory Report for Handeni Hill Forest Reserve in Handeni District, Tanzania for the Tanga Catchment Forest Project*, FORCONSULT, Faculty of Forestry and Nature Conservation, Sokoine University of Agriculture, Morogoro, Tanzania, 2002.
[26] H. Nuru, C. D. K. Rubanza, and C. B. Nezia, "Governance of key players at district and village levels on health improvement of Urumwa Forest reserve, Tabora: ten years of Joint Forest Management," in *Proceedings of the 1st Participatory Forest Management Research Workshop: Participatory Forest Management for Improved Forest Quality, Livelihood and Governance (PFM '09)*, pp. 111–122, 2009.
[27] A. De Gier, "Woody biomass assessment in woodland and shrub-lands," in *Off-Forest Tree Resources of Africa*, Proceedings of a Workshop, pp. 89–98, Arusha, Tanzania, 1999.
[28] S. Brown, "Measuring, monitoring and verification of carbon benefits for forest based projects," in *Capturing Carbon and Conserving Biodiversity: The Market Approach*, I. R. Swingland, Ed., pp. 118–133, Earthscan Publications, London, UK, 2003.
[29] K. MacDicken, "A guide to monitoring carbon storage in forestry and agroforestry projects," Tech. Rep. 1, Winrock International Institute for Agricultural Development, Arlington, Tex, USA, 1997.
[30] A. Martin, "Lessons for REDD from PES research," in *REDD, Forest Governance and Rural Livelihoods: The Emerging Agenda*, O. Springate-Baginski and E. Wollenberg, Eds., pp. 36–39, CIFOR, Bogor, Indonesia, 2010.
[31] I. Mustalahti, "Msitu wa Angai: haraka, haraka, haina baraka! Why does handing over the Angai forest to local villages proceed so slowly?" in *Anomalies of Aid. A Festschrift for Juhani Koponen*, J. Gould and L. Siitonen, Eds., vol. 15, pp. 168–186, Institute of Development Studies, University of Helsinki, Helsinki, Finland, 2007.
[32] United Republic of Tanzania (URT), *National Adaptation Programme of Action (NAPA)*, Vice President's Office, Division of Environment, 2007.
[33] FAO, *Global Forest Resources Assessment 2005*, Food and Agriculture Organization of the United Nations, Rome, Italy, 2006.
[34] B. K. Kaale and A. L. Simula, *Contribution of Participatory Forest Management to Poverty Eradication in Liwale District, Lindi Region*, Ministry of Natural Resources and Tourism. National Forest Programme Coordination Unit Support Programme, Dar es Salaam, Tanzania, 2004.
[35] T. Veltheim and M. Kijazi, "Lessons learned on participatory forest management," Tech. Rep. 61, East Usambara Conservation Area Management Programme, 2002.
[36] T. Blomley and H. Ramadhani, "Lessons learned from Participatory Forest Management in Tanzania (PFM)," *SLSA Newsletter*, no. 17, 2005.
[37] D. Murdiyarso and M. Skutsch, *Community Forest Management as a Carbon Mitigation Option: Case Studies*, CIFOR, Bogor, Indonesia, 2006.
[38] M. Skutsch, B. Karky, E. Zahabu, M. McCall, and G. Peters-Guarin, *Community Measurement of Carbon Stock Change for REDD*, Special study on forest degradation. Working paper 156, FAO, Rome, Italy, 2009.
[39] R. E. Malimbwi, J. Kielland-Lund, J. Nduwamung, and A. O. A. Chamshama, "Species diversity and standing crop development in four miombo vegetation communities," in *Proceedings of the First Annual Forest Research Workshop*, vol. 67, pp. 201–212, Falcuty of Forest, 1998.
[40] E. J. Luoga, *The effect of human disturbances on diversity and dynamics of Eastern Tanzania Miom¬bo arborescent species*, Ph.D. thesis, Faculty of Science, University of the Witwatersrand, Johannesburg, South Africa, 2000.
[41] J. M. Abdallah, *Assessment of the impact of non-timber forest products utilization on sustainable management of Miombo woodlands in Urumwa Forest Reserve, Tabora, Tanzania*, M.S. thesis, Sokoine University of Agriculture, Morogoro, Tanzania, 2001.
[42] I. Backéus, B. Pettersson, L. Strömquist, and C. Ruffo, "Tree communities and structural dynamics in miombo (Brachystegia-Julbernardia) woodland, Tanzania," *Forest Ecology and Management*, vol. 230, no. 1–3, pp. 171–178, 2006.

[43] J. Isango, M. Varmola, S. Valkonen, and S. Tapaninen, "Stand structure and tree species composition of Tanzania miombo woodlands: a case study from miombo woodlands of community-based forest management in Iringa District," in *Proceedings of the 1st MITIMIOMBO Project Workshop*, pp. 43–56, Finnish Forest Research Institute, Morogoro, Tanzania, 2007.

# 4

# Forest Succession and Maternity Day Roost Selection by *Myotis septentrionalis* in a Mesophytic Hardwood Forest

**Alexander Silvis,[1] W. Mark Ford,[1,2] Eric R. Britzke,[3] Nathan R. Beane,[3] and Joshua B. Johnson[4]**

[1] *Department of Fish and Wildlife Conservation, Virginia Polytechnic Institute and State University, Blacksburg, VA 24061, USA*
[2] *US Geological Survey, Virginia Cooperative Fish and Wildlife Research Unit, Blacksburg, VA 24061, USA*
[3] *Environmental Laboratory, US Army Engineer Research and Development Center, 3909 Halls Ferry Road, Vicksburg, MS 39180, USA*
[4] *Pennsylvania Game Commission, 2001 Elmerton Avenue, Harrisburg, PA 17110, USA*

Correspondence should be addressed to Alexander Silvis, silvis@vt.edu

Academic Editor: Brian C. McCarthy

Conservation of summer maternity roosts is considered critical for bat management in North America, yet many aspects of the physical and environmental factors that drive roost selection are poorly understood. We tracked 58 female northern bats (*Myotis septentrionalis*) to 105 roost trees of 21 species on the Fort Knox military reservation in north-central Kentucky during the summer of 2011. Sassafras (*Sassafras albidum*) was used as a day roost more than expected based on forest stand-level availability and accounted for 48.6% of all observed day roosts. Using logistic regression and an information theoretic approach, we were unable to reliably differentiate between sassafras and other roost species or between day roosts used during different maternity periods using models representative of individual tree metrics, site metrics, topographic location, or combinations of these factors. For northern bats, we suggest that day-roost selection is not a function of differences between individual tree species *per se*, but rather of forest successional patterns, stand and tree structure. Present successional trajectories may not provide this particular selected structure again without management intervention, thereby suggesting that resource managers take a relatively long retrospective view to manage current and future forest conditions for bats.

## 1. Introduction

Prior to the onset of white-nose syndrome (WNS) in North America [1], northern bats (*Myotis septentrionalis*) were common in most forest types in the eastern United States and southern Canada [2–9]. Northern bat foraging activity consistently has been greatest in closed-canopy forests [5, 10–13] and maternity roosts and roost areas of live trees and/or snags are typically located in upland forests [4, 7, 11, 14, 15].

Management for Myotine bats in North America often is based on the conservation of summer maternity roosts and winter hibernacula [16, 17]. In particular, summer maternity roosts are widely assumed to be critical, and possibly limiting, environmental features for bats roosting in forested or formerly forested landscapes [16, 18]. Although northern bats have been the focus of several recent studies [9, 19, 20], patterns of maternity roost selection at the forest stand and landscape scale are not well understood. There is considerable variation in forest conditions and roost tree species preference and differences (or lack thereof) between roosts and other available trees across the northern bat's distribution [4, 5, 21]. Nonetheless, consensus is that northern bat maternity colonies typically use snags or decaying live trees with cavities or loose/exfoliating bark and that management efforts should focus on ensuring that suitable roosts are maintained on the landscape long term [4, 7, 11].

Although conservation of individual roosts may provide local or individual benefits, bat habitat conservation at this scale is unfeasible in the context of other forest stand and landscape management objectives. However, recent research suggests that nonrandom assorting social groups should be the focal point of roost conservation and that there is a need to better quantify the multiscale habitat features necessary to preserve and maintain maternity colonies [22]. In the central Appalachians, Johnson et al. [9] identified such nonrandom assorting social groups in female northern bats and further illustrated how these groups form scale-free networks of dayroosts on the landscape. Within this context, roost networks are the units of biological relevance that may allow management of forest bats to occur at the appropriate scales within the framework of conventional forestry [9, 22]. Relating forest processes such as disturbance and establishment to creation of suitable roost structure and conditions over areas relevant to social networks therefore would be useful in developing continually adaptive landscape management plans that consider bats among other resource objectives. We collected day roost data as part of a larger long-term study on northern bat social ecology on the Fort Knox military reservation in north-central Kentucky, USA; our overall objective was to document patterns in northern bat summer maternity roost selection at Fort Knox. Because we documented strong selection of a single species as a day roost that was not historically prevalent in presettlement closed canopy forests, we attempt to describe this pattern in the larger context of forest establishment and disturbance processes related to land-use history using our data and examples from previous northern bat research.

## 2. Methods

*2.1. Study Site Description.* We conducted our northern bat day roost study on the Fort Knox military reservation in Meade, Bullitt, and Hardin counties, Kentucky, USA. Fort Knox lies within the Western Pennyroyal subregion of the Mississippian portion of the Interior Low Plateau physiographic province [23]. Topography in the region consists of dissected rolling plateaus, narrow valleys, and entrenched streams. Much of the area is underlain by karst formations and winter cave hibernacula are abundant. Elevations range from 116 m above sea level (asl) along the Ohio River to 323 m asl, though most uplands generally are between 180 and 275 m asl [23]. Forest cover is predominantly a western mixed-mesophytic association [24], with second- and third-growth forests dominated by white oak (*Quercus alba*), black oak (*Q. velutina*), chinkapin oak (*Q. muehlenbergii*), shagbark hickory (*Carya ovata*), yellow poplar (*Liriodendron tulipifera*), white ash (*Fraxinus americana*), and American beech (*Fagus grandifolia*) in the overstory, and sassafras (*Sassafras albidum*), redbud (*Cercis canadensis*) and sugar maple (*Acer saccharum*) in the understory [25]. Climate at Fort Knox is warm temperate with an average temperature between 18 and 25°C during the growing season and an average annual precipitation of 113 cm.

*2.2. Data Collection.* We captured northern bats using mist nets erected across closed forest corridors, streams, standing water, or in close proximity to previously documented trees currently being used as maternity roosts. For all northern bats netted, we determined age (by degree of epiphyseal-diaphyseal fusion [26]), mass, forearm length, sex, and reproductive condition [27] and placed uniquely numbered lipped aluminum bands on the forearms. We also attached LB-2 radio transmitters (0.46–0.54 g: Holohil Systems Ltd. Woodlawn, ON, Canada) between the scapulae of female northern bats using Perma-Type Surgical Cement (Perma-Type Company Inc., Plainville, CT, USA). We released tagged bats near net sites within a few minutes of capture. We followed the guidelines of Virginia Polytechnic Institute and State University Institutional Animal Care and Use Committee permit 11-040-FIW.

Using TRX-1000S receivers and folding three-element Yagi antennas (Wildlife Materials Inc., Carbondale, IL, USA), we located northern bat day roosts every day for the life of the transmitter or until the unit dropped from the bat. We georeferenced all day roosts located using a Garmin GPSmap 60CSx global positioning system (Olathe, KS, USA). At each female northern bat day roost, we recorded roost species, diameter at breast height (dbh), height, crown class ([28]; i.e., 1 = suppressed, 2 = intermediate, 3 = codominant, 4 = dominant), and decay class ([29]; 1 = live, 2 = declining, 3 = recent dead, 4 = loose bark, 5 = no bark, 6 = broken top, 7 = broken bole) and visually estimated percent remaining bark. Additionally, we measured the nearest four trees using the point-quarter system [30]; for each of these trees we determined species and measured distance to roost, dbh, decay class, and crown class. To assess canopy cover and canopy gap characteristics of day roosts we measured gap fraction, leaf area index (LAI), and total below canopy photosynthetically active photon flux density (PPFD) using WinSCANOPY and XLScanopy softwares (Régent Instruments Inc., Canada). Forest canopy photos were collected using a Nikon Coolpix 8400 camera and FC-E9 fisheye lens (Melville, NY, USA).

To calculate percent slope, elevation, and aspect at each day roost, we input geo-referenced locations into ArcMap 9.3 (ESRI Inc., Redlands, CA, USA) and calculated position metrics with the Spatial Analyst extension. To compare distribution of used day roost species versus availability of other potential roost species in the forest stands containing day roosts, we recorded potential roosts (i.e., boles with loose/exfoliating bark, visible cavities, or other defects) by species along randomly directed $20 \times 100$ m belt transects from the periphery of each point-quarter plot [15].

*2.3. Data Analysis.* We used two-sample Wilcoxon tests for simple comparisons of individual day roost metrics between day roosts and nearest neighbor trees as well as for comparisons of live and snag day roosts. We used chi-square goodness-of-fit tests to determine if day roosts were equitably distributed by live/dead status and by species. Although not commonly addressed in habitat preference studies of bats, the chi-square goodness-of-fit test assumes independence of relocations among individuals and temporal independence of relocations of individuals [31, 32]. By pooling use data across individuals and counting each tree as a single location rather than each use, we assumed independence of individual

TABLE 1: Candidate model sets used to compare female *Myotis septentrionalis* day-roost species selection and day-roost selection by maternity status on the Fort Knox military reservation in Hardin, Bullitt, and Meade counties, Kentucky, USA, 2011.

| Model | Parameters |
|---|---|
| 1 | dbh + height + Decay + percent bark + percent slope + sin (aspect) + cos (aspect) + elevation + gap fraction + LAI + PPFD |
| 2 | dbh + height + Decay + percent bark |
| 3 | dbh + height |
| 4 | Decay + percent bark |
| 5 | Decay + percent bark + gap fraction + LAI + PPFD |
| 6 | Gap fraction + LAI + PPFD |
| 7 | Percent + sin (aspect) + cos (aspect) + elevation + gap fraction + LAI + PPFD |
| 8 | Percent slope + sin (aspect) + cos (aspect) + elevation |
| 9 | Null model |

trees. We used an information theoretic approach (IT) to compare day roosts used by bats during pregnancy to those used by bats during lactation and nonlactation by examining a set of 9 candidate models representing tree-specific characteristics, topological characteristics, micro-site characteristics, combinations of those characteristics, and a null (Table 1). Although we measured canopy position, we removed it from all models due to unacceptably high-standard errors. We followed Garroway and Broders [33] in separating day roost maternity use status (lactation or non-lactation) by date of capture of the first lactating female and volant juvenile. We used the same IT approach and candidate model set to compare the most selected day roost species to the collective of all other species. We ranked models using Akaike's criteria (AIC), the difference between the model with the lowest AIC and the AIC of the *i*th model ($\Delta_i$) and Akaike's weights ($w_i$) [34]. We assessed significance of individual parameters within the best supported model using Wald's $X^2$ test and overall fit of the model using the log-likelihood ratio test against a null model, area under the receiver operating characteristic (AUC), and percent correct classification. To avoid problems of circularity when analyzing aspect, we used sine and cosine transformations of aspect for logistic analysis. All tests were performed using the R statistical program (version 2.14) [35] with significance for all tests accepted at $\alpha \leq 0.05$.

## 3. Results

We mist-netted on 33 nights between 24 May and 17 July 2011 and captured fifty-eight adult females, two juvenile females and 16 adult male northern bats. Eighteen of the adult females were pregnant, 29 were lactating, 3 had ceased lactation, and 8 were nonreproductive. Transmitters were attached to all 58 adult female northern bats; mean transmitter retention time was 4.1 days. Overall, we tracked these bats to 105 day roosts that comprised 21 tree species (Table 2) for a total of 270 relocation events. Number of tagged bats within a single day roost ranged from 1 to 15, with a mean of $1.55 \pm 0.16$ bats/tree. The number of uses of an individual day roost by tagged bats ranged from 1 to 84, with a mean of $2.55 \pm 0.79$, where use is defined as the sum of presences of all bats using that day roost.

Spatially, most day roosts were located near ridge tops or plateaus with a mean elevation of $217.95 \pm 18.26$ m and a mean slope of $13.46 \pm 11.10\%$ (Table 3). Day roosts not on plateaus tended to be located on south-facing slopes (mean aspect = $222.90 \pm 90.84$ degrees) when topography permitted an aspect value to be meaningfully discerned. Day roosts had a larger dbh ($30.19 \pm 18.59$ cm; $W = 28443$, $P < 0.001$) and were in later stages of decay (mean = $3.53 \pm 1.59$; $W = 36504$, $P < 0.001$) than neighboring trees. Of the located trees, 71.4% were in the suppressed canopy class, 16.3% were intermediate, 5.7% were codominant, and 6.6% were dominant. Mean day roost height was $14.56 \pm 7.07$ m. Day roost sites had a mean gap fraction of $8.37 \pm 2.26\%$, a mean leaf area index of $2.55 \pm 0.37\ m^2 * m^{-2}$, and a mean below canopy photosynthetic photon flux density of $1.65 \pm 0.76\ Mol * m^{-2}$. Bark retention was low (64.57%) across day roosts.

Seventy of the day roosts we observed were in snags whereas 35 were in live trees with visible cavities. Cavities accounted for 104 of 105 roosts. The remaining day roost was located under exfoliating bark of a snag. Snags were used as day roosts more than expected based on availability (67%; $W = 19$, $P < 0.001$). We detected no difference in dbh ($W = 1334$, $P = 0.46$), gap fraction ($W = 1062$, $P = 0.27$), LAI ($W = 1273$, $P = 0.75$) PPFD ($W = 1250$, $P = 0.87$), slope position ($W = 1034$, $P = 0.20$), aspect ($W = 1175$, $P = 0.74$) or elevation ($W = 1312$, $P = 0.55$) between snag and live day roosts. Snags were significantly shorter ($W = 1576$, $P = 0.02$) and had significantly less remaining bark than live trees ($W = 2205$, $P < 0.001$; Table 3).

Our best supported model differentiating day roosts by maternity status was a site-specific canopy condition model containing gap fraction, LAI, and PPFD (Table 4). Under this model probability of a day roost being used during the lactation period increased with gap fraction and LAI, but decreased with PPFD (Table 5). This model provided a better fit than a null model (log-likelihood = $-60.16$, $P = 0.002$), but nonetheless had poor predictive power (71% correct classification rate, AUC = 0.72).

The most commonly observed day roost species, sassafras, sugar maple, and white oak accounted for 48.6%, 9.5%, and 7.6% of the total recorded day roosts, while the remaining species each accounted for $\leq$6.0% of day roosts used (Table 2). Sassafras was the most commonly observed potential roost species (34.6%) and was used more than

TABLE 2: Female *Myotis septentrionalis* day roosts by species and the number of corresponding potential day roosts (%) in a mesophytic forest on the Fort Knox military reservation in Hardin, Bullitt and Meade counties, Kentucky, USA, 2011. Available roosts were those trees with visible cavities or exfoliating bark found on 20 × 100 m belt transects oriented at random azimuths from each day roost.

| Species | Available (%) | Day roosts (%) |
|---|---|---|
| Sassafras (*Sassafras albidum*) | 847 (34.6) | 51 (48.6) |
| Sugar maple (*Acer saccharum*) | 357 (14.6) | 10 (9.5) |
| White oak (*Quercus alba*) | 132 (5.4) | 8 (7.6) |
| White ash (*Fraxinus americana*) | 166 (6.8) | 6 (5.7) |
| Eastern redbud (*Cercis canadensis*) | 99 (4.0) | 4 (3.8) |
| Winged elm (*Ulmus alata*) | 32 (1.3) | 4 (3.8) |
| American beech (*Fagus grandifolia*) | 7 (0.3) | 3 (2.9) |
| Black locust (*Robinia pseudoacacia*) | 117 (4.8) | 3 (2.9) |
| Shagbark hickory (*Carya ovata*) | 122 (5.0) | 2 (1.9) |
| Black walnut (*Juglans nigra*) | 56 (2.3) | 2 (1.9) |
| Chinkapin oak (*Quercus muhlenbergii*) | 108 (4.4) | 2 (1.9) |
| Slippery elm (*Ulmus rubra*) | 33 (1.3) | 1 (1.0) |
| Boxelder (*Acer negundo*) | 5 (0.2) | 1 (1.0) |
| Pignut hickory (*Carya glabra*) | 48 (2.0) | 1 (1.0) |
| Hackberry (*Celtis occidentalis*) | 9 (0.4) | 1 (1.0) |
| Blue Ash (*Fraxinus quadrangulata*) | 29 (1.2) | 1 (1.0) |
| Eastern redcedar (*Juniperus virginiana*) | 52 (2.1) | 1 (1.0) |
| Blackgum (*Nyssa sylvatica*) | 6 (0.2) | 1 (1.0) |
| American sycamore (*Planatus occidentalis*) | 11 (0.4) | 1 (1.0) |
| Northern red oak (*Quercus rubra*) | 64 (2.6) | 1 (1.0) |
| Black oak (*Quercus velutina*) | 44 (1.8) | 1 (1.0) |
| American hornbeam (*Carpinus caroliniana*) | 1 (0.04) | 0 (0.0) |
| Tree of heaven (*Ailanthus altissima*) | 3 (0.1) | 0 (0.0) |
| Shellbark hickory (*Carya laciniosa*) | 2 (0.1) | 0 (0.0) |
| Flowering dogwood (*Cornus florida*) | 19 (0.8) | 0 (0.0) |
| Common persimmon (*Diospyros virginiana*) | 24 (1.0) | 0 (0.0) |
| Sweet gum (*Liquidambar styraciflua*) | 5 (0.2) | 0 (0.0) |
| Yellow poplar (*Liriodendron tulipifera*) | 47 (1.9) | 0 (0.0) |
| Red mulberry (*Morus rubra*) | 2 (0.1) | 0 (0.0) |
| Pitch pine (*Pinus rigida*) | 1 (0.04) | 0 (0.0) |
| Loblolly pine (*Pinus taeda*) | 2 (0.1) | 0 (0.0) |

expected based on availability on the landscape ($\chi^2 = 6.8$, $d.f. = 1$, $P = 0.009$). Sugar maple was the second most commonly observed potential roost species (14.3%), but was used in proportion to its availability in the surrounding forest stand ($\chi^2 = 1.5$, $d.f. = 1$, $P = 0.22$). In general, the ranked order of abundance of species most used as day roosts was equivalent to the ranking of species deemed potential roosts. Our best supported model differentiating sassafras day roosts from other species was a tree-specific model containing dbh and height, decay status, and percent bark (Table 6). Within this model, probability of a roost being sassafras decreased with increasing dbh, percent bark, when decay stage was 2 or 4, but increased with height and when decay stage was 3, 5, or 6 (Table 7). This model provided a better fit than a null model (log-likelihood = −49.3, $P < 0.001$), but had a low AUC (0.84) and poor correct classification rate (72%).

## 4. Discussion

The limited day roost documentation of northern bats mostly has shown maternity colony use of snags and live trees larger and more decadent than neighboring boles. It is widely accepted that increased solar exposure at female day roost sites provides important thermal benefits to temperate bat species and that larger trees under open canopies receive more solar radiation and presumably provide better day roost structures [18, 36], particularly during lactation [34]. Despite this speculation, no studies have yet reported direct measures of solar radiation such as photosynthetic flux density. Similarly, measures of canopy structure/complexity such as leaf area index that may affect the amount of solar radiation reaching day roosts are lacking [36]. Use of direct measures of solar radiation and canopy complexity, as well as more accurate measurement of gap fraction, may improve the understanding of the effects of solar radiation on roost selection and identify previously unexplored commonalities between day roost sites across latitudinal and elevation gradients. Compared to LAI values reported for other deciduous forests (range $2–10\,m^2{*}m^{-2}$, mean = $5.41\,m^2{*}m^{-2}$) our observed LAI was low [37]. Nonetheless, our LAI value is greater than those reported for nonshrubby forest understory conditions (range $0.2–13.3\,m^2{*}m^{-2}$, mean $1.81\,m^2{*}m^{-2}$; [37]) supporting the supposition that northern bats roost in relatively dense forest at Fort Knox.

Many studies have used canopy closure as a surrogate measure for solar radiation. We found that our canopy closure values calculated using hemispherical photographs were similar to those reported in the central Appalachians in both Kentucky and West Virginia by Lacki and Schwierjohann [4] and Menzel et al. [5] as well as to Johnson et al. [36]. However, canopy closures at our day roosts were substantially greater than those reported by Garroway and Broders [33] in Nova Scotia, Canada, and Carter and Feldhamer [38] in southern Illinois bottomland hardwoods. Additionally, while the use of day roosts that were larger than surrounding trees in our study would generally support the solar radiation hypothesis, the majority of day roosts (71.4%) we observed were suppressed and under substantial canopy cover.

Our analysis of day roost selection by maternity status indicates that gap fraction and solar radiation best differentiate day roosts used during lactation and nonlactation periods. However, the overall poor performance of the model suggests that female northern bats at our study sites did not strongly differentiate between day roosts used during different reproductive conditions. This contrasts with Garroway and Broders [33] who found that female northern bats selected larger day roosts during the lactation period relative to non-lactation periods and suggests that selection of larger day roosts is a function of increased solar radiation.

TABLE 3: Mean ± SD values of day-roost characteristics for all trees, live and snag trees, and the two tree species most commonly used by female *Mytois septentrionalis* in a mesophytic forest on the Fort Knox military reservation in Hardin, Bullitt, and Meade counties, Kentucky, USA, 2011.

| | All trees | Live | Snag | *Sassafras albidum* | *Acer saccharum* |
|---|---|---|---|---|---|
| *N* | 105 | 35 | 70 | 51 | 10 |
| dbh (cm) | 30.19 ± 18.59 | 31.69 ± 18.06 | 29.44 ± 18.94 | 20.96 ± 7.69 | 26.09 ± 10.92 |
| Height (m) | 14.56 ± 7.07 | 17.39 ± 8.14 | 13.15 ± 6.05 | 11.86 ± 4.65 | 13.48 ± 5.79 |
| Decay (stage) | 3.53 ± 1.59 | 1.71 ± 0.46 | 4.44 ± 1.10 | 4.02 ± 1.48 | 3.6 ± 1.65 |
| Live (*N*) | 33 | NA | NA | 11 | 4 |
| Remaining bark (%) | 64.57 ± 36.11 | 93.00 ± 4.47 | 50.36 ± 36.61 | 52.82 ± 39.10 | 76.00 ± 28.36 |
| Suppressed (%) | 71.4 | 51.4 | 81.4 | 92.2 | 70.0 |
| Gap fraction (%) | 8.37 ± 2.26 | 7.98 ± 1.96 | 8.57 ± 2.38 | 8.13 ± 2.25 | 9.42 ± 1.81 |
| Slope (%) | 13.46 ± 11.10 | 11.67 ± 9.42 | 14.35 ± 11.81 | 12.05 ± 10.65 | 16.68 ± 11.35 |
| Aspect (°) | 222.90 ± 90.84 | 211.08 ± 93.09 | 227.51 ± 90.34 | 223.80 ± 92.15 | 241.91 ± 79.22 |
| Elevation (m) | 217 ± 18.26 | 219.92 ± 16.38 | 216.96 ± 19.17 | 219.35 ± 16.96 | 216.92 ± 11.78 |
| LAI ($m^2{*}m^{-2}$) | 2.55 ± 0.37 | 2.74 ± 0.34 | 2.71 ± 0.43 | 2.71 ± 0.38 | 2.68 ± 0.34 |
| PPFD ($\mu mol{*}m^{-2}{*}s^{-1}$) | 1.65 ± 0.76 | 1.68 ± 0.82 | 1.63 ± 0.72 | 1.65 ± 0.89 | 1.71 ± 0.59 |

TABLE 4: Rankings of models used to compare female *Myotis septentrionalis* day roosts used during lactation and non-lactation periods on the Fort Knox military reservation in Hardin, Bullitt, and Meade counties, Kentucky, USA, 2011. Model parameters are given as well as number of parameters ($K$), Akaike's information criteria (AIC) value, difference in AIC value between top model and $i$th model ($\Delta_i$), and model support ($w_i$).

| Model | $K$ | AIC | $\Delta_i$ | $w_i$ |
|---|---|---|---|---|
| Gap fraction + LAI + PPFD | 5 | 128.33 | 0 | 0.94 |
| Percent + sin (aspect) + cos (aspect) + elevation + gap Fraction + LAI + PPFD | 9 | 135.21 | 6.89 | 0.03 |
| Null model | 2 | 137.01 | 8.69 | 0.01 |
| Decay + percent bark + gap fraction + LAI + PPFD | 11 | 137.95 | 9.62 | 0.01 |
| dbh + height | 4 | 138.95 | 10.63 | 0.00 |
| Percent slope + sin (aspect) + cos (aspect) + elevation | 6 | 140.78 | 12.46 | 0.00 |
| dbh + height + decay + percent bark | 10 | 143.14 | 14.81 | 0.00 |
| dbh + height + decay + percent bark + percent Slope + sin (aspect) + cos (aspect) + elevation + gap Fraction + LAI + PPFD | 17 | 145.15 | 16.83 | 0.00 |
| Decay + percent bark | 8 | 146.13 | 17.81 | 0.00 |

TABLE 5: Parameter summary of the best supported model comparing female *Myotis septentrionalis* day roosts used during lactation and non-lactation periods on the Fort Knox military reservation in Hardin, Bullitt, and Meade counties, Kentucky, USA, 2011.

| Variable | Parameter estimate | SE | Wald $\chi^2$ | $P > \chi^2$ | Odds ratio |
|---|---|---|---|---|---|
| Intercept | −4.32 | 4.12 | 1.1 | 0.29 | — |
| Gap fraction | 0.56 | 0.22 | 6.4 | 0.01 | 1.75 |
| LAI | 0.9 | 1.09 | 0.69 | 0.41 | 2.47 |
| PPFD | −1.15 | 0.38 | 9.3 | 0.002 | 0.32 |

Assuming solar radiation is important in roost temperature regulation and selection, latitudinal temperature gradients likely impose different restrictions on roost selection. If this is the case, greater canopy cover in our sites relative to generally cooler northern sites [33] is not surprising, nor is comparable canopy cover between our sites and sites along similar latitudinal gradients [4, 5, 36]. A latitudinal temperature gradient would explain differences between our analysis of day roost selection by maternity status and that of Garroway and Broders [33].

We found that day roosts were consistently located on ridge tops and plateaus primarily on south facing aspects similar to the observations of other researchers working with northern bats [4, 36, 39]. Although it may be that such positions increase solar radiation at roost sites, upper slopes and ridges also have the highest natural disturbance frequency and severity [40]. Historic cycles of repeated natural stand disturbance and increased snag presence rather than increased solar radiation might be the primary influence on northern bat selection locally. Currently, it is unclear to what extent roost solar exposure, roost availability due to increased disturbance, or some interaction of these factors play a role in northern bat day roost selection.

Table 6: Rankings of models used to compare *Myotis septentrionalis* sassafras day roosts to day roosts of other species on the Fort Knox military reservation in Hardin, Bullitt, and Meade counties, Kentucky, USA, 2011. Model parameters are given as well as number of parameters ($K$), Akaike's information criteria (AIC) value, difference in AIC value between top model and $i$th model ($\Delta_i$), and model support ($w_i$).

| Model | $K$ | AIC | $\Delta_i$ | $w_i$ |
|---|---|---|---|---|
| dbh + height + decay + percent bark | 10 | 116.67 | 0.00 | 0.65 |
| dbh + height | 4 | 118.12 | 1.44 | 0.31 |
| dbh + height + decay + percent bark + percent Slope + sin (aspect) + cos (aspect) + elevation + gap Fraction + LAI + PPFD | 17 | 122.38 | 5.71 | 0.04 |
| Decay + percent bark + gap fraction + LAI + PPFD | 11 | 136.32 | 19.65 | <0.001 |
| Decay + percent bark | 8 | 141.22 | 24.55 | <0.001 |
| Gap fraction + LAI + PPFD | 5 | 146.88 | 30.20 | <0.001 |
| Null model | 2 | 147.48 | 30.80 | <0.001 |
| Percent + sin (aspect) + cos (aspect) + elevation + gap Fraction + LAI + PPFD | 9 | 152.20 | 35.53 | <0.001 |
| Percent slope + sin (aspect) + cos (aspect) + elevation | 6 | 153.56 | 36.88 | <0.001 |

Table 7: Parameter summary of the best supported model comparing *Myotis septentrionalis* sassafras day-roosts to day-roosts of other species on the Fort Knox military reservation in Hardin, Bullitt and Meade counties, Kentucky, USA, 2011.

| Variable | Parameter Estimate | SE | Wald $\chi^2$ | $P > \chi^2$ | Odds ratio |
|---|---|---|---|---|---|
| Intercept | 3.31 | 2.13 | — | — | — |
| dbh | −0.12 | 0.03 | 13.00 | <0.001 | 0.88 |
| Height | 0.06 | 0.06 | 0.97 | 0.32 | 1.06 |
| Decay stage 2 | −0.05 | 1.33 | <0.001 | 0.97 | 0.95 |
| Decay stage 3 | 0.8 | 1.33 | 0.36 | 0.55 | 2.23 |
| Decay stage 4 | −0.01 | 1.43 | <0.001 | 0.99 | 0.99 |
| Decay stage 5 | 0.91 | 1.82 | 0.25 | 0.62 | 2.49 |
| Decay stage 6 | 1.01 | 1.57 | 0.41 | 0.52 | 2.75 |
| Percent bark | −0.02 | 0.01 | 1.70 | 0.20 | 0.98 |

We documented a wider use of tree species by northern bats than reported elsewhere [4, 5, 7, 36], including the first recorded use of eastern redcedar (*Juniperus virginiana*). Northern bats in our study displayed a marked preference for sassafras as day roosts. Black locust, a preferred day roost species by both male and female northern bats in the central Appalachians of West Virginia [5, 15, 36], was rarely used at Fort Knox. Despite overwhelming selection by northern bats, our inability to reliably differentiate sassafras day roosts from other day roost using logistic regression was sur prising. As suggested by Ford et al. [15], and following classical use-availability theory, roost selection probably is a function of the abundance of individual species, rather than differences between species, assuming the desired physical characteristics are present. Although sassafras at our sites clearly provided appropriate roosting structure, preference for this tree species is probably ecologically novel at the scale we observed. In the context of forest succession, this fast-growing shade-intolerant species [41] would not historically have been a large component of closed canopy forests in our study region under small disturbance gap-phase dynamics [42–45]. Much of the extant sassafras at Fort Knox is almost certainly a product of extensive timber harvest from the late 1700s to late 1800s followed by decades of agricultural use through the early 1900s and subsequent abandonment of agricultural areas following acquisition by the United States Army beginning in 1919 and ending in 1942. On some portions of the installation, escaped fires from weapons ranges constitute an important recurring disturbance that may mimic historic fire return intervals appropriate to the area [46, 47].

Given the "aberrant" nature of sassafras as a major component of forests within mixed-mesophytic forests, the wide range of tree species used as day roosts by northern bats [4, 5, 17, 33, 36] and the lack of differences between species used as day roosts in this study, we believe that ecological processes of forests may play a greater role in bat day roost selection than is currently recognized. Because our original intent was not to analyze the effects of forest succession on day roost selection, we are unable to test this hypothesis directly using our data. However, we believe it is appropriate to reinterpret day roost selection by forest bats in other studies in the context of unique forest disturbances and establishment conditions. For example, shortleaf pines (*Pinus echinata*) selected as day roosts by northern bats in Arkansas [7] largely were snags created following pine beetle (*Ips* spp.) outbreaks (R. Perry, personal communication). Ford et al. [15] suggested that northern bat use of black locusts *(Robinia pseudoacacia)* as day roosts in West Virginia was a recent ecological phenomenon directly related to disturbances from exploitative logging in the early- to mid-1900s whereby widespread landscape-level clearcutting favored regeneration and growth of black locust over historically more prevalent species. There, black locust day roosts were in early stages of decay that were comparable to the decay stage of sassafras we observed at Fort Knox. In the absence of suppression by competitors, the observed high rates of decadence in West

Virginia and at Fort Knox in black locust and sassafras, respectively, are unlikely to occur within the first several decades after establishment [48].

In the context of day roost spatial networks it is important to move beyond individual tree concepts and incorporate larger forest establishment conditions that create and maintain suitable long-term roosting opportunities and networks. We are aware of no comprehensive attempts to relate past land use or forest development to patterns of day roost selection by bats, yet understanding these relationships should be invaluable for managers to relate current day roost conditions and availability with necessary future conditions. We believe the following questions should be considered relative to understanding day roost ecology of tree-roosting bats such as the northern bat: (1) is current forest composition largely a result of historical or anthropogenically disturbed conditions rather than natural processes?, (2) are the species used as day roosts typical of the regional and local forest type?, (3) are bats adapting to novel conditions related to anthropogenic or stochastic natural events that have drastically altered forest structure and composition?, (4) are the conditions in place for creation of suitable day roosts into the future, or will the creation of suitable day roosts be dependent on management activities?

## 5. Conclusions

Differences in establishment history and disturbance processes can lead to a myriad of alternative stable or dynamic states of forest communities that vary in their successional trajectories and long-term composition and structure [49]. In the context of bats, day roost species selection may be a function of the regional species candidate pool and successional processes. The particular anthropogenic and stochastic forest disturbance processes that shaped present forest conditions [40, 49] across the range of the northern bat may not be feasibly recreated by managers, or even desired in the light of other stewardship needs. Furthermore, current conditions do not necessarily represent desired future conditions for managers. By linking forest successional and disturbance processes to bat day roost networks, bat habitat may be managed at spatial and temporal scales compatible with larger forest management objectives for a fuller compliment of desired natural resource outcomes, however, further work directly addressing this topic across a wider range of bat species and forest types is needed.

## Acknowledgments

This paper was supported by the US Army Environmental Quality and Installation Basic Research 6.1 program. The authors thank Jimmy Watkins, Mike Brandenberg, and Charlie Logsdon for their assistance in supporting this project. The Kentucky Department of Fish and Wildlife Resources graciously provided field housing for this project. Meryl Friedrich and Mark Lawrence provided invaluable field assistance on this project. Use of trade, product, or firm names does not imply endorsement by the US government.

## References

[1] D. S. Blehert, A. C. Hicks, M. Behr et al., "Bat white-nose syndrome: an emerging fungal pathogen?" *Science*, vol. 323, no. 5911, article 227, 2009.

[2] R. W. Foster and A. Kurta, "Roosting ecology of the Northern bat (*Myotis septentrionalis*) and comparisons with the endangered Indiana bat (*Myotis sodalis*)," *Journal of Mammalogy*, vol. 80, no. 2, pp. 659–672, 1999.

[3] M. C. Caceres and R. M. R. Barclay, "*Myotis septentrionalis*," *Mammalian Species*, no. 634, pp. 1–4, 2000.

[4] M. J. Lacki and J. H. Schwierjohann, "Day-roost characteristics of Northern bats in mixed mesophytic forest," *The Journal of Wildlife Management*, vol. 65, no. 3, pp. 482–488, 2001.

[5] M. A. Menzel, S. F. Owen, W. M. Ford et al., "Roost tree selection by Northern long-eared bat (*Myotis septentrionalis*) maternity colonies in an industrial forest of the central Appalachian mountains," *Forest Ecology and Management*, vol. 155, no. 1–3, pp. 107–114, 2002.

[6] H. G. Broders, G. J. Forbes, S. Woodley, and I. D. Thompson, "Range extent and stand selection for roosting and foraging in forest-dwelling Northern long-eared bats and little brown bats in the Greater Fundy Ecosystem, New Brunswick," *The Journal of Wildlife Management*, vol. 70, no. 5, pp. 1174–1184, 2006.

[7] R. W. Perry and R. E. Thill, "Roost selection by male and female Northern long-eared bats in a pine-dominated landscape," *Forest Ecology and Management*, vol. 247, no. 1–3, pp. 220–226, 2007.

[8] A. D. Morris, D. A. Miller, and M. C. Kalcounis-Rueppell, "Use of forest edges by bats in a managed pine forest landscape," *The Journal of Wildlife Management*, vol. 74, no. 1, pp. 26–34, 2010.

[9] J. B. Johnson, W. Mark Ford, and J. W. Edwards, "Roost networks of Northern myotis (*Myotis septentrionalis*) in a managed landscape," *Forest Ecology and Management*, vol. 266, pp. 223–231, 2012.

[10] T. S. Jung, I. D. Thompson, R. D. Titman, and A. P. Applejohn, "Habitat selection by forest bats in relation to mixed-wood stand types and structure in central Ontario," *The Journal of Wildlife Management*, vol. 63, no. 4, pp. 1306–1319, 1999.

[11] S. F. Owen, M. A. Menzel, J. W. Edwards et al., "Bat activity in harvested and intact forest stands in the allegheny mountains," *Northern Journal of Applied Forestry*, vol. 21, no. 3, pp. 154–159, 2004.

[12] R. T. Brooks and W. M. Ford, "Bat activity in a forest landscape of central Massachusetts," *Northeastern Naturalist*, vol. 12, no. 4, pp. 447–462, 2005.

[13] S. C. Loeb and J. M. O'Keefe, "Habitat use by forests bats in South Carolina in relation to local, stand, and landscape characteristics," *The Journal of Wildlife Management*, vol. 70, no. 5, pp. 1210–1218, 2006.

[14] M. A. Menzel, T. C. Carter, J. M. Menzel, W. Mark Ford, and B. R. Chapman, "Effects of group selection silviculture in bottomland hardwoods on the spatial activity patterns of bats," *Forest Ecology and Management*, vol. 162, no. 2-3, pp. 209–218, 2002.

[15] W. M. Ford, S. F. Owen, J. W. Edwards, and J. L. Rodrigue, "*Robinia pseudoacacia* (black locust) as day-roosts of male *Myotis septentrionalis* (Northern bats) on the fernow experimental forest, West Virginia," *Northeastern Naturalist*, vol. 13, no. 1, pp. 15–24, 2006.

[16] M. B. Fenton, "Science and the conservation of bats," *Journal of Mammalogy*, vol. 78, no. 1, pp. 1–14, 1997.

[17] J. M. Psyllakis and R. M. Brigham, "Characteristics of diurnal roosts used by female Myotis bats in sub-boreal forests," *Forest Ecology and Management*, vol. 223, no. 1–3, pp. 93–102, 2006.

[18] T. H. Kunz and L. F. Lumsden, "Ecology of cavity and foliage roosting bats," in *Bat Ecology*, T. H. Kunz and M. B. Fenton, Eds., pp. 2–90, University of Chicago Press, Chicago, Ill, USA, 2003.

[19] C. J. Garroway and H. G. Broders, "Nonrandom association patterns at Northern long-eared bat maternity roosts," *Canadian Journal of Zoology*, vol. 85, no. 9, pp. 956–964, 2007.

[20] K. J. Patriquin, M. L. Leonard, H. G. Broders, and C. J. Garroway, "Do social networks of female Northern long-eared bats vary with reproductive period and age?" *Behavioral Ecology and Sociobiology*, vol. 64, no. 6, pp. 899–913, 2010.

[21] C. L. Lausen, T. S. Jung, and J. M. Talerico, "Range extension of the Northern long-eared bat (*Myotis septentrionalis*) in the Yukon," *Northwestern Naturalist*, vol. 89, no. 2, pp. 115–117, 2008.

[22] M. Rhodes, "Roost fidelity and fission-fusion dynamics of white-striped free-tailed bats (*Tadarida australis*)," *Journal of Mammalogy*, vol. 88, no. 5, pp. 1252–1260, 2007.

[23] E. L. P. Arms, M. J. Mitchell, F. C. Watts, and B. L. Wilson, *Soil Survey of Hardin and Larue Counties, Kentucky*, USDA Soil Conservation Service, 1979.

[24] E. L. Braun, *Deciduous Forests of Eastern North America*, Blakiston Company, Philadelphia, Pa, USA, 1950.

[25] R. Cranfill, "Flora of Hardin County, Kentucky," *Castanea*, vol. 56, no. 4, pp. 228–267, 1991.

[26] E. L. P. Anthony, "Age determination in bats," in *Ecological and Behavioral Methods for the Study of Bats*, T. H. Kunz, Ed., pp. 47–58, Smithsonian Institution Press, Washington, DC, USA, 1988.

[27] M. A. Menzel, J. M. Menzel, S. B. Castleberry, J. Ozier, W. M. Ford, and J. W. Edwards, "Illustrated key to skins and skulls of bats in the Southeastern and mid-Atlantic states," Research Note NE-376, USDA Forest Service, Newton Square, Pa, USA, 2002.

[28] R. D. Nyland, *Silviculture: Concepts and Applications*, McGraw-Hill, New York, NY, USA, 1996.

[29] S. P. Cline, A. B. Berg, and H. M. Wight, "Snag characteristics and dynamics in Douglas-Fir forests, Western Oregon," *The Journal of Wildlife Management*, vol. 44, no. 4, pp. 773–786, 1980.

[30] J. E. Brower and J. H. Zar, *Field and Laboratory Methods for General Ecology*, W. C. Brown, Dubuque, Iowa, USA, 1984.

[31] J. R. Alldredge and J. Griswold, "Design and analysis of resource selection studies for categorical resource variables," *The Journal of Wildlife Management*, vol. 70, no. 2, pp. 337–346, 2006.

[32] D. L. Thomas and E. J. Taylor, "Study designs and tests for comparing resource use and availability II," *The Journal of Wildlife Management*, vol. 70, no. 2, pp. 324–336, 2006.

[33] C. J. Garroway and H. G. Broders, "Day roost characteristics of Northern long-eared bats (*Myotis septentrionalis*) in relation to female reproductive status," *Ecoscience*, vol. 15, no. 1, pp. 89–93, 2008.

[34] K. P. Burnham and D. R. Anderson, *Model Selection and Multimodel Inference: A Practical Information-Theoretic Approach*, Springer, New York, NY, USA, 2002.

[35] R. D. C. Team, *R: A Language and Environment for Statistical Computing*, R Foundation for Statistical Computing, Vienna, Austria, 2011.

[36] J. B. Johnson, J. W. Edwards, W. M. Ford, and J. E. Gates, "Roost tree selection by Northern myotis (*Myotis septentrionalis*) maternity colonies following prescribed fire in a Central Appalachian Mountains hardwood forest," *Forest Ecology and Management*, vol. 258, no. 3, pp. 233–242, 2009.

[37] L. Breuer, K. Eckhardt, and H. G. Frede, "Plant parameter values for models in temperate climates," *Ecological Modelling*, vol. 169, no. 2-3, pp. 237–293, 2003.

[38] T. C. Carter and G. A. Feldhamer, "Roost tree use by maternity colonies of Indiana bats and Northern long-eared bats in Southern Illinois," *Forest Ecology and Management*, vol. 219, no. 2-3, pp. 259–268, 2005.

[39] T. S. Jung, I. D. Thompson, and R. D. Titman, "Roost site selection by forest-dwelling male Myotis in central Ontario, Canada," *Forest Ecology and Management*, vol. 202, no. 1–3, pp. 325–335, 2004.

[40] C. G. Lorimer and A. S. White, "Scale and frequency of natural disturbances in the Northeastern US: implications for early successional forest habitats and regional age distributions," *Forest Ecology and Management*, vol. 185, no. 1-2, pp. 41–64, 2003.

[41] R. M. Burns and B. H. Honkala, *Silvics of North America*, vol. 2 of *Hardwoods, Agriculture Handbook*, no. 654, USDA Forest Service, Washington, DC, USA, 1984.

[42] M. K. Trani, R. T. Brooks, T. L. Schmidt, V. A. Rudis, and C. M. Gabbard, "Patterns and trends of early successional forests in the Eastern United States," *Wildlife Society Bulletin*, vol. 29, no. 2, pp. 413–424, 2001.

[43] M. Lemenih and D. Teketay, "Effect of prior land use on the recolonization of native woody species under plantation forests in the highlands of Ethiopia," *Forest Ecology and Management*, vol. 218, no. 1–3, pp. 60–73, 2005.

[44] M. A. Albrecht and B. C. McCarthy, "Effects of prescribed fire and thinning on tree recruitment patterns in central hardwood forests," *Forest Ecology and Management*, vol. 226, no. 1–3, pp. 88–103, 2006.

[45] C. J. Schweitzer and D. C. Dey, "Forest structure, composition, and tree diversity response to a gradient of regeneration harvests in the mid-Cumberland Plateau escarpment region, USA," *Forest Ecology and Management*, vol. 262, no. 9, pp. 1729–1741, 2011.

[46] C. C. Frost, "Presettlement fire frequency regimes of the United States: a first approximation," in *Fire in Ecosystem Management: Shifting the Paradigm from Suppression to Prescription: Proceedings, 20th Tall Timbers Fire Ecology Conference*, pp. 70–81, Tall Timbers Research Station, Tallahassee, Fla, USA, 1998.

[47] R. W. McEwan, T. F. Hutchinson, R. P. Long, D. R. Ford, and B. C. McCarthy, "Temporal and spatial patterns in fire occurrence during the establishment of mixed-oak forests in Eastern North America," *Journal of Vegetation Science*, vol. 18, no. 5, pp. 655–664, 2007.

[48] J. S. Ward and G. R. Stephens, "Influence of crown class and shade tolerance on individual tree development during deciduous forest succession in Connecticut, USA," *Forest Ecology and Management*, vol. 60, no. 3-4, pp. 207–236, 1993.

[49] G. J. Nowacki and M. D. Abrams, "The demise of fire and "mesophication" of forests in the Eastern United States," *BioScience*, vol. 58, no. 2, pp. 123–138, 2008.

# 5

# Yield Responses of Black Spruce to Forest Vegetation Management Treatments: Initial Responses and Rotational Projections

**Peter F. Newton**

*Canadian Wood Fibre Centre, Canadian Forest Service, Natural Resources Canada, Sault Ste. Marie, ON, Canada P6A 2E5*

Correspondence should be addressed to Peter F. Newton, pnewton@nrcan.gc.ca

Academic Editor: John Sessions

The objectives of this study were to (1) quantitatively summarize the early yield responses of black spruce (*Picea mariana* (Mill.) B.S.P.) to forest vegetation management (FVM) treatments through a meta-analytical review of the scientific literature, and (2) given (1), estimate the rotational consequences of these responses through model simulation. Based on a fixed-effects meta-analytic approach using 44 treated-control yield pairs derived from 12 experiments situated throughout the Great Lakes—St. Lawrence and Canadian Boreal Forest Regions, the resultant mean effect size (response ratio) and associated 95% confidence interval for basal diameter, total height, stem volume, and survival responses, were respectively: 54.7% (95% confidence limits (lower/upper): 34.8/77.6), 27.3% (15.7/40.0), 198.7% (70.3/423.5), and 2.9% (−5.5/11.8). The results also indicated that early and repeated treatments will yield the largest gains in terms of mean tree size and survival. Rotational simulations indicated that FVM treatments resulted in gains in stand-level operability (e.g., reductions of 9 and 5 yr for plantations established on poor-medium and good-excellent site qualities, resp.). The challenge of maintaining coniferous forest cover on recently disturbed sites, attaining statutory-defined free-to-grow status, and ensuring long-term productivity, suggest that FVM will continue to be an essential silvicultural treatment option when managing black spruce plantations.

## 1. Introduction

The underlying objective of intensive forest management (IFM) is to increase the intrinsic productivity of the forest land base via the application of an integrated silvicultural regime involving the application of various temporal-spatial-specific treatment matrices encompassing intensive site preparation (e.g., mechanical scarification), plantation establishment including the use of genetically improved stock, eliminating competing vegetation via mechanical, chemical, or biological control mechanisms, continuous and active protection from insects and pathogens, and density management (e.g., maximizing product quality and quantity via initial spacing, precommercial thinning, and/or commercial thinning). At the forest level, these stand-level increases in productivity when translated into accelerated rates of development can be used to offset future wood supply deficits or increase the annual allowable cut through the annual allowable cut effect [1]. Although numerous case studies of individual experiments have clearly demonstrated the benefits of IFM treatments, collective summaries documenting the response of boreal species are limited. Consequently, as part of a larger quantitative synthesis regarding yield responses of boreal conifers to IFM treatments (e.g., tree improvement [2], forest fertilization [3], density management [4]), the objectives of this study were to (1) quantitatively summarize the early yield responses of black spruce (*Picea mariana* (Mill.) BSP) to forest vegetation management (FVM) treatments through a meta-analytical review of the scientific literature, and (2) given (1), project rotational outcomes of these responses employing a modified variant of the comprehensive growth, yield, and wood quality simulator, CROPLANNER [5, 6].

FVM has been defined as the practice of preferentially allocating the finite environmental resources (e.g., solar radiation, nutrients, and water) of a given site towards the selected crop tree species and away from noncrop species via the suppression, reduction, or elimination of the unwanted competitors, in order to realize specified silviculture and/or forest management objectives (sensu Walstad and Kuch [7];

Wagner et al. [8]; Bell et al. [9]). FVM can increase the likelihood of successful plantation establishment and assist in attaining statutory-defined free-to-grow status. Conversely, without FVM, forest managers may have difficulty in realizing their stand-level timber management objectives given the negative effect of herbaceous and woody competitors on the growth and survival rates of various crop species. Historically, the dominant means for controlling competing vegetation in Canada has been through the use of herbicides (e.g., aerial applications of glyphosate [10]). However, concern regarding the potential negative impacts of herbicide applications on wildlife habitat and biodiversity has resulted in numerous studies exploring various alternatives (e.g., [11–16]). Furthermore, the use of chemical herbicides has been prohibited in some regions (e.g., Canadian province of Quebec), and resistance to their use has been growing elsewhere (e.g., Europe [17]). Consequently, the focus of this study was to document the yield consequences of reducing or eliminating interspecific competitors on black spruce, irrespective of the actual control mechanism employed.

## 2. Method

*2.1. Quantifying Early Yield Responses to FVM Treatments.* Based on a meta-analytical approach, the analysis consisted of 4 sequential steps. Firstly, electronic databases were systematically searched for relevant publications using various keywords and phrases (e.g., combinations and permutations of the following terms: vegetation management treatments; black spruce, plantations; mechanical, chemical and biological release; brushing; herbicide, Vision; Roundup; glyphosate; Canadian Boreal and Great Lakes—St. Lawrence Forest Regions). These databases consisted of (1) SilverPlatter WebSPIRS (Ovid Technologies Inc., USA) which included Agricola, Biological Abstracts, CAB Abstracts, and TreeCD, (2) NRCan Forestry Library Catalogue (Natural Resources Canada), and (3) Science Direct Database and Scopus (Elsevier Science B.V.). Furthermore, the World Wide Web was similarly searched using Google Web Search and Google Scholar (Google Inc., CA, USA). Secondly, the identified studies were assessed for their specific applicability in terms of locality (Great Lakes—St. Lawrence, and/or Boreal Forest Regions of Canada), stand-type (upland black spruce), treatments (mechanical, chemical, or biological), yield parameters assessed (basal stem diameter, total stem height, total stem volume, and survival), and publication date (<2011). Thirdly, a subset of applicable studies was selected and their results summarized via the calculation of the grand mean effect size (response ratio) and associated 95% confidence intervals for each yield variate (Table 1). The responses associated with the maximum dosage of a specific chemical treatment (e.g., glyphosate, hexazinone, sulfonylurea, combination of 2,4-D, mecoprop, and dicamba) or the maximum removal for a specific mechanical treatment (e.g., brushing, clipping, extraction) were used. Fourthly, the linear association between effect size and the individual experimental factors, that is, number of annual sequential treatments ($T_N$), initial year of treatment application relative to the year of seedling establishment ($T_I$ (yr)), and length of the observation period relative to the year of seedling establishment ($T_L$ (yr)), was examined through graphical and correlation analyses.

Computational formulae based on a meta-analytic approach were used throughout (Hedges and Olkin [31]; Hedges et al. [32]). Specifically, the grand mean response ratio ($\overline{L}_j^*$) of the $j$th yield variate ($j$ = 1 (basal diameter), 2 (total height), 3 (stem volume), and 4 (survival)) was calculated using the mean logarithmic response ratio of the $j$th yield variate within the $k$th experiment ($\overline{L}_{j(k)}$) along with the reciprocal of the total unconditional variance estimate of $L_{j(k)}$ ($w_{j(k)}^*$), according to (1).

$$\overline{L}_j^* = \text{EXP}\left[\frac{\sum_{k=1}^{n_{j(k)}} w_{j(k)}^* \cdot \overline{L}_{j(k)}}{\sum_{k=1}^{n_{j(k)}} w_{j(k)}^*}\right], \quad (1)$$

where

$$\overline{L}_{j(k)} = \frac{\sum_{i=1}^{n_{j(k)}} L_{jk(i)}}{n_{j(k)}} \quad \text{where } L_{jk(i)} = \log_e\left(R_{jk(i)}\right),\ R_{jk(i)} = 100 \cdot \left(\frac{\overline{T}_{jk(i)} - \overline{C}_{jk}}{\overline{C}_{jk}}\right),$$

$$w_{j(k)}^* = \left(v_{j(k)} + \hat{\sigma}_j^2\right), \quad \text{where}\begin{cases} \hat{\sigma}_j^2 = \dfrac{\left[\sum_{k=1}^{n_{j(k)}} w_{j(k)}\left(L_{j(k)}\right)^2 - \left(\sum_{k=1}^{n_{j(k)}} w_{j(k)}L_{j(k)}\right)^2 / \sum_{k=1}^{n_{j(k)}} w_{j(k)}\right] - \left(n_{j(k)} - 1\right)}{\sum_{k=1}^{n_{j(k)}} w_{j(k)} - \sum_{k=1}^{n_{j(k)}} w_{j(k)}^2 / \sum_{k=1}^{n_{j(k)}} w_{j(k)}}, \\ v_{j(k)} = \dfrac{\sum_{i=1}^{n_{j(k)}} \left(L_{jk(i)} - \overline{L}_{j(k)}\right)^2}{n_{j(k)} - 1}, \end{cases} \quad (2)$$

where $L_{jk(i)}$ and $R_{jk(i)}$ are the logarithmic ($\log_e$(%)) and arithmetic (%) relative response of the $j$th yield variate within the $k$th experiment associated with the $i$th experimental factor set (time of treatment, number of treatments applied, and type of treatment), respectively, $n_{j(k)}$ is the total number of response ratios observed for the $j$th yield variate within the $k$th experiment, $\overline{T}_{jk(i)}$ is the mean value of the $j$th yield variate within the treated plots within the $k$th experiment associated with the $i$th experimental factor set, $\overline{C}_{jk}$ is the mean value of the $j$th yield variate within the untreated

Table 1: Relative yield responses and associated experimental details ordered by number of sequential treatments ($T_N$), initial year of treatment application relative to the year of seedling establishment ($T_I$ (yr)); and length of the observation period relative to the year of seedling establishment ($T_L$ (yr)).

| $T_N^a$ (n) | $T_I^a$ (yr) | $T_L^a$ (yr) | Relative response[b] $R_D$ (%) | $R_H$ (%) | $R_V$ (%) | $R_S$ (%) | Source and treatment details |
|---|---|---|---|---|---|---|---|
| 1 | −1 | 4 | 73 | 43 | 339 | 10 | Tables 1 and 2 and Figures 1, 4(b), and 6(b) as reported by Pitt et al. [18]; mean responses based on bareroot and container stock-types combined results; treatment = hexazinone (Velpar L) at 4 kg ai/ha (liquid formulation); effectiveness = reduction of 20% and 1% in herbaceous and woody vegetation cover, respectively, relative to untreated control plots, at time of last remeasurement. |
| 1 | −1 | 4 | 67 | 60 | — | 8 | Tables 1 and 2 and Figures 1(a,b) and 4(b,d) as reported by Reynolds and Roden [19]; mean responses based on spring and fall combined results; treatment = hexazinone (Velpar L) at 2 kg ai/ha (liquid formulation); effectiveness = reduction of 44% in red raspberry (*Rubus idaeus* L. var. strigosus (Michx.) Maxim.) (major competitor) cover, relative to untreated control plots, at time of last remeasurement. |
| 1 | −1 | 4 | 57 | 48 | — | 29 | Tables 1 and 2 and Figures 1(a,b), and 4(b,d) as reported by Reynolds and Roden [19]; mean responses based on spring and fall combined results; treatment = hexazinone (PRONONE 10 G) at 2 kg ai/ha (granular formulation); effectiveness = reduction of 38% in red raspberry (major competitor) cover, relative to untreated control plots, at time of last remeasurement. |
| 1 | −1 | 4 | 31 | 43 | — | 26 | Tables 1 and 2 and Figures 1(a,b) as reported by Reynolds and Roden [20]; mean responses based on spring and fall combined results; treatment = sulfonylurea (Metsulfuron (ESCORT)) at 72 g ai/ha (liquid formulation); effectiveness = reduction of 47% in red raspberry (major competitor) cover, relative to untreated control plots, at time of last remeasurement. |
| 1 | −1 | 5 | 100 | 63 | — | 48 | Tables 2 and 3 as reported by Wood and von Althen [21]; mean responses based on bareroot and container stock-types combined results; treatment = glyphosate (Roundup) at 2 kg ai/ha (liquid formulation); effectiveness = reductions of 20% and 25% in herbaceous and woody vegetation cover, respectively, relative to the untreated control plots, at time of last remeasurement. |
| 1 | 0 | 5 | 87 | 35 | — | −5 | Tables 1 and 2 and Figures 2(b,d) as reported by Reynolds and Roden [19]; mean responses based on spring and fall combined results; treatment = hexazinone (Velpar L) at 2 kg ai/ha (liquid formulation); effectiveness = reduction of 44% in red raspberry (major competitor) cover, relative to untreated control plots, at time of last remeasurement. |
| 1 | 0 | 5 | 53 | 25 | — | −27 | Tables 1 and 2 and Figures 2(a,c) as reported by Reynolds and Roden [19]; mean responses based on spring and fall combined results; treatment = hexazinone (PRONONE 10 G) at 2 kg ai/ha (granular formulation); effectiveness = reduction of 38% in red raspberry (major competitor) cover, relative to untreated control plots, at time of last remeasurement. |
| 1 | 0 | 5 | 47 | 23 | — | −38 | Tables 1 and 2 and Figures 2(a,b) as reported by Reynolds and Roden [20]; treatment = sulfonylurea (Sulfometuron (OUST) at 300 g ai/ha (liquid formulation); effectiveness = reduction of 57% in red raspberry (major competitor) cover, relative to untreated control plots, at time of last remeasurement. |
| 1 | 0 | 5 | 20 | — | — | −15 | Tables 1 and 2 and Figure 1(a) as reported by Reynolds and Roden [20]; treatment = sulfonylurea (Metsulfuron (ESCORT) at 72 g ai/ha (liquid formulation); effectiveness = reduction of 47% in red raspberry (major competitor) cover, relative to untreated control plots, at time of last remeasurement. |
| 1 | 0 | 5 | 35 | 15 | 112 | — | Table 1 as reported by Sutherland et al. [22] and Figure 1(d) as reported by Sutherland and Foreman [23]; treatment = hexazinone (Velpar L) at 3.1 kg ai/ha (liquid formulation); effectiveness = initial reduction of competing vegetation at time of treatment; however, based on a vegetation index metric (height × cover), competition (red raspberry and trembling aspen (*Populus tremuloides* Michx.) increased 40%, relative to the untreated control plots, at the time of last remeasurement. |

Table 1: Continued.

| $T_N^a$ (n) | $T_I^a$ (yr) | $T_L^a$ (yr) | Relative response[b] $R_D$ (%) | $R_H$ (%) | $R_V$ (%) | $R_S$ (%) | Source and treatment details |
|---|---|---|---|---|---|---|---|
| 1 | 0 | 5 | 26 | 19 | 109 | −9 | Tables 1, 2, 3, and 4 as reported by Jobidon et al. [24]; results for chemical treatment at the Flynn experimental site; treatment = glyphosate (Vision) at 1.5 kg ai/ha (liquid formulation); untreated control plots dominated by red raspberry (63–75% cover) at time of establishment. |
| 1 | 0 | 5 | 42 | 27 | 204 | −4 | Tables 1, 2, 3, and 4 as reported by Jobidon et al. [24]: results for manual treatment at the Flynn experimental site; treatment = physical removal via hedge clipper; untreated control plots dominated by red raspberry (63–75% cover) at time of establishment. |
| 1 | 0 | 5 | 27 | 13 | 93 | 0 | Tables 1, 2, 3, and 4 as presented Jobidon et al. [24]; results for chemical treatment at the Joncas experimental site; treatment = glyphosate (Vision) at 1.6 kg ai/ha (liquid formulation); untreated control plots dominated by red raspberry (24–45% cover) at time of establishment. |
| 1 | 0 | 5 | 24 | 14 | 86 | 3 | Tables 1, 2, 3, and 4 as reported by Jobidon et al. [24]: results for manual treatment at the Joncas experimental site; treatment = physical removal via machete; untreated control plots dominated by red raspberry (24–45% cover) at time of establishment. |
| 1 | 0 | 5 | 45 | 20 | 104 | 2 | Tables 1, 2, 3, and 4 as reported by Jobidon et al. [24]; results for chemical treatment at the Pilote experimental site; treatment = glyphosate (Vision) at 1.6 kg ai/ha (liquid formulation); untreated control plots dominated by red raspberry (71–79% cover) at time of establishment. |
| 1 | 0 | 5 | 40 | 23 | 99 | 13 | Tables 1, 2, 3, and 4 as reported by Jobidon et al. [24]: results for manual treatment at the Pilote experimental site; treatment = physical removal via brush cutter; untreated control plots dominated by red raspberry (71–79% cover) at time of establishment. |
| 1 | 0 | 5 | 67 | 41 | — | 12 | Tables 2 and 3 as reported by Wood and von Althen [21]; mean responses based on bareroot and container stock-types combined results; treatment = glyphosate (Roundup) at 2 kg ai/ha (liquid formulation); effectiveness = reductions of 10% and 42.5% in herbaceous and woody vegetation cover, respectively, relative to untreated control plots, at time of last remeasurement. |
| 1 | 0 | 5 | 68 | 38 | 273 | 5 | Table 2 and Figures 1, 4(a), and 6(a) as reported by Pitt et al. [18]; mean responses based on bareroot and container stock-types combined results; treatment = hexazinone (Velpar L) at 4 kg ai/ha (liquid formulation); effectiveness = reduction of 20% and 1% in herbaceous and woody vegetation cover, respectively, relative to untreated control plots, at time of last remeasurement. |
| 1 | 0 | 5 | 17 | 13 | 69 | 2 | Table 1 and Figure 2 as reported by Wagner et al. [25]; treatment = glyphosate (Vision) at 4 kg ai/ha (liquid formulation); effectiveness = reduction of 10% in herbaceous vegetation cover relative to the untreated control plots at the time of last remeasurement (*n*., nil competition from woody species). |
| 1 | 0 | 10 | 15 | 7 | — | 7 | Table 2 as reported by Jobidon and Charette [26]; results for Squatec experimental site; treatment = manual removable via pulling of all vegetation within 60 cm of subject trees; principal competitor species were removed (red raspberry and fireweed). |
| 1 | 0 | 10 | 49 | 40 | — | 20 | Table 4 as reported by Jobidon and Charette [26]; results for Lake Anna experimental site; treatment = mechanized manual removable via brush cutter of all vegetation within 60 cm of subject trees; principal competitor species were removed (shade intolerant deciduous trees). |
| 1 | 0 | 10 | 79 | 52 | — | 11 | Table 4 as reported by Jobidon and Charette [26]; results for Lake Anna; treatment = manual removable via pulling of all competing vegetation; principal competitor species were removed (shade intolerant deciduous trees). |

Table 1: Continued.

| $T_N^a$ ($n$) | $T_I^a$ (yr) | $T_L^a$ (yr) | Relative response[b] $R_D$ (%) | $R_H$ (%) | $R_V$ (%) | $R_S$ (%) | Source and treatment details |
|---|---|---|---|---|---|---|---|
| 1 | 0 | 11 | 78 | 18 | — | −10 | Figures 3, 5, and 7 as reported by Wood and Mitchell [27]; results for Bragg experiment; mean responses based on bareroot and container stock-types combined results; treatment = glyphosate (Roundup) at 2.5 kg ai/ha (liquid formulation); effectiveness = reductions of 97% in competing vegetation (principally trembling aspen), relative to the untreated control plots, at time of last remeasurement. |
| 1 | 0 | 15 | 67 | 48 | 240 | — | Figure 1 as reported by Robinson et al. [28]; mean response of brush control treatment relative to untreated control plots; treatment = glyphosate (Roundup) at 2.2 kg acid equivalent/ha (liquid formulation) on bladed scarified sites; effectiveness: at the time of the 15 yr post-treatment assessment, the treatment had eliminated the noncrop hardwood species. |
| 1 | 1 | 5 | 31 | 10 | — | −11 | Tables 2 and 3 as reported by Wood and von Althen [21]; mean responses based on bareroot and container stock-types combined results; treatment = glyphosate (Roundup) at 2 kg ai/ha (liquid formulation); effectiveness = reductions of 5% and 35% in herbaceous and woody vegetation cover, respectively, relative to untreated control plots, at time of last remeasurement. |
| 1 | 1 | 11 | 51 | 8 | — | −7 | Figures 3, 5, and 7 as reported by Wood and Mitchell [27]; results for Bragg experiment; mean responses based on bareroot and container stock-types and spring and summer seasonal results combined; treatment = glyphosate (Roundup) at 2.5 kg ai/ha (liquid formulation); effectiveness = reductions of 98% in competing vegetation (principally trembling aspen), relative to the untreated control plots, at time of last remeasurement. |
| 1 | 2 | 10 | 87 | 46 | 246 | — | Textual description and Figure 1 as reported in Pitt et al. [29] for Corrigal experimental site; annual directed foliar application of 1.58 or 2% liquid solution of glyphosate (Vision); effectiveness: at the time of the 10 yr remeasurement the average of all treatments had reduced the cover of deciduous woody tree species to less than 10%, tall shrub species to 2%, and low shrub species to 20%. |
| 1 | 2 | 11 | 38 | 25 | — | 5 | Figures 3, 5, and 7 as reported by Wood and Mitchell [27]; results for Kenogaming experiment; mean responses based on bareroot and container stock-types and spring and summer seasonal results combined; treatment = glyphosate (Roundup) at 2.14 kg ai/ha (liquid formulation); effectiveness = reductions of 76% in competing vegetation (principally trembling aspen), relative to the untreated control plots, at time of last remeasurement. |
| 1 | 2 | 11 | 33 | 11 | — | −19 | Figures 3, 5, and 7 as reported by Wood and Mitchell [27]; results for Lampugh experiment; mean responses based on bareroot and container stock-types and spring and summer seasonal results combined; treatment = glyphosate (Roundup) at 2.14 kg ai/ha (liquid formulation); effectiveness = reductions of 94% in competing vegetation (principally trembling aspen), relative to the untreated control plots, at time of last remeasurement. |
| 1 | 3 | 10 | 87 | 8 | 126 | — | Textual description and Figure 1 as reported in Pitt et al. [29] for Hele experimental site; treatment = glyphosate (Vision) at 1.7 kg ai/ha (liquid formulation); effectiveness: at the time of the 10 yr remeasurement the average of all treatments had reduced the cover of deciduous woody tree species to less than 10%, tall shrub species to 2%, and low shrub species to 20%. |
| 1 | 4 | 5 | 24 | 3 | 69 | −1 | Table 1 and Figure 2 as reported by Wagner et al. [25]; treatment = glyphosate (Vision) at 4 kg ai/ha (liquid formulation); effectiveness = reduction of 20% in herbaceous vegetation cover, relative to the untreated control plots, at the time of last remeasurement (*n.*, nil competition from woody species). |
| 2 | 0, 1 | 5 | 59 | 21 | 228 | −6 | Table 1 and Figure 2 as reported by Wagner et al. [25]; treatment = glyphosate (Vision) at 4 kg ai/ha (liquid formulation); effectiveness = reduction of 21% in herbaceous vegetation cover, relative to the untreated control plots, at the time of last remeasurement (*n.*, nil competition from woody species). |

TABLE 1: Continued.

| $T_N^a$ (n) | $T_I^a$ (yr) | $T_L^a$ (yr) | Relative response[b] $R_D$ (%) | $R_H$ (%) | $R_V$ (%) | $R_S$ (%) | Source and treatment details |
|---|---|---|---|---|---|---|---|
| 2 | 3, 4 | 5 | 40 | −1 | 117 | 1 | Table 1 and Figure 2 as reported by Wagner et al. [25]; treatment = glyphosate (Vision) at 4 kg ai/ha (liquid formulation); effectiveness = reduction of 92% in herbaceous vegetation cover, relative to the untreated control plots, at the time of last remeasurement (n., nil competition from woody species). |
| 3 | 0, 1, 2 | 5 | 82 | 31 | 336 | −4 | Table 1 and Figure 2 as reported by Wagner et al. [25]; treatment = glyphosate (Vision) at 4 kg ai/ha (liquid formulation); effectiveness = reduction of 44% in herbaceous vegetation cover, relative to the untreated control plots, at the time of last remeasurement (n., nil competition from woody species). |
| 3 | 2, 3, 4 | 5 | 79 | −2 | 270 | −8 | Table 1 and Figure 2 as reported by Wagner et al. [25]; treatment = glyphosate (Vision) at 4 kg ai/ha (liquid formulation); effectiveness = reduction of 93% in herbaceous vegetation cover, relative to the untreated control plots, at the time of last remeasurement (n., nil competition from woody species). |
| 3 | 0, 2, 4 | 5 | 126 | 26 | 471 | — | Table 1 as reported by Sutherland et al. [22] and Figure 1(d) as presented in Sutherland and Foreman [23]; treatment = hexazinone (Velpar L) at 3.1 kg ai/ha (liquid formulation) in year 0 followed by chemical tending in years 2 and 4 using glyphosate (Vision) at 1.78 kg ai/ha (liquid formulation); effectiveness = reduction of 96% in combined herbaceous and woody vegetation (principally red raspberry and trembling aspen; vegetation measure based on height × cover index), relative to untreated control plots, at time of last remeasurement. |
| 4 | 0, 1, 2, 3 | 5 | 109 | 22 | 426 | −8 | Table 1 and Figure 2 as reported by Wagner et al. [25]; treatment = glyphosate (Vision) at 4 kg ai/ha (liquid formulation); effectiveness = reduction of 57% in herbaceous vegetation cover, relative to the untreated control plots, at the time of last remeasurement (n., nil competition from woody species). |
| 4 | 1, 2, 3, 4 | 5 | 79 | 15 | 270 | −5 | Table 1 and Figure 2 as reported by Wagner et al. [25]; treatment = glyphosate (Vision) at 4 kg ai/ha (liquid formulation); effectiveness = reduction of 94% in herbaceous vegetation cover, relative to the untreated control plots, at the time of last remeasurement (n., nil competition from woody species). |
| 4 | −1, 0, 1, 2 | 14 | 93 | 43 | 125 | 12 | Figures 13, 14, 15, and 16 as reported by Robinson et al. [28]; results for the scarified but unfertilized plots only; treatment = glyphosate (Vision) at 2 kg ai/ha (liquid formulation); effectiveness = approximate reduction of 100% in herbaceous and woody vegetation cover, relative to the untreated control plots, 3 years post-planting. |
| 5 | 0, 1, 2, 3, 4 | 5 | 213 | 96 | — | 20 | Tables 2 and 3 as reported by Wood and von Althen [21]; mean responses based on bareroot and container stock-types results combined; treatment = annual application of glyphosate (Roundup) at 2 kg ai/ha (liquid formulation); effectiveness = reductions of 90% and 100% in herbaceous and woody vegetation cover, respectively, relative to the untreated control plots, at time of last remeasurement. |
| 5 | 0, 1, 2, 3, 4 | 5 | 108 | 22 | 420 | 12 | Table 1 and Figure 2 as reported by Wagner et al. [25]; treatment = glyphosate (Vision) at 4 kg ai/ha (liquid formulation); effectiveness = reduction of 96% in herbaceous vegetation cover, relative to the untreated control plots, at the time of last remeasurement (n., nil competition from woody species). |
| 5 | 2, 3, 4, 5, 6 | 10 | 87 | 46 | 246 | — | Textual description and Figure 1 as reported in Pitt et al. [29] for Corrigal experimental site; annual directed foliar application of 1.58 or 2% liquid solution of glyphosate (Vision); effectiveness: at the time of the 10 yr remeasurement the average of all treatments had reduced the cover of deciduous woody tree species to less than 10%, tall shrub species to 8%, and low shrub species to 20%. |

Table 1: Continued.

| $T_N^a$ ($n$) | $T_I^a$ (yr) | $T_L^a$ (yr) | Relative response[b] $R_D$ (%) | $R_H$ (%) | $R_V$ (%) | $R_S$ (%) | Source and treatment details |
|---|---|---|---|---|---|---|---|
| 5 | 3, 4, 5, 6, 7 | 10 | 87 | 8 | 126 | — | Textual description and Figure 1 as reported in Pitt et al. [29] for Hele experimental site; annual directed foliar application of 1.58 or 2% liquid solution of glyphosate (Vision); effectiveness: at the time of the 10 yr remeasurement the average of all treatments had reduced the cover of deciduous woody tree species to less than 10%, tall shrub species to 2%, and low shrub species to 20%. |
| 6 | −1, 0, 1, 2, 3, 4 | 5 | 221 | 95 | — | 22 | Tables 2 and 3 as reported by Wood and von Althen [21]; mean responses based on bareroot and container stock-types results combined; treatment = annual application glyphosate (Roundup) at 2 kg ai/ha (liquid formulation) including pre-planting treatment; effectiveness = reductions of 90% and 100% in herbaceous and woody vegetation cover, respectively, relative to the untreated control plots, at time of last remeasurement. |

Note, all selected studies were located within the Great Lakes - St. Lawrence or Boreal Forest Regions [30].

[a]Number of sequential treatments ($T_N$); Initial year of treatment application relative to the year of seedling establishment (−1 = year before planting (e.g., chemical site preparation); 0 = year of seedling establishment; 1, 2, . . . , 7 years after seedling establishment, resp.; ($T_I$)); and Length of the observation period relative to the year of seedling establishment ($T_L$).

[b]$R_k = 100 \cdot (\overline{T}_k - \overline{C}_k)/\overline{C}_k$ where $R_j$ is the relative response (%) of the $k$th yield variate ($D$ = basal stem diameter; $H$ = total stem height; $V$ = total stem volume; and $S$ = survival), $\overline{T}_k$ is the mean value of the $k$th yield variate within the treated population, and $\overline{C}_k$ is the mean value of the $k$th yield variate within the untreated control population.

plots within the $k$th experiment, $v_{j(k)}$ is variation due to sampling of the $j$th yield variate within the $k$th experiment, $w_{j(k)}$ is equal to the inverse of $v_{j(k)}$, and $\hat{\sigma}_j^2$ is the among-study variation associated with the $j$th yield variate. The 95% confidence intervals for $\overline{L}_j^*$ were first calculated according to (3) and then reexpressed in arithmetic terms:

$$\overline{L}_j^* - z_{\alpha/2} \cdot \mathrm{SE}\left(\overline{L}_j^*\right) \le \mu \le \overline{L}_j^* + z_{\alpha/2} \cdot \mathrm{SE}\left(\overline{L}_j^*\right), \quad \text{where } \mathrm{SE}\left(\overline{L}_j^*\right) = \sqrt{\frac{1}{\sum_{k=1}^{n_{j(k)}} w_{j(k)}^*}}, \tag{3}$$

where $z_{\alpha/2}$ is the 97.5% point of the standard normal distribution, and $\mathrm{SE}(\overline{L}_j^*)$ is the standard error of the weighted mean associated with the $j$th yield variate. Note that the variance in response among treatments within an individual experiment was used as a surrogate measure of sample variation ($v_{j(k)}$) given the lack of studies reporting treatment-specific sample variation. Consequently, the resultant mean values and the associated confidence limits represent the average collective response of black spruce over a range of FVM treatments.

*2.2. Modeling Long-Term Yield Responses.* Responses to FVM treatments were assumed to follow a Type I growth response pattern (sensu [33]): a temporary increase in growth rate that advances the stage of stand development but does not change the inherent productivity of a site. Consequently, an approach analogous to the technique used to account for genetic worth effects in growth and yield projections was employed (i.e., genetic worth effect is quantified as the temporary and instantaneous increase in mean dominant height growth at the specified selection age which is thereafter proportionally discounted until rotation age [34]). Thus the FVM treatment effect was defined as the temporary and instantaneous increase in mean dominant height growth ($V_E$; %) at a specified effect age ($V_A$; yr) which thereafter proportionally declined until rotation age. Mathematically, the following script describes the computation:

$$\begin{aligned} &\text{If } V_E > 0, \quad \text{then for } t = V_A, \\ &\hat{H}'_{d(t)} = \hat{H}_{d(t)} + \left(\frac{V_E}{100}\right)\hat{H}_{d(t)}, \\ &\text{otherwise when } t > V_A, \\ &\hat{H}'_{d(t)} = \hat{H}_{d(t)} + \left(\frac{V_E}{100} - \left(\left(\frac{V_E/100}{R_A - V_A}\right)(t - V_A)\right)\right)\hat{H}_{d(t)}, \end{aligned} \tag{4}$$

where $\hat{H}_{d(t)}$ is the dominant height (m) at time $t$ as predicted by the site-based height-age function, $\hat{H}'_{d(t)}$ is the treatment-adjusted dominant height (m) at time $t$, and $R_A$ is the rotation age (yr). Furthermore, in order to account for the effect of FVM treatments on survival, the annual mortality rate during the pre-crown closure period within the untreated stand was adjusted according to the following computational script:

$$\begin{aligned} &\text{If } S_E > 0 \text{ then for } t < T_{CC}, \\ &M_{(t)} = \hat{N}_{(t)}\left(1 - \left(\frac{1.0}{1 + S_E/100S_A}\right)\right), \end{aligned} \tag{5}$$

where $S_E$ is the grand mean survival response (%) occurring at a specified effect age ($S_A$; yr), $T_{CC}$ is the age of the untreated stand at crown closure, $\hat{N}_{(t)}$ is the model-based predicted density (stems/ha) at time $t$ within the control stand, and $M_{(t)}$ is the annual increase in the mortality rate (stems/ha/yr) at time $t$.

These modifications to the height and survival computations were then incorporated within the CROPLANNER decision-support model [5, 6]. Briefly, CROPLANNER is the algorithmic analogue of the structural stand density management model (SSDMM) which was developed for black spruce and jack pine stand-types through the expansion of the dynamic stand density management diagram modelling framework. Structurally, the model consists of a number of functional and empirical quantitative relationships, which collectively represent the cumulative effect of various underlying competition processes on tree and stand yield parameters. The temporal dependency of these processes is governed by the intensity of competition and site quality as expressed by relative density index and site index, respectively. Hence, the site-specific mean dominant height-age function largely governs the rate of stand development. Thus embedding the anticipated Type I response pattern within this function was considered a logical approach to modeling the effects of FVM treatments on growth. For a complete analytical description of the approach used in the development and calibration of the modular-based SSDMMs, refer to [5, 6].

## 3. Results

*3.1. Early Responses and Their Correlation to Experimental Variables.* Results from 12 experiments were used in the calculation of the grand mean response ratios and associated 95% confidence intervals. The author(s) of these studies were as follows: Wood and von Althen [21], Reynolds and Roden [19, 20], Wood and Mitchell [27], Jobidon and Charette [26], Jobidon et al. [24], Pitt et al. [18], Wagner et al. [25], Sutherland et al. [22], Sutherland and Foreman [23], Robinson et al. [28], Pitt et al. [29], and Fu et al. [35]. Experimentally, the studies differed in terms of the number of sequential treatments applied, initial year of treatment application relative to the year of seedling establishment, and length of the observation period relative to the year of seedling establishment: $\overline{T}_N$ = 1.9 (min/max = 1/6), $\overline{T}_I$ = 0.4 yr (−1/4), and $\overline{T}_L$ = 6.7 yr (4/15). The mean percentage response for basal diameter, total height, stem volume, and survival across all experiments was, respectively, 66.9 (min/max/$n$ = 14.9/220.9/44), 29.3 (−1.6/96.0/43), 208.1 (68.6/470.7/25), and 2.5 (−37.7/47.6/37). Table 1 summarizes the characteristics of the selected studies in terms of their experimental design, type and number of sequential treatments applied, time of initial treatment relative to seedling establishment, duration of response period, effectiveness of the treatments, and mean responses observed.

The mean effect size and associated 95% confidence intervals for basal diameter, total height, stem volume, and survival were, respectively, 54.7% (95% confidence limits (lower/upper): 34.8/77.6), 27.3% (15.7/40.0), 198.7% (70.3/423.5), and 2.9% (−5.5/11.8). Hence, FVM treatments resulted in significant ($P \leq 0.05$) early gains in mean tree size. Linear correlation analysis revealed the following significant ($P \leq 0.05$) associations: (1) mean effect size for diameter (product-moment correlation coefficient ($r$) = 0.74) and volume ($r$ = 0.48) increased with increasing $T_N$, and (2) mean effect size for height ($r = -0.53$) and survival ($r = -0.38$) decreased with increasing $T_I$. These results suggest that early and repeated FVM treatments will yield the largest gains in terms of mean tree size and survival.

*3.2. Long-Term Effects on Yield Outcomes and Stand Operability Status.* The lower and upper limits of the 95% confidence interval for the mean percent gain in height growth ($V_E$) and the corresponding mean effect age ($V_A$), along with the mean percent gain in survival ($S_E$) and the associated mean effect age ($S_A$), were used as input to the modified CROPLANNER model. A conventional silvicultural regime applicable to the upland black spruce stand-type was simulated: planting 2100 seedlings per hectare on a scarified site was modeled over a 75-year rotation for a poor-medium quality site (site index = 14 [36]), and a 50-year rotation for a good-excellent (site index = 18 [36]) quality site. The objective underlying the FVM treatment was to reduce the time to operability status as defined by site-specific piece-size and merchantable volume productivity thresholds: 15 stems/m$^3$ and 150 m$^3$/ha for the lower site quality, and 10 stems/m$^3$ and 200 m$^3$/ha for the higher site quality. Three regimes were assessed on each site: (1) Regime 1 consisted of an untreated control stand in which the short-lived competitors (e.g., ericaceae plants, herbaceous forbs, shrubs, and small trees) were assumed to be eventually shaded out once crown closure status was achieved and hence the post-closure succession pattern followed that of monospecific plantation; (2) Regime 2 was subjected to an FVM treatment in which the response was assumed to follow the patterns described by the aforementioned scripts using the results from the meta-analysis; specifically, the (i) height growth increase was defined by the lower limit of the 95% confidence limit, that is, $V_E$ = 15.7%, which took effect at the specified $V_A$ of 7 yr, and (ii) survival increase was defined by the mean value derived from all the studies ($S_E$ = 2.9%; $S_A$ = 7 yr); (3) Regime 3, similar to Regime 2, was subjected to an FVM treatment in which the (i) height growth response was defined by the upper limit of the 95% confidence limit, $V_E$ = 40.0%, taking effect at the specified $V_A$ of 7 yr, and (ii) survival increase was defined by the mean value derived from the studies ($S_E$ = 2.9%; $S_A$ = 7 yr).

The resultant mean volume-density trajectories for each regime within the context of the CROPLANNER-based graphic are illustrated in Figures 1 and 2 for plantations situated on poor-medium and good-excellent site qualities, respectively. Although the graphics can be used to derive estimates of density, mean volume, total volume, quadratic mean diameter, basal area, and mean live crown ratio, for any age on a given trajectory via graphical interpolation, they are presented so that the general stand development implications of FVM treatments can be visualized and interpreted within the context of the widely used stand density management diagram. Contrasting the two diagrams reveals that the trajectories progressed through the size-density space at a greater rate on the good-excellent site than on the poor-medium site, irrespectively of treatments. The divergent mean volume-density trajectories among the regimes during the pre-crown closure period (i.e., portion of the trajectories that are below the crown closure line) are indicative of the greater mortality rate arising from interspecific competition effects within the control stands. Similarly, this differential in survival rate is carried forward until rotation age as evident from the separation of the control and treated trajectories

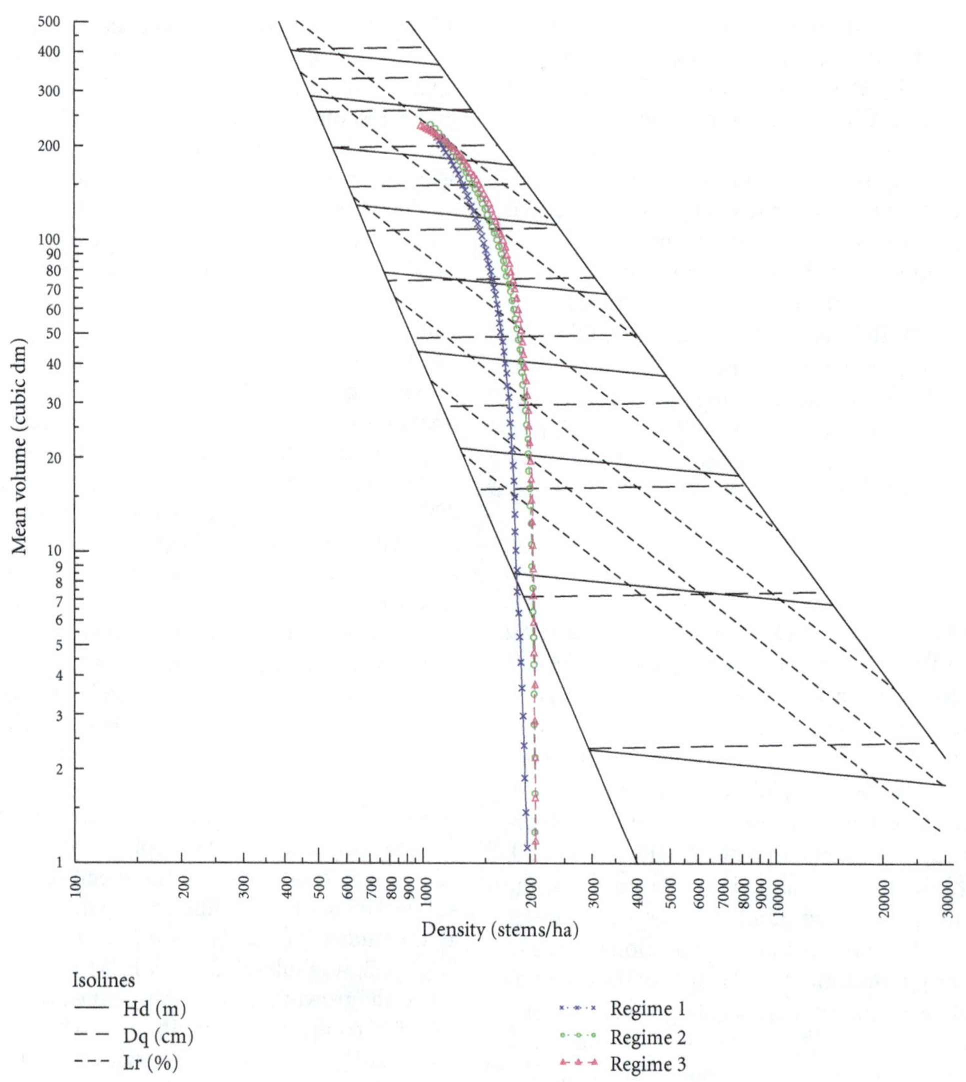


FIGURE 1: Seventy-five year temporal size-density trajectories for upland black spruce plantations situated on low-medium quality sites (site index = 14): Regime 1—no FVM treatment applied; Regime 2—FVM treatment with an expected effect of a 15.7% height gain at age 7; and Regime 3—FVM applied with an expected effect of a 40.0% height gain at age 7. Graphically, illustrating the (1) isolines for mean dominant height (Hd; 4–20 m by 2 m intervals), quadratic mean diameter (Dq; 4–26 cm by 2 cm intervals), and mean live crown ratio (Lr; 35, 40, 50, ..., 80%), and (2) crown closure line (lower boundary of the size-density space as delineated by the lower left diagonal solid line) and the self-thinning rule (upper boundary of the size-density space as delineated by the upper right diagonal solid line).

during the post-crown closure period. In accordance with the Type I response pattern assumed in the model, all three regimes approached equivalence in terms of their size-density trajectories and associated yield-based productivity indices at rotation, as inferred from the approximate convergence of the size-density trajectories. However, it was evident from the annual progression within the trajectories that the treated regimes experienced an acceleration in the rate of their development during the post-treatment period.

Figure 3 illustrates the predicted mean dominant height development with the FVM treatment effect incorporated. Based on the Type I response pattern, (1) height growth underwent an instantaneous acceleration at the specified effect age (7 yr) which then declined for the remaining portion of the rotation, and (2) from the point of treatment until rotation age, the productivity as measured by site height was always greater within the treated stands than that within the untreated stands. This increase in the rate of stand development resulted in crown closure occurring approximately 5 yr earlier within the treated stands, than within the control stands. Over the rotation, the effect of the increase in the rate of stand development and survival was to increase site occupancy and volumetric production, which translated into a substantial reduction in the time to operability status. Specifically, examining the temporal dynamics of the piece size-age (Figure 4) and the merchantable volume-age (Figure 5) relationships derived from the modified CROPLANNER model, indicated that the operability thresholds for the treated and control regimes were attained at (1) ages 39 and 45 for the 15 stems/m$^3$ piece size threshold, respectively, and

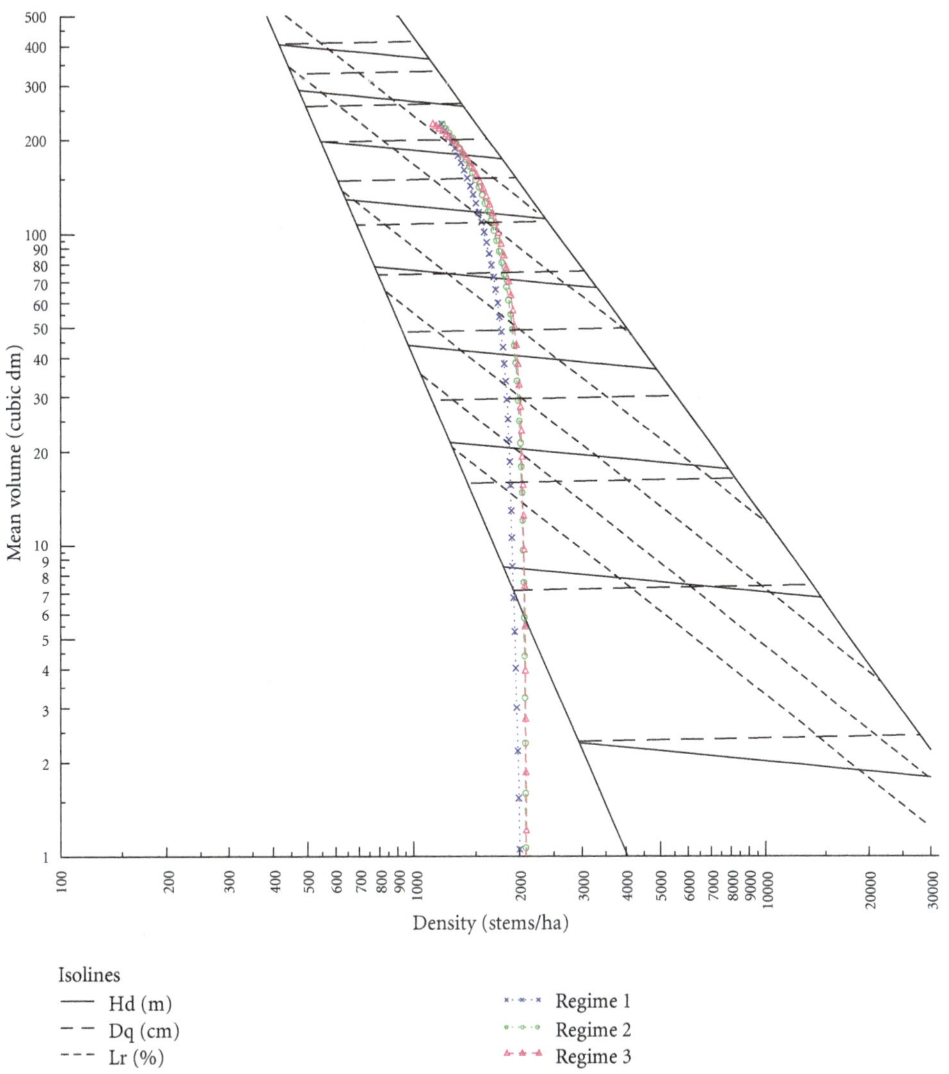


FIGURE 2: Fifty-year temporal size-density trajectories for upland black spruce plantations situated on good-excellent quality sites (site index = 18): Regime 1: no FVM treatment applied; Regime 2: FVM treatment with an expected effect of a 15.7% height gain at age 7; and Regime 3: FVM applied with an expected effect of a 40.0% height gain at age 7. Graphical denotations are given in Figure 1.

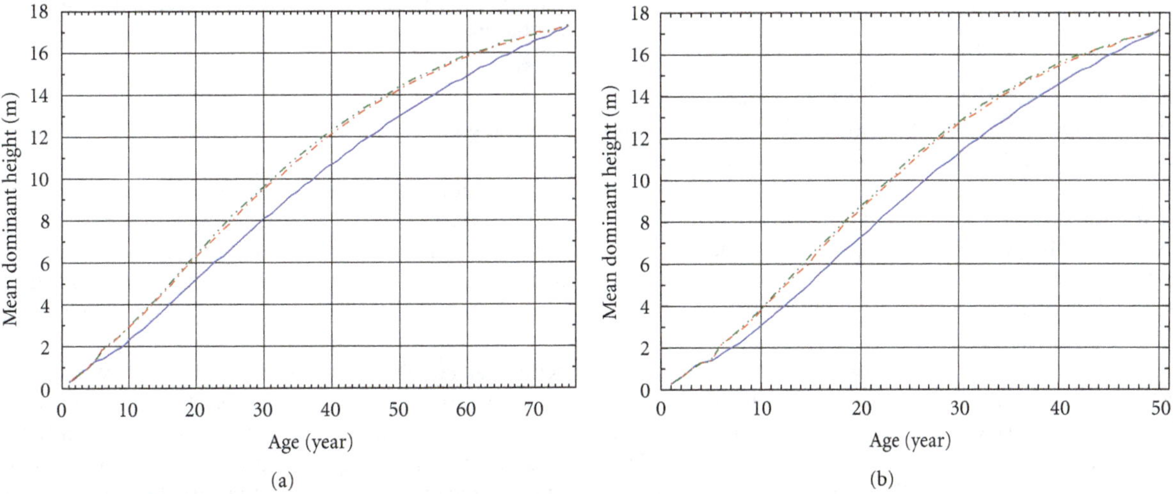


FIGURE 3: Type I response pattern as reflected in the mean dominant height-age relationship for poor-medium (a) and good-excellent (b) site qualities: Regime 1: no FVM treatment applied (solid line); Regime 2: FVM treatment with an expected effect of a 15.7% gain at age 7 (long-short dash line); and Regime 3: FVM applied with an expected effect of a 40.0% gain at age 7 (long-short-short dash line).

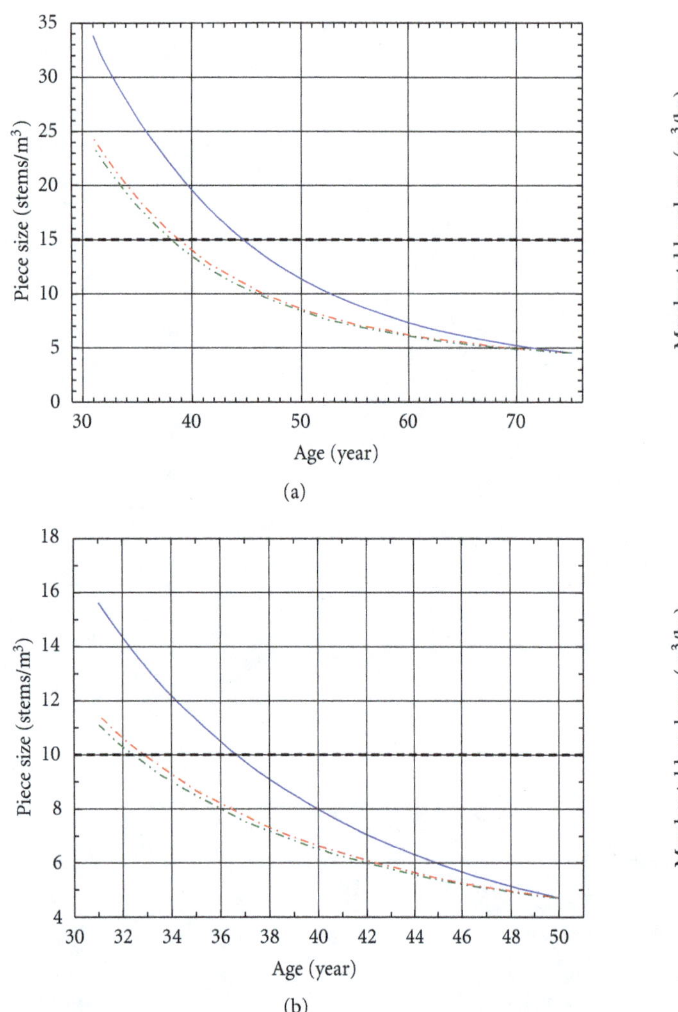


FIGURE 4: Type I response pattern as reflected in the piece size-age relationship for poor-medium (a) and good-excellent (b) site qualities. Regime denotations are given in Figure 3. Note that the intersection of the horizontal dotted line denoting the specified operability criterion with a production curve indicates the earliest age at which the piece size operability criterion is attained.

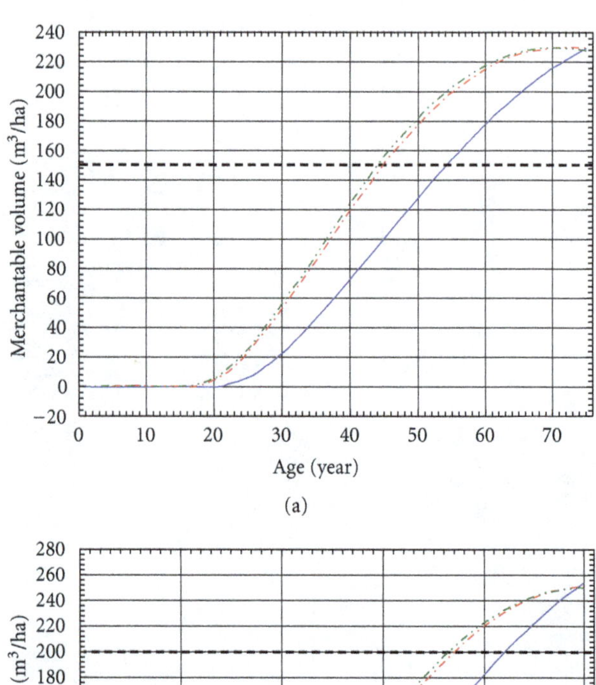


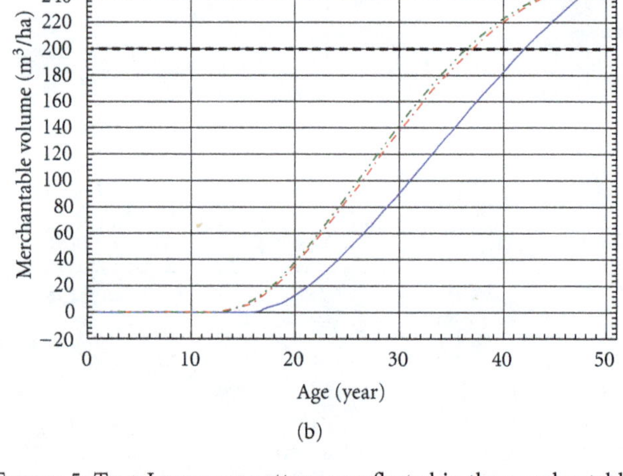


FIGURE 5: Type I response pattern as reflected in the merchantable volume-age relationship for poor-medium (a) and good-excellent (b) site qualities. Regime denotations are given in Figure 3. Note that the intersection of the horizontal dotted line denoting the specified operability criterion with a production curve indicates the earliest age at which the merchantable volume operability criterion is attained.

at ages 45 and 54 for the 150 $m^3$/ha merchantable volume thresholds, respectively, on the poor-medium quality site (Figures 4(a) and 5(a), resp.), and (2) ages 33 and 37 for the 10 stems/$m^3$ piece size threshold, respectively, and at ages 37 and 42 for the 200 $m^3$/ha merchantable volume thresholds, respectively, on the good-excellent quality site (Figures 4(b) and 5(b), resp.). Thus relative to the untreated control stands, the earliest time that the treated stands reached the specified piece size and merchantable volume thresholds was approximately 9 yr less on the poor-medium site (earliest time at which both operability criteria were realized: 45 yr for the treated stands versus 54 yr for the control stand), and approximately 5 yr less on the good-excellent site (37 yr for the treated stands versus 42 yr for the control stand). These operability improvements suggest that substantial reductions in economic rotation lengths are possible with the application of early and repeated FVM treatments.

## 4. Discussion

Competition from interspecific herbaceous and woody competitors during the early stages of plantation development may involve a continuum ranging from a resource depletion to a preemption process (sensu [37]). The resource depletion process is a symmetrical relationship in which all species compete for belowground resources (e.g., all competitors acquire an equal share of the available nutrients and moisture on a per-unit size basis). Conversely, the resource preemption process is an asymmetrical relationship in which the larger-sized competitors acquire a greater size-proportional share of the aboveground resources than smaller-sized competitors (e.g., larger-sized competitors acquire a greater share of the solar-based resources than smaller-sized competitors on a per-unit size basis). FVM treatments are an effective method of minimizing the detrimental effects of these competition processes on the crop tree population via the selective elimination of interspecific competitors. However, the duration of the effect is dependent on the effectiveness of the FVM treatment applied in terms of reducing both the herbaceous and noncrop woody competitors. Based on a collective review of studies in the Pacific Northwest, Lake States, and Ontario, Wagner [38] suggested that removing interspecific woody competitors may only secure an increase in the order of 20–30% of early crop tree volume growth potential and hence achieving the remaining 70–80% will also require effective control of the herbaceous vegetation.

The results of this study indicated that relative to untreated control stands, mean basal diameter, total height, stem volume, and survival within treated stands, increased by 54.7%, 27.3%, 198.7%, and 2.9%, respectively. These responses correspond to the maximal treatment reported in the selected studies and consequently represent the best-case scenario, irrespective of the control mechanism employed. These empirical results are in accordance with the scientific consensus regarding the general effectiveness of FVM treatments on conifers [8, 39]: effective removable of competing vegetation results in initial increases in growth. The results also indicated that early and repeated treatments yielded the largest gains. The generalized responses are consistent with those reported in individual experiments which did not meet the selection criteria employed in this study. For example, Hoepting et al. [40] reported on 15 yr results for a set of Ontario-based black spruce plantations which were repeatedly treated from age 1 to 5 yr following planting. The best treatment, consisting of a single direct foliar application of 2 kg ai/ha of glyphosate, resulted in significant ($P \leq 0.05$) gains in basal diameter and total height of 50% and 30%, respectively, over untreated control plantations.

Although results from long-term empirical experiments have yet to provide a clear picture in regards to the permanency of FVM gains, the modelling approach as applied in this study, provided insight into a plausible range of yield outcomes at rotation under a conventional silvicultural regime. Results from some of the older black spruce experiments are starting to provide a preliminary perspective on the longer-term consequences of FVM treatments. Cyr and Thiffault [41] reported 22 yr post-treatment results for black spruce plantations which were mechanically released in the second year of their development. Results from this experiment indicated that the initial diameter and height gains observed at 5 yr post-treatment were still significant ($P \leq 0.05$), although diminished in magnitude, by age 22. Significant ($P \leq 0.05$) survival gains (21.5%) were also evident; however they were not detected at the 5 yr post-treatment assessment. A similar pattern of decline was also reported by Hoepting et al. [40]. Collectively, these results suggest that untreated and treated plantations follow different developmental trajectories and thus support the modelling assumptions employed in this study. Specifically, in contrast to untreated control stands, FVM-treated stands exhibit an initial acceleration in growth which dissipates in magnitude over time.

Modelling has been used to project the consequences of yield gains arising from FVM treatments given the general lack of long-term experimental response data. The efforts have involved the employment of process-based and empirical models (e.g., benchmark yield curves [42]; forest vegetation simulator [43]). The approach utilized in this study consisted of modifying the CROPLANNER simulation model in order to account for changes in height growth and survival arising from FVM treatments through adjustments to the height growth and mortality elements of the model. These modifications were consistent with the postulated long-term effects of FVM treatments: a temporary increased in the rate of stand development and a reduction in the occurrence of suppression-related mortality during the pre-crown closure stage of stand development. Specifically, under the assumption that the response in height growth follows a Type I pattern after FVM treatments are applied, the site-based height-age function was modified accordingly. Although empirical-based results from long-term experiments for boreal species are currently insufficient in regards to empirically testing this assumption, early remeasurement data [41] and rotation length observations from other forest regions [44] do not contradict the hypothesis that FVM treatments result in an instantaneous increase in growth which dissipates over time.

The modeling approach proposed in this study is similar to procedure employed by South et al. [45] to project long-term yield gains of loblolly pine (*Pinus taeda* L.) arising from FVM treatments. A graphical age-shift adjustment was made to the volume-age production curve in which the treated stand artificially advanced along the function for a specified number of years. This adjustment reflected the FVM treatment-induced acceleration in the rate of stand development which effectively resulted in a reduction in the time required to attain a specified volume threshold. Supporting empirical data consisting of 20 yr remeasurement data derived from 11 FVM studies revealed that the percentage difference in merchantable volume production between control and treated stands declined with age.

In addition to the Type I response in which the carrying capacity of a site is not increased due to treatment, Type II and III responses patterns in which carrying capacity increases and decreases, respectively, have been postulated for some specific species and site conditions. Richardson [44] hypothesized that radiata pine (*Pinus radiata* D. Don) may exhibit a Type II pattern in situations where a competitor which is tolerant to low light conditions and competes aggressively for below ground resources (i.e., symmetric competition for moisture and nutrients) remains in the control stand until rotation. This ultimately could result in the degradation of site quality and result in a decline in the carrying capacity of the untreated control stand, relative to the treated stand. Conversely, a Type III pattern could conceivably arise in cases where a light intolerant competitor rapidly establishes itself on a control site and acquires the below ground resources that would normally be lost to leaching. Subsequently at the time of crown closure, the competitor could release these conserved nutrients upon decomposition, resulting in an increase in the rate of stand development. Thus in this situation, productivity may actually increase in the untreated stand relative to the treated stand. However, irrespective of these theoretical alternative response patterns, the existing empirical evidence largely supports the applicability of the Type I response pattern for modeling the yield consequences of FVM treatments (e.g., [41, 44]).

The modelled increase in survival until the point of crown closure for the treated stands assumes that the benefit of avoiding mortality of the crop tree species will only last until crown closure occurs. Considering that the principal competition on boreal sites has been characterized as consisting of short-lived ericaceae species (e.g., *Kalmia angustrifolia* L., Labrador tea (*Rhododendron groenlandicum* (Oeder) K. A. Kron and Judd), and blueberries (*Vaccinium* spp.)) which compete symmetrically for below-ground resources, and herbaceous forbs (e.g., fireweed (*Epilobium angustifolium* L.), small shrubs (e.g., red raspberry (*Rubus idaeus* L.), and small trees (e.g., mountain maple (*Acer spicatum* Lamb.), beaked hazel (*Corylus cornuta* Marsh.), and pin cherry (*Prunus pensylvanica* L. F.)) which compete asymmetrically for above-ground resources [16, 46], this was a reasonable assumption. Given that long-term experimental results are lacking to test this assumption and early survival gains reported during the pre-crown closure period are often characterized by considerable variation in terms of their magnitude and duration (e.g., [29, 41, 47]), the use of a constant but developmental-limited mortality rate (until crown closure) to account for survival differences, was considered a prudent approach.

The CROPLANNER simulation model used in this study does not explicitly account for the potential effect of competing vegetation on succession patterns within untreated stands. Basically, the model assumes that once crown closure is achieved, the competitors will be eventually shaded out and the untreated stand will return to a mostly monospecific condition (e.g., >90% black spruce in terms of basal area). Consequently, this assumption is only realistic if the population of competitors is comprised of short-lived ericaceous species, herbaceous forbs, small shrubs, and (or) small trees. Conversely, if woody competitors remained within the control plantations following crown closure (e.g., ≥10% of the basal area composed of interspecific woody competitors, such as trembling aspen (*Populus tremuloides* (Michx.)) and white birch (*Betula papyifera* (Marsh.)), then the uses of the model would result in an under estimation of the effects the FVM treatments. Bell et al. [42] attempted to project the long-term effects of FVM treatments on black spruce volume productivity using observed differences in net merchantable volume, diameter, density, and species composition between treated and control stands at 10 yrs post-treatment, in combination with a set of empirical benchmark yield curves. The assumption that species composition remained static from age 10 to a rotation age (60 yr) was implemented. The results suggested that the large differences in species composition at year 10 greatly influence the merchantable yield outcomes at year 60: the preferred net merchantable conifer volume on the treated plots was 152% (minimum–maximum: −55 to 670%) greater than that projected for the control plots (*n*, values derived from the Nipigon Hele and Nipigon Corrigal experiments as reported in Table 7 of Bell et al. [42]). Similarly, Homagain et al. [43] employed the forest vegetation simulator in combination with a bucking algorithm to project 70 yr rotational effects of FMV treatments, in order to conduct a cost and benefit analysis for these same two experiments. Although the projection period differed by only 10 years between the studies (60 yr in Bell et al. [42] versus 70 yr in Homagain et al. [43]), the estimated merchantable yields were different. Homagain et al. [43] reported that the net merchantable preferred conifer volume on the treated plots was only 29% (minimum–maximum: 22–41%) greater than that projected for the control plots (*n*, values derived from the Nipigon Hele and Nipigon Corrigal experiments as reported in Table 3 of Homagain et al. [43]). These differences highlight the variation in outcomes that can arise when using different modeling approaches and assumptions. The approach utilized in this study, in which FVM treatment effects are explicitly modelled in terms of height and survival gains based on a Type I response pattern, represents an alternative approach to modeling the long-term effects of FVM treatments. Nevertheless, although modeling provides a plausible range of outcomes to FVM treatments, the rotational consequences of FVM treatments will not be fully ascertained until remeasurement data from long-term experiments become available.

## 5. Conclusion

The results and associated inferences of this study clearly illustrated the important role FVM treatments can play in minimizing the negative yield consequences of interspecific competition on black spruce. Consequential gains in basal diameter, total height, stem volume, and survival were evident within the initial post-treatment period and when, projected over the rotation, resulted in gains in stand-level operability. However, the modelling assumptions used in these long-term projections still require empirical validation and hence these rotational effects should be considered tentative. Remeasurement data derived from long-term experiments are required to fully understand the growth response pattern and the magnitude of the yield gains possible through FVM treatments. Irrespective of the evolving shift from phytocidal chemicals to various nonchemical treatment alternatives [17, 47], the challenge of maintaining coniferous forest cover on recently distributed sites, attaining statutory-defined free-to-grow status, and ensuring long-term productivity, suggests that FVM will continue to be an essential silvicultural treatment option.

## Acknowledgments

The author expresses his appreciation to the Forestry Research Partnership and the Ontario Living Legacy Trust, for fiscal support.

## References

[1] T. Erdle, “Forest level effects of stand level treatments: using silviculture to control the AAC via the allowable cut effect,” in

*Expert Workshop on the Impact of Intensive Forest Management on the Allowable Cut*, P. F. Newton, Ed., pp. 19–30, Forestry Research Partnership, Canadian Ecology Centre, Mattawa, Ontario, Canada, 2001.

[2] P. F. Newton, "Systematic review of yield responses of four North American conifers to forest tree improvement practices," *Forest Ecology and Management*, vol. 172, no. 1, pp. 29–51, 2003.

[3] P. F. Newton and I. G. Amponsah, "Systematic review of short-term growth responses of semi-mature black spruce and jack pine stands to nitrogen-based fertilization treatments," *Forest Ecology and Management*, vol. 237, no. 1–3, pp. 1–14, 2006.

[4] P. F. Newton, "Meta-analytical trends in diameter response of black spruce and jack pine to pre-commercial thinning," in *Proceedings of the 2nd International Conference on Forest Measurements and Quantitative Methods and Management & The 2004 Southern Mensurationists Meeting*, C. J. Cieszewski and M. Strub, Eds., pp. 217–223, Fiber Supply Assessment, Center for Forest Business, Warnell School of Forestry and Natural Resources, University of Georgia, Athens, Ga, USA, 2006.

[5] P. F. Newton, "Development of an integrated decision-support model for density management within jack pine stand-types," *Ecological Modelling*, vol. 220, no. 23, pp. 3301–3324, 2009.

[6] P. F. Newton, "A decision-support system for density management within upland black spruce stand-types," *Environmental Modelling & Software*. In revision.

[7] J. D. Walstad and P. J. Kuch, "Introduction to forest vegetation management," in *Vegetation Management For Conifer Production*, J. D. Walstad and P. J. Kuch, Eds., pp. 3–14, John Wiley & Sons, Toronto, Canada, 1987.

[8] R. G. Wagner, K. M. Little, B. Richardson, and K. McNabb, "The role of vegetation management for enhancing productivity of the world's forests," *Forestry*, vol. 79, no. 1, pp. 57–79, 2006.

[9] F. W. Bell, N. Thiffault, K. Szuba, N. J. Luckai, and A. Stinson, "Synthesis of silviculture options, costs, and consequences of alternative vegetation management practices relevant to boreal and temperate conifer forests: introduction," *Forestry Chronicle*, vol. 87, no. 2, pp. 155–160, 2011.

[10] D. G. Thompson and D. G. Pitt, "A review of Canadian forest vegetation management research and practice," *Annals of Forest Science*, vol. 60, no. 7, pp. 559–572, 2003.

[11] J. D. Walstad, M. Newton, and D. H. Gjerstad, "Overview of vegetation management alternatives," in *Forest Vegetation Management For Conifer Production*, J. D.and Walstad and P. J. Kuch, Eds., pp. 157–200, John Wiley & Sons, Toronto, Ontario, Canada, 1987.

[12] F. W. Bell, K. R. Ride, M. L. St-Amour, and M. Ryans, "Productivity, cost, efficacy and cost effectiveness of motor-manual, mechanical, and herbicide release of boreal spruce plantations," *Forestry Chronicle*, vol. 73, no. 1, pp. 39–46, 1997.

[13] F. W. Bell, R. A. Lautenschlager, R. G. Wagner, D. G. Pitt, J. W. Hawkins, and K. R. Ride, "Motor-manual, mechanical, and herbicide release affect early successional vegetation in northwestern Ontario," *Forestry Chronicle*, vol. 73, no. 1, pp. 61–68, 1997.

[14] J. A. Matarczyk, A. J. Willis, J. A. Vranjic, and J. E. Ash, "Herbicides, weeds and endangered species: management of bitou bush (*Chrysanthemoides monilifera* ssp. rotundata) with glyphosate and impacts on the endangered shrub, *Pimelea spicata*," *Biological Conservation*, vol. 108, no. 2, pp. 133–141, 2002.

[15] V. Roy, N. Thiffault, and R. Jobidon, "Maîtrise intégrée de la végétation au Québec (Canada) : une alternative efficace aux phytocides chimiques," Note de recherche forestière n∘ 123, Direction de la recherche forestière, Gouvernement du Québec, Québec, Canada, 2003.

[16] N. Thiffault and V. Roy, "Living without herbicides in Québec (Canada): historical context, current strategy, research and challenges in forest vegetation management," *European Journal of Forest Research*, vol. 130, no. 1, pp. 117–133, 2011.

[17] N. McCarthy, N. S. Bentsen, I. Willoughby, and P. Balandier, "The state of forest vegetation management in Europe in the 21st century," *European Journal of Forest Research*, vol. 130, no. 1, pp. 7–16, 2011.

[18] D. G. Pitt, C. S. Krishka, F. W. Bell, and A. Lehela, "Five-year performance of three conifers stock types on fine sandy loam soils treated with hexazinone," *Northern Journal of Applied Forestry*, vol. 16, pp. 72–81, 1999.

[19] P. E. Reynolds and M. J. Roden, "Hexazinone site preparation improves black spruce seedling survival and growth," *Forestry Chronicle*, vol. 71, pp. 426–433, 1995.

[20] P. E. Reynolds and M. J. Roden, "Site preparation with sulfonylurea herbicides improves black spruce seedling growth," *Forestry Chronicle*, vol. 72, no. 1, pp. 80–85, 1996.

[21] J. E. Wood and F. W. von Althen, "Establishment of white spruce and black spruce in boreal Ontario: effects of chemical site preparation and post-planting weed control," *Forestry Chronicle*, vol. 69, no. 5, pp. 554–560, 1993.

[22] B. Sutherland, I. K. Morrison, F. F. Foreman, and P. E. Reynolds, "Response of black spruce seedlings and competitive vegetation following chemical and mechanical site preparation on a boreal mixedwood site in northern Ontario," in *Proceedings of the 3th International Conference on Forest Vegetation Management*, G. Wagner and D. G. Thompson, Eds., pp. 326–328, Ontario Forest Research Institute, Ministry of Natural Resources, Government of Ontario, Ontario, Canada, 1998, Forest Research Information Paper no. 141.

[23] B. Sutherland and F. R. Foreman, "Black spruce and vegetation response to chemical and mechanical site preparation on a boreal mixedwood site," *Canadian Journal of Forest Research*, vol. 30, no. 10, pp. 1561–1570, 2000.

[24] R. Jobidon, F. Trottier, and L. Charette, "Dégagement chimique ou manuel de plantations de l'épinette noire? Étude de cas dans la domaine de la sapinière à bouleau blanc au Québec," *Forestry Chronicle*, vol. 75, no. 6, pp. 973–979, 1999.

[25] R. G. Wagner, G. H. Mohammed, and T. L. Noland, "Critical period of interspecific competition for northern conifers associated with herbaceous vegetation," *Canadian Journal of Forest Research*, vol. 29, no. 7, pp. 890–897, 1999.

[26] R. Jobidon and L. Charette, "Effets, après 10 ans, du dégagement manuel simple ou répété et de al période de coupe de la végétation de compétition sur la croissance de l'épinette noire en plantation," *Canadian Journal of Forest Research*, vol. 27, no. 12, pp. 1979–1991, 1997.

[27] J. E. Wood and E. G. Mitchell, "Silvicultural treatments for black spruce establishment in boreal Ontario: effects of weed control, stock type, and planting season," NODA/NFP Technical Report TR-10, Great Lakes Forestry Centre, Canadian Forest Service number, Department of Natural Resources, Government of Canada, Sault Ste. Marie, Ontario, Canada, 1995.

[28] G. Robinson, S. Wetzel, and D. Burgess, "Remeasurement of Cartier Lake and Foleyet site preparation experiments," Interim Report, Forestry Research Partnership, Canadian Ecology Centre, Ontario, Canada, 2001.

[29] D. G. Pitt, R. G. Wagner, and W. D. Towill, "Ten years of vegetation succession following ground-applied release treatments in young black spruce plantations," *Northern Journal of Applied Forestry*, vol. 21, no. 3, pp. 123–134, 2004.

[30] J. S. Rowe, *Forest Regions of Canada*, Canadian Forestry Service, Department Environment, Government of Canada, Ottawa, Ontario, Canada, 1972.

[31] L. V. Hedges and I. Olkin, *Statistical Methods for Meta-Analysis*, Academic Press, New York, NY, USA, 1995.

[32] L. V. Hedges, J. Gurevitch, and P. S. Curtis, "The meta-analysis of response ratios in experimental ecology," *Ecology*, vol. 80, no. 4, pp. 1150–1156, 1999.

[33] P. Snowdon, "Modeling type 1 and type 2 growth responses in plantations after application of fertilizer or other silvicultural treatments," *Forest Ecology and Management*, vol. 163, no. 1–3, pp. 229–244, 2002.

[34] C. Y. Xie and A. D. Yanchuk, "Breeding values of parental trees, genetic worth of seed orchard seedlots, and yields of improved stocks in British Columbia," *Western Journal of Applied Forestry*, vol. 18, no. 2, pp. 88–100, 2003.

[35] S. Fu, F. W. Bell, and H. Y. H. Chen, "Long-term effects of intensive silvicultural practices on productivity, composition, and structure of northern temperate and boreal plantations in Ontario, Canada," *Forest Ecology and Management*, vol. 241, no. 1–3, pp. 115–126, 2007.

[36] W. H. Carmean, G. Hazenberg, and K. C. Deschamps, "Polymorphic site index curves for black spruce and trembling aspen in northwest Ontario," *Forestry Chronicle*, vol. 82, no. 2, pp. 231–242, 2006.

[37] P. F. Newton and P. A. Jolliffe, "Aboveground dry matter partitioning responses of black spruce to directional-specific indices of local competition," *Canadian Journal of Forest Research*, vol. 33, no. 10, pp. 1832–1845, 2003.

[38] R. G. Wagner, "Competition and critical-period thresholds for vegetation management decisions in young conifer stands," *Forestry Chronicle*, vol. 76, no. 6, pp. 961–968, 2000.

[39] R. E. Stewart, L. L. Gross, and B. H. Honkola, "Effects of competing vegetation on forest trees: a bibliography with abstracts," General Technical Report WO-43, Department of Agriculture Forest Service, Washington, DC, USA, 1984.

[40] M. K. Hoepting, R. G. Wagner, J. McLaughlin, and D. G. Pitt, "Timing and duration of herbaceous vegetation control in northern conifer plantations: 15th-year tree growth and soil nutrient effects," *Forestry Chronicle*, vol. 87, no. 3, pp. 398–413, 2011.

[41] G. Cyr and N. Thiffault, "Long-term black spruce plantation growth and structure after release and juvenile cleaning: a 24-year study," *Forestry Chronicle*, vol. 85, no. 3, pp. 417–426, 2009.

[42] F. W. Bell, J. Dacosta, M. Penner et al., "Longer-term volume trade-offs in spruce and jack pine plantations following various conifer release treatments," *Forestry Chronicle*, vol. 87, no. 2, pp. 235–250, 2011.

[43] K. Homagain, C. K. Shahi, N. J. Luckai, M. Leitch, and F. W. Bell, "Benefit-cost analysis of vegetation management alternatives: an Ontario case study," *Forestry Chronicle*, vol. 87, no. 2, pp. 260–273, 2011.

[44] B. Richardson, "Vegetation management practices in plantation forests of Australia and New Zealand," *Canadian Journal of Forest Research*, vol. 23, no. 10, pp. 1989–2005, 1993.

[45] D. B. South, J. H. Miller, M. O. Kimberley, and C. L. Vanderschaaf, "Determining productivity gains from herbaceous vegetation management with "age-shift" calculations," *Forestry*, vol. 79, no. 1, pp. 43–56, 2006.

[46] F. W. Bell, M. Kershaw, I. Aubin, N. Thiffault, J. Dacosta, and A. Wiensczyk, "Ecology and traits of plant species that compete with boreal and temperate forest conifers: an overview of available information and its use in forest management in Canada," *Forestry Chronicle*, vol. 87, no. 2, pp. 161–174, 2011.

[47] B. S. Biring, P. G. Comeau, and P. Fielder, "Long-term effects of vegetation control treatments for release of Engelmann spruce from a mixed-shrub community in Southern British Columbia," *Annals of Forest Science*, vol. 60, no. 7, pp. 681–690, 2003.

6

# The Effects of Selective Logging Behaviors on Forest Fragmentation and Recovery

**Xanic J. Rondon,[1,2] Graeme S. Cumming,[1] Rosa E. Cossío,[3] and Jane Southworth[4]**

[1] *Percy FitzPatrick Institute, DST-NRF Centre of Excellence, University of Cape Town, Rondebosch, Cape Town 7701, South Africa*
[2] *Center for Latin American Studies, University of Florida, Gainesville, FL 32611, USA*
[3] *School of Natural Resources and Environment, University of Florida, Gainesville, FL 32611, USA*
[4] *Department of Geography, University of Florida, Gainesville, FL 32611, USA*

Correspondence should be addressed to Xanic J. Rondon, rondonxj@gmail.com

Academic Editor: Todd S. Fredericksen

To study the impacts of selective logging behaviors on a forest landscape, we developed an intermediate-scale spatial model to link cross-scale interactions of timber harvesting, a fine-scale human activity, with coarse-scale landscape impacts. We used the Lotka-Volterra predator-prey model with Holling's functional response II to simulate selective logging, coupled with a cellular automaton model to simulate logger mobility and forest fragmentation. Three logging scenarios were simulated, each varying in timber harvesting preference and logger mobility. We quantified forest resilience by evaluating (1) the spatial patterns of forest fragmentation, (2) the time until the system crossed a threshold into a deforested state, and (3) recovery time. Our simulations showed that logging behaviors involving decisions made about harvesting timber and mobility can lead to different spatial patterns of forest fragmentation. They can, together with forest management practices, significantly delay or accelerate the transition of a forest landscape to a deforested state and its return to a recovered state. Intermediate-scale models emerge as useful tools for understanding cross-scale interactions between human activities and the spatial patterns that are created by anthropogenic land use.

## 1. Introduction

Humans both create and respond to spatial patterns across a range of spatial and temporal scales [1–3]. Although the real world is multiscale [4, 5], most models of land use and land cover change are built at a single spatio-temporal scale. Social-ecological dynamics tend to be most predictable at broader analytical scales (i.e., broad extent and coarse grain), in part because analysis that uses a higher level of data aggregation obscures the variability of processes (such as idiosyncratic decisions by people) that occur at finer scales [4, 6–8]. Broad-scale models, however, often lack important elements of complex processes that can be modeled using a multi-scale approach [9–11].

Land use is both a response to socioeconomic driving factors (e.g., the price of beef) and a cause of changes in socioeconomic systems (e.g., forest clearing for increased cattle production leads to an increased supply of beef, reducing prices) [12, 13]. Influences on land use occur at many different scales, and their interactions and feedbacks can create nonlinear dynamics and the potential for alternate stable states [14–16]. Cross-scale interactions can have important influences on fine-scale processes, or vice versa [17, 18]. In this context, intermediates or mesoscale models, which focus on connecting fine- and broad-scale pattern-process relationships [19], have an important role to play because they are well suited to capturing human agency, an important element that many models of land-use ignore [8, 10, 20, 21].

We used intermediate-scale models to evaluate the broad-scale impacts of selective logging, a fine-scale process, on a simulated forest landscape. In the Amazon basin, selective logging has great economic importance but is also a large-scale driver of forest fragmentation [22–24]. Although timber loggers do not clear-cut the forest and burn it for land conversion, they thin the forest by harvesting marketable tree species. In the process they typically degrade the forest, damaging both the canopy and the understory [25–27]. One of the current concerns in the Amazon is that timber harvesting may degrade a forest to the point of passing one or more key thresholds, beyond which the forest can lose its ability (at time scales of >50 years) to both sustain biodiversity and provide important ecosystem goods and services [28–30]. Such changes can be considered a regime shift: a substantial reorganization of a complex system with prolonged or irreversible consequences [31–33].

Thus far, most studies on selective logging have focused on quantifying the extent, distribution, and rate of selective logging in the forest landscape [24, 26, 38]. Such studies have found that selective logging can leave a complex array of canopy gaps caused by tree falls, roads, skid trails and log decks [27, 39, 40]. Selective logging has also been found to cause alterations in forest biophysical properties (e.g., water and wind stress and changes in micro-meteorological systems [41]), which could lead to forest fires [25, 42] and changes in forest structure and composition [43]. Although these studies are important, they have not linked the direct effects of different logging approaches to their coarse-scale landscape impacts.

The purpose of this study was to determine whether different selective logging behaviors could influence the resilience of a simulated forest landscape. We constructed three scenarios, representative of logging behaviors in the Peruvian Amazon [44]. In scenario 1, we simulated a null model, where selective logging occurred randomly and there was no timber harvesting preference. In scenario 2, we simulated the behavior of timber loggers that harvest valuable timber species at a high mobility cost [44]. In scenario 3, we simulated the behavior of loggers that harvest timber species of low value at a low mobility cost, because species of high value have already been harvested [44]. Note that the focus of this analysis was on the spatial patterns of impacts, rather than on logging intensity. For each scenario, we quantified forest resilience by evaluating: (1) spatial patterns of forest fragmentation and the transition time to reaching a deforested state; and (2) the time taken to return to a forested state through management. We defined social-ecological regime shifts as being the transitions between (1) an old-growth forest state (timber is abundant) and a deforested state (entire forest landscape has been logged); (2) a deforested state and a recovered (forested, but not old growth) state [45]. In the recovered state, timber trees reach a minimum commercial volume in the short term and in the long term, more complete ecological function may return. In both cases, the states under consideration are social-ecological rather than purely ecological states; the stability or instability of patterns on the landscape is contingent on human agency and decisions as well as on the ecology of the system.

## 2. Methods

We developed an intermediate-scale approach to model timber and logger dynamics. Published data were used to parameterize the model when they were available (Table 1). The model consisted of three main parts: (1) a module with timber density, volume and distribution; (2) a module simulating logger-timber dynamics, (3) a cellular automaton module simulating logger mobility and fragmentation. The simulated forest landscape consisted of a two-dimensional space of 65 × 65 1-ha cells (4225 cells or ha), the area of a small forest concession in southwestern Amazonia [44]. Following Peters et al. [19], our modeling approach incorporated three scales. The finest scale (single cell = 1 ha) was the scale at which individual timber trees were found, selective timber logging occurred, and logged forest patches began forming. Key pattern-process relationships at this scale included timber harvesting that influence the distribution and abundance of timber trees in the landscape. The intermediate-scale (>1 cell or 1 ha) was the scale at which loggers dispersed to other forest cells and the number of logged forest patches increased. Key pattern-process relationships at this scale included the spatial patterns of loggers and their mobility processes. The coarse scale is the scale at which forest fragmentation occurs. Key pattern-process relationships at this scale consisted of the spatial patterns of logged forest patches and the transition to the forest degradation state, in which the provision of timber was exhausted.

*2.1. Timber Density, Volume, and Distribution.* In southwestern Amazonia, the density of timber species of low and high value can show considerable variation. Highly valuable timber species as mahogany (*Swietenia macrophylla*) can have densities >0.03 trees ha$^{-1}$ [36], whereas the density of Spanish cedar (*Cedrela odorata*) can range from 0.17 to 0.35 trees ha$^{-1}$ [46]. Timber species of much lower value as *Cedrelinga catenaeformis* can be found at much higher densities (e.g., 0.8 trees ha$^{-1}$ [46]). Furthermore, logging intensity in this region varies from 1 to 6 trees ha$^{-1}$ [47].

We assumed that the forested landscape consisted of an old-growth forest, where timber trees of high and low value coexisted, regardless of species. The density of timber trees of high value in the simulated landscape was set at 0.5 trees ha$^{-1}$, which is within the range (0.3 to 2 tree ha$^{-1}$) found in Verissimo et al. [48]. In the Peruvian Amazon, the minimum cutting diameter (MCD) of trees of high value such as mahogany is ≥75 cm dbh or ≥ 4.5 m$^3$ per tree ("real volume"), but they can also reach volumes as high as 21–27 m$^3$ (150–190 cm dbh) [36, 49]. We set the volume of timber trees of high value to range from 5 to 26 m$^3$. Thus, the volume density for timber trees of high value in the simulated forest landscape ranged from 2.5 to 13 m$^3$ ha$^{-1}$, which is similar to the range of mahogany volumes (1–11 m$^3$ ha$^{-1}$) found in Verissimo et al. [48]. The density of timber trees of low value was set at 4 trees ha$^{-1}$, within the range of the extraction rate in southwestern Amazonia [47]. The volume of timber trees of low value ranged from 12 to 47 m$^3$ ha$^{-1}$; this is equivalent to 3–12 m$^3$ per tree [37]. The volumes of both types of timber trees were drawn randomly from a normal distribution and were placed randomly within the cells of the forest landscape at the start of each simulation.

Table 1: Summary of parameters for the Lotka-Volterra component of the model.

| Parameter | High-value tree | Low-value tree | Citation |
|---|---|---|---|
| $r$, intrinsic growth rate for trees (timber volume) | $0.02\,m^3\,yr^{-1}$ | $0.04\,m^3\,yr^{-1}$ | [34, 35] |
| $a$, harvesting volume per time step (harvesting rate) | $8\,m^3$ | $4\,m^3$ | [36, 37] |
| $b$, conversion efficiency | 20% | 20% | |
| $d$, declining rate of logger population in absence of local timber | 2% | 2% | |
| $c$, maximum harvesting rate | $20\,m^3\,ha^{-1}$ | $40\,m^3\,ha^{-1}$ | [25] |
| $K$, carrying capacity (maximum volume) for a timber tree | $28\,m^3$ | $12.5\,m^3\,yr^{-1}$ | [36, 37] |

*2.2. Timber-Logger Dynamics Module.* We used an adaptation of the Lotka-Volterra predator-prey equations [50, 51] to simulate the harvesting of timber resources ($N_1$) by a mobile predator, the logger ($N_2$). Timber volume growth was modeled using the logistic growth function with Holling's functional response II; that is, the timber harvesting rate (predation attack rate) increased in a decelerating fashion, reaching a maximum harvesting rate ($c$) at high timber volumes. The Lotka-Volterra equations are:

$$\frac{dN_1}{dt} = \frac{rN_1(K - N_1)}{K - c(1 - 1/e^{aN_1/c})N_2} \quad \text{timber},$$
$$\frac{dN_2}{dt} = \mathrm{bc}\left(1 - 1/e^{aN_1/c}\right)N_2 - dN_2 \quad \text{logger}, \tag{1}$$

where $r$ is the intrinsic growth rate for timber volume growth, $a$ is the harvesting rate, $b$ is the conversion efficiency, $d$ is the declining rate for loggers, $c$ is the maximum harvesting rate, and $K$ is the carrying capacity for a timber tree.

Equation (1) were modified to introduce one logger and timber trees of low and high value. Situations with one predator and two prey items typically use a competition coefficient for prey 1 on prey 2 ($\alpha_{12}$) and for prey 2 on prey 1 ($\alpha_{21}$). However, we assumed that there was no competition between timber trees of low and high value, and hence competition coefficients ($\alpha_{12}$ and $\alpha_{21}$) were equal to zero. The modified Lotka-Volterra equations were thus:

$$\frac{dN_1}{dt} = \frac{r_1N_1(K_1 - N_1)}{K_1} - c_1\left(1 - \frac{1}{e^{a_1N_1/c_1}}\right)N_3 \quad \text{low-value timber trees},$$
$$\frac{dN_2}{dt} = \frac{r_2N_2(K_2 - N_2)}{K_2} - c_2\left(1 - \frac{1}{e^{a_2N_2/c_2}}\right)N_3 \quad \text{high-value timber trees},$$
$$\frac{dN_3}{dt} = bc_1\left(1 - \frac{1}{e^{a_1N_1/c_1}}\right)N_3 + bc_3\left(1 - \frac{1}{e^{a_2N_2/c_2}}\right)N_3 - dN_3 \quad \text{logger}. \tag{2}$$

We obtained parameter values from the literature, making some assumptions (as explained below) due to lack of data. A mean tree growth rate of $0.03\,m^3\,yr^{-1}$ has been reported for mahogany trees [34]. We took a lower volume growth rate because tree growth is typically non-linear, having low growth rates for small and large-sized individuals and high growth rates for the intermediate sizes (e.g., [35, 52]). We set the growth rates of timber trees of high value ($r_2$) at $0.02\,m^3\,yr^{-1}$ and the average growth rate for timber trees of low value was set at $0.01\,m^3\,yr^{-1}$ for a total value of ($r_1$) $0.04\,m^3\,yr^{-1}$. Carrying capacity was defined as the maximum volume that timber of low and high value can be achieved in an old-growth forest. The carrying capacity for a timber tree of high value ($K_2$) was set at $28\,m^3$, the maximum real volume for mahogany trees in an old-growth forest [36], and that of low value was reached at about $12.5\,m^3$ [37] for a maximum volume ($K_1$) of about $50\,m^3$ for 4 trees.

Where timber trees of both low and high value were found, the minimum commercial harvesting volumes for each were set $8\,m^3$ ($a_1$) and $4\,m^3$ ($a_2$). These volumes were equivalent to harvesting low-and high-value timber trees above their MCD [36, 37]. In terms of low-value timber trees, the harvesting of $8\,m^3$ could be compared to harvesting 2 trees of *Cedrelinga catenaeformis*, each one ≥61 cm dbh (MCD) [37]. Timber trees were only harvested when they were greater than or equal to their minimum harvesting or commercial volume. Due to previous logging, some forest concessions in southwestern Amazonia have very low numbers of high value timber species; thus, timber loggers in these concessions have a preference for harvesting timber species of low value at very high rates [44]. The maximum harvesting rate for timber trees of low value ($c_1$) was set at $40\,m^3\,ha^{-1}$ and for high value was set at $20\,m^3\,ha^{-1}$ ($c_2$), a low intensity extraction [25]. Each logging team consisted of 3 loggers [53]. A 20% conversion efficiency ($b_1$ and $b_2$) was assumed to be associated with harvesting timber and logger population growth. If there was no timber to harvest, we assumed that the population of loggers declined ($d_1$ and $d_2$) by 2% per year.

*2.3. Cellular Automaton Module.* We coupled the Lotka-Volterra predator-prey model with a cellular automaton model [54, 55]. In this study we assumed that timber logging was associated with the opening of the forest canopy. Each forest cell contained timber trees of either low or high value and in some cases both types. Before logging, all forest cells in the landscape were in the unlogged state (cells = 1) and they transitioned into the logged state (cell = 0) when timber from a cell was harvested. The cell remained in the logged state for the duration of the simulation ($t = 80$ years), although timber volume kept on growing. A logger dispersed into a new forest cell when timber volume was $< a_1$ or $a_2$, the threshold volume. A logger could return to a previously logged cell if there was timber Larger than or equal to threshold volume. The simulation model started with 89 teams of three loggers per cell, placed in the forest cells closest to a hypothetical road in the forest landscape. We do not know the actual number of loggers that work in the forest concessions in southwestern Amazonia, but experienced foresters who work in the region indicate that after the first forest concession contract of 40 years, most of the high value timber species are depleted, particularly in the Tambopata and Manu regions of Madre de Dios, Peru [pers. comm. Cossío 2010]. In our simulations this situation occurred when there were about 89 teams of loggers.

*2.4. Selective Logging Scenarios and Data Analysis.* We contrasted three different scenarios to explore the long-term effects of different selective logging behaviors on forest fragmentation (Figure 1(a)). The differential equations used for are presented in the Appendix. All modeling was done using MATLAB 7.6. In *scenario 1*, logger mobility was random within the forest landscape and there was no timber harvesting preference. Loggers had a 50% chance of harvesting timber trees of either low or high value when both types of timber were present in a forest cell (Figure 1(a)). This scenario was used as a null model for comparison purposes, since we simulated an expected behavior in the absence of any specific processes [56, 57].

In scenarios 2 and 3, loggers had a timber harvesting preference and were assumed to access timber sequentially, along an accessibility gradient, starting from a hypothetical road and using a mobility cost function. In both scenarios, loggers preferred to disperse to forest cells closer to the road to reduce transportation costs. In *scenario 2*, loggers harvested timber trees of high value and when both timber trees were present there was a 75% chance of harvesting timber of high value (Figure 1(a)). In this scenario, a mobility cost was introduced using linear relation between distance to the road (line-haul cost) and transport cost [58]. Scenario 2 simulated the behavior of timber loggers in forest concessions that harvest highly valuable timber species such as mahogany and Spanish cedar and depend heavily on roads for timber transport [44]. In *scenario 3*, loggers harvested timber trees of low value and when both timber trees were present there was a 75% chance of harvesting timber of low value (Figure 1(a)). The mobility cost in this scenario was based on a logistic function between distance to the road and transportation cost. Scenario 3 simulated the behavior of timber loggers in forest concessions that harvest timber species of low value and have much less financial capital to invest in road construction [44]. These loggers tend to use a combination of roads and rivers to reduce transport costs [44].

To estimate the time until a regime shift to forest degradation occurred, each simulation model was run for 80 years for a total of 50 times. By this time the entire forest landscape was logged and very little timber was left. In scenario 2, the harvesting of high-value timber trees ended 36 years after the initiation of logging; thus, for the remaining time of the simulations we assumed that loggers changed their harvesting preference to timber trees of low value, which is very common [44]. We analyzed the spatial patterns of forest fragmentation for each scenario at the scale of the concession landscape (4225 cells or has). Landscape pattern metrics were calculated in Fragstats 3.3.1 using the 8-neighbor-cell rule [59]. We quantified the number of logged forest patches, length of edge, and area logged in the concession landscape for each year of simulated selective timber logging.

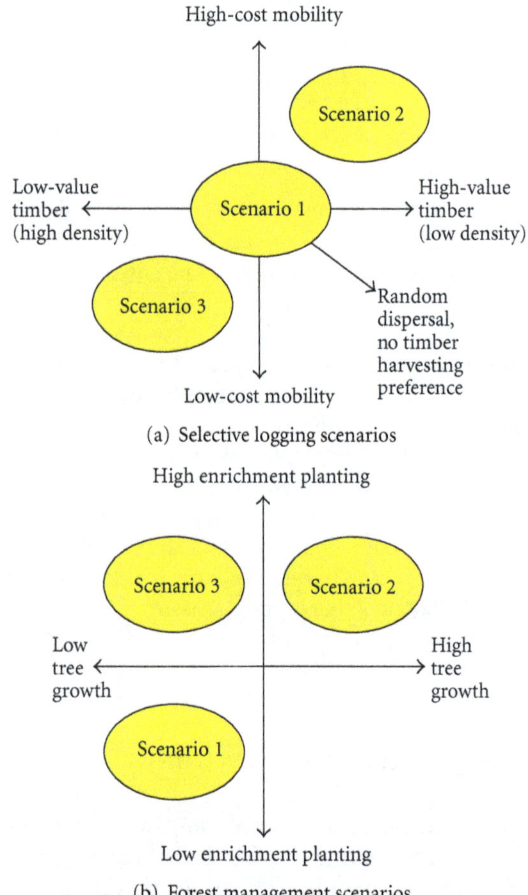


FIGURE 1: Three selective logging scenarios, based on timber harvesting preference, mobility cost, and forest management practices, were simulated in a forest landscape.

*2.5. Forest Management Scenarios and Data Analysis.* Once the forest landscape was in the deforested state, we explored the transition of the forest landscape to a recovered state by associating logging behaviors with different forest management practices (Figure 1(b)). We estimated the time of reaching a recovered state, defined as the system recovery after a perturbation where timber in the forest cells grew to a minimum commercial volume. The recovered state was assumed to be an alternative state [45] because these forests are assumed to be actively managed for the long-term production of timber and often exhibit a different floristic composition and structure from the prelogged state. Due to the low timber yields at the end of 80 years of logging simulation (<0.1 m$^3$/ha), we enhanced timber recruitment by assuming the use of enrichment planting of timber saplings during harvesting. In all scenarios we used higher tree growth rates to simulate silvicultural thinning and exposure to high light intensities due to canopy opening from logging.

In *scenario 1*, we simulated the lowest levels of forest management practices (Figure 1(b)). We assumed that timber trees of high value could reach a volume ranging of 0.5-1 m$^3$ at the end of logging simulations, equivalent to mahogany trees of ~30–40 cm dbh [36]. We assumed that timber trees of low value could reach a volume ranging 1.5–2 m$^3$, the equivalent to two trees per cell >30 cm dbh. This was calculated with a diameter-volume equation for trees of low commercial value (Vol = 9.1405 dbh$^{2.1382}$, $r^2$ = 0.95, $n$ = 38), using the data from Lombardi et al. [37]. In this scenario timber growth rates were set for $r_1$ and $r_2$ at 0.03 and 0.06 m$^3$ yr$^{-1}$ [34]. In *scenario 2*, we assumed that there were higher levels of forest management practices (Figure 1(b)). We assumed that timber trees of high and low value could reach 1–1.5 m$^3$ (a mahogany tree of 40–60 cm dbh [36]) and 2–2.5 m$^3$ (corresponding to two timber trees of low value >35 cm dbh using the equation above). Timber growth rates for low ($r_1$) and high ($r_2$) value timber were 0.04 and 0.08 m$^3$ yr$^{-1}$ [34]. In *scenario 3*, there were intermediate levels of forest management practices. Enrichment planting levels were as high as the values in scenario 2, but timber growth rates were as low as the values in scenario 1 (Figure 1(b)).

## 3. Results

The amount of timber harvested in each selective logging scenario varied through time. Selective logging under scenarios 1 and 2 harvested the lowest amount of timber, about 0.80 m$^3$ ha$^{-1}$ and 1 m$^3$ ha$^{-1}$, respectively (Figure 2). Selective logging under scenario 3 harvested the highest amount of timber, reaching rates >35 m$^3$ ha$^{-1}$. Out of the three scenarios, scenario 2 harvested the lowest amount of timber (0.40 m$^3$ ha$^{-1}$) at the initiation of logging (Figure 2). In this scenario there was a harvesting preference for timber trees of high value, which were exhausted when about 50% of the landscape had been logged. Once the high-value timber trees were exhausted (below commercial volumes), loggers switched their harvesting preference to timber trees of low value, harvesting them at much higher rates.

The three selective logging behaviors or scenarios created different spatial patterns of forest fragmentation. In all three logging scenarios, the number of logged forest patches varied nonlinearly through time (Figure 3). Scenarios 1 and 2 created unimodal trajectories in the number of logged forest patches, whereas scenario 3 created a slightly bimodal trajectory (Figure 3). Throughout the simulations, scenario 2 produced the lowest number of logged forest patches (<50 logged forest patches), but their sizes were much greater than those of scenarios 1 and 3 (Figure 3). The number of logged forest patches in scenario 2 peaked when about 50–60% of the forest landscape (4225 ha) had been logged, ~30 to 40 years after the initiation of timber harvesting (Figure 3). Scenarios 1 and 3 produced >300 and >250 logged forest patches and their numbers peaked when about 20% of the forest landscape had been logged, ~10 to 11 years after the initiation of timber harvesting (Figure 3).

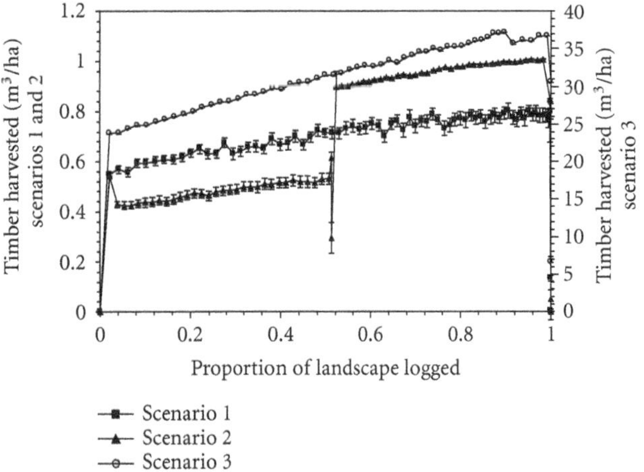

FIGURE 2: Timber harvested rates over 80 years of selective logging simulations.

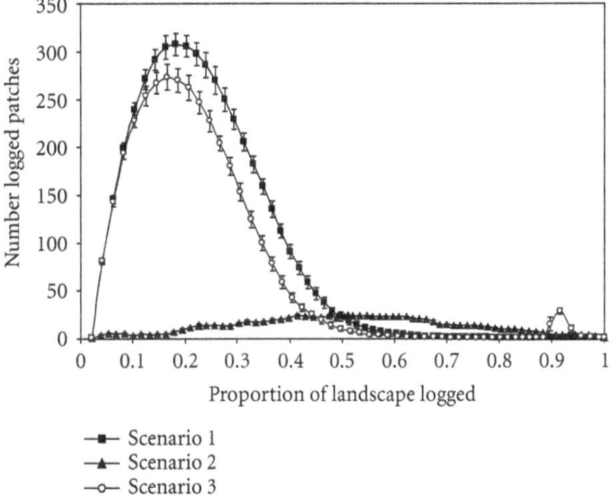

FIGURE 3: Number of logged forest patches over 80 years of selective logging simulations in a forest landscape (4225 ha).

The three logging scenarios also produced non-linear edge-length trajectories of quadratic form (Figure 4). Scenarios 1 and 2 produced unimodal edge-length trajectories, whereas scenario 3 produced a slightly bimodal trajectory. During the simulations, scenarios 1 and 2 produced the highest edge-length peak at about 400 km, when about 50% of the landscape had been logged, ~28 and 33 years after the initiation of logging. Scenario 3 produced the lowest edge-length peak at about 350 km, when >40% of the landscape had been logged, ~23 years after the initiation of logging.

The three selective logging behaviors also delayed or accelerated the transition of the forest landscape to a deforested state. Both scenarios 1 and 2 took >66 years to transition to the deforested state, but scenario 3 transitioned

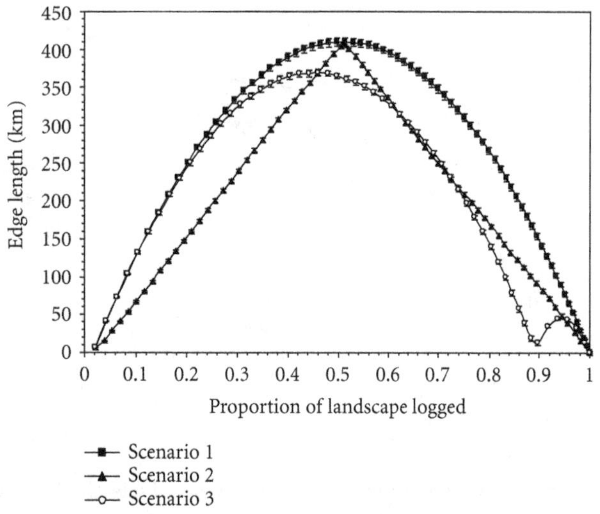


Figure 4: Edge length over 80 years of selective logging simulations in a forest landscape (4225 ha).

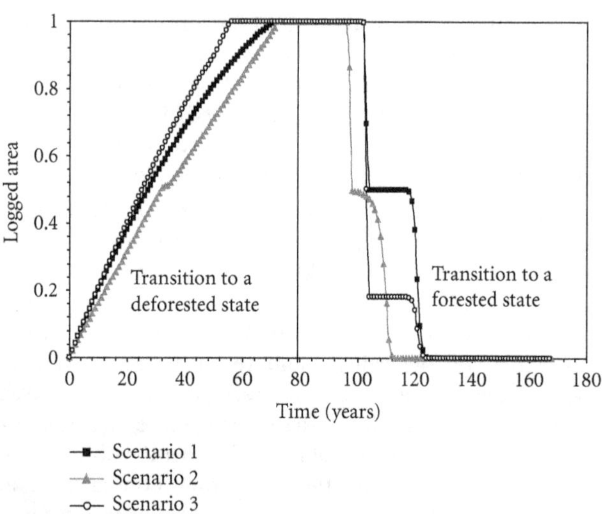


Figure 5: Regime shifts into a deforested state and recovered state in a forest landscape, under three different scenarios of selective logging and forest management practices.

10 years earlier (>51 years) (Figure 5). Forest management practices also delayed or accelerated the transition to a recovered state (Figure 5). When higher levels of enrichment planting and more intensive silvicultural thinning practices were used (e.g., higher timber growth rates), scenario 2 reached the restoration state ~32 years after entering the deforested state, but when lower levels of these factors were used, scenarios 1 and 3 took a longer time to reach the recovered state (>32 years) (Figure 5).

## 4. Discussion

Selective timber logging is not often considered as a source of forest fragmentation; however, depending on the harvesting intensity, logging operations can greatly reduce canopy cover (up to 60% [39]), resulting in extensive forest fragmentation and edge effects. When selective logging takes place, the transition from a forest to a logged forest is not as abrupt as that of forest to land uses such as pastures or agricultural areas. Logging disturbances also create edges. The difference is that logging disturbances adjoining forested areas create soft edges whereas deforested or clear-cut areas adjoining forested areas create hard edges [24]. Soft edges can eventually recover in time through regrowth, reducing their overall influence because the transition becomes less severe [60], but soft edges can transition to hard edges if the logged patch is clear-cut.

Selective logging behaviors can create different spatial patterns of forest fragmentation. Several studies have found that timber harvesting regimes or behaviors can influence the patterns of forest fragmentation in a landscape [61]. In this study, scenario 2, the selective logging behavior associated with harvesting timber trees of high value at a high mobility cost, led to the harvesting of lower timber volumes and the creation of a lower number of logged forest patches in the landscape. Scenario 3, the selective logging behavior involving harvesting timber trees of low value at a low mobility cost, led to the harvesting of high timber volumes and to the creation of a higher number of logged forest patches. As in other studies, the three selective logging scenarios showed non-linear forest fragmentation and edge-length trajectories [57, 61–63]. Scenario 3 showed slightly bimodal fragmentation and edge-length trajectories due to the crossing of a distance threshold of the logistic function of distance to mobility cost. This allowed dispersal into more distant areas for the same mobility cost, increasing the number of logged forest patches and edge length (Figures 3 and 4)

Selective logging behaviors can have different impacts on forest resilience. Other studies have found that selective logging behaviors have highly variable impacts on forests, depending on the use of conventional logging versus reduced impact logging techniques [64–66]. In this study we found that logging behavior can affect the speed of the fragmentation process, accelerating or delaying the transition into a deforested state. Any additional logging after reaching the peak level of fragmentation increased the size of the logged forest patches until they coalesced and declined in number to the point that the entire landscape became one logged forest patch (Figures 3 and 5). In the Amazon basin, harvesting timber species of high value (e.g., mahogany) may have a smaller impact on forest structure and function than harvesting timber species of low value, which are more abundant and are extracted at much higher volumes [48]. Sawmills in the Amazon are involved in the processing of more then 100 tree species [67, 68] and the volume of timber species of high value makes up only a small percentage of the total production (e.g., mahogany is 5% of the total timber production) [48].

We associated the harvesting of timber trees of high value at a high mobility cost (scenario 2) with intensive forest management practices. Harvesting timber of high value is usually associated with timber enterprises that have a high financial capital to invest in infrastructure (e.g., roads, feeder roads, etc.) and machinery, and a high human capital (e.g., education in forestry, experience in logging, skills and knowledge) to implement forest management plans [44, 48, 68].

Lotka-Volterra predator-prey equations represent one of the simplest dynamic consumer-resource systems [69]. Although real consumer-resource systems are more complex, these equations have been previously used for modeling renewable resources and population dynamics, with man as a predator and the resource base as the prey [70–74]. The main assumption in these studies is that consumer or logger dynamics, as in the case of this study, depend solely on a particular resource. However, it is well known that agents of resource use in the Amazon pursue a diverse portfolio of subsistence and market-oriented activities (e.g., hunting, harvesting of NTFP, shifting cultivation, agroforestry, cattle ranching, and agriculture) in order to spread their risks in an extremely variable environment [75, 76].

This framework to connect cross-scale interactions through the transfer of processes at intermediate scales has been used to study several systems, but not in the analyses of forest fragmentation and degradation by human agents [77–80, among others]. We used logger mobility as the transfer process for linking fine-scale processes with coarse-scale pattern as well as the propagation of selective logging in the simulated landscape. To make our model more realistic, we introduced decision making in loggers based on timber harvesting preference (timber of low value versus high value) and logger mobility (high cost versus low cost) from a hypothetical road. However, logging behavior is more complex and depends on other factors such as timber prices, land tenure, forest policy, and accessibility to local markets, among other factors [81, 82]. To evaluate whether different spatial patterns of forest fragmentation could arise when making simple decisions, we also had to ignore the potential effects of other drivers of landscape change such as forest fires, road network expansion, and land conversion after selective logging [25, 81, 83]. We also assumed that there was no tree mortality, tree competition or further tree recruitment during the tree growth simulations.

Despite some of the obvious limitations of the model, our analysis demonstrates that intermediate-scale models can serve as a useful tool for understanding cross-scale interactions between human activities and impacts to the landscape. Although the drivers of tropical deforestation are very well studied, far less is known about the spatial patterns of forest fragmentation and degradation that agents of different land-use systems in the Amazon create [81, 84]. We focused on selective logging behaviors, but intermediate-scale models could be used to study other land-use agents, their interaction and impacts to the landscape. Such studies can offer some insights into the mechanisms that give rise to the patterns of forest fragmentation found in the Amazon. These models can also contribute to the designing of more sustainable developmental or forestry policies, through their assessment of potential consequences or coarse-scale impacts to the landscape. As we seek to balance societal benefits with the cost of ecological degradation, intermediate-scale models have a potential role in generating perspectives on the causes and consequences of fine-scale actions of agents involved in land use, forest management and policy.

## Appendix

We developed the following differential equations based on the harvesting preference.

(a) Logger prefers to harvest timber of low value. This case occurred when timber of low value ($N_1$) was ≥ threshold volume and it was either the only species present in a cell or timber of high value ($N_2$) < threshold volume. If present $N_2$ (<threshold volume) experienced logistic growth and the growth of the logger ($N_3$) was a function of low quality timber species:

$$\begin{aligned}\frac{dN_1}{dt} &= \frac{r_1N_1(k_1 - N_1)}{k_1} - c_1\left(1 - e^{-a_1N_1/c_1}\right)N_3 \quad \text{timber of low value,}\\ \frac{dN_2}{dt} &= \frac{r_2N_2(k_2 - N_2)}{k_2} \quad \text{timber of high value,}\\ \frac{dP}{dt} &= bc_1\left(1 - e^{-a_1N_1/c_1}\right)N_3 - dN_3 \quad \text{logger.}\end{aligned} \tag{A.1}$$

(b) Logger prefers to harvest timber of high value. This case occurred when timber of high value ($N_2$) was ≥ its threshold volume and it was either the only species present in a cell or timber of low value ($N_1$) < threshold volume. If $N_1$ was present, it experienced logistic growth and the growth of the logger ($N_3$) was a function of high-value timber:

$$\begin{aligned}\frac{dN_1}{dt} &= \frac{r_1N_1(k_1 - N_1)}{k_1} \quad \text{timber of low value,}\\ \frac{dN_2}{dt} &= \frac{r_2N_2(K_2 - N_2 - \alpha_{21}N_1)}{K_2} - c_2\left(1 - e^{-a_2N_2/c_2}\right)N_3 \quad \text{timber of high value,}\\ \frac{dN_3}{dt} &= bc_2\left(1 - e^{-a_2N_2/c_2}\right)N_3 - dN_3 \quad \text{logger.}\end{aligned} \tag{A.2}$$

(c) Loggers ($N_3$) "declined" because there was no timber of either high or low value above their respective threshold volume:

$$\begin{aligned}\frac{dN_1}{dt} &= \frac{r_1N_1(k_1 - N_1)}{k_1} \quad \text{timber of low value,}\\ \frac{dN_2}{dt} &= \frac{r_2N_2(k_2 - N_2)}{k_2} \quad \text{timber of high value,}\\ \frac{dP}{dt} &= -dN_3 \quad \text{logger.}\end{aligned} \tag{A.3}$$

## References

[1] A. Veldkamp and L. O. Fresco, "CLUE: a conceptual model to study the conversion of land use and its effects," *Ecological Modelling*, vol. 85, no. 2-3, pp. 253–270, 1996.

[2] E. F. X. Lambin, N. Baulies, G. Bockstael et al., "Land-use and land-cover change implementation plan," Tech. Rep., LUCC International Project Office, University of Louvain, Louvain-la-Neuve, Belgium, 1999.

[3] C. H. Wood and R. Porro, Eds., *Deforestation and Land Use in the Amazon*, University Press of Florida, Gainesville, Fla, USA, 2002.

[4] S. A. Levin, "The problem of pattern and scale in ecology," *Ecology*, vol. 73, no. 6, pp. 1943–1967, 1992.

[5] S. A. Levin, "Multiple scales and the maintenance of biodiversity," *Ecosystems*, vol. 3, no. 6, pp. 498–506, 2000.

[6] E. Lambin, "Modelling deforestation processes: a review," Research Report 1, European Commission, Office for Official Publications of the European Community, European Commission Joint Research Centre and the European Space Agency, Luxembourg, 1994.

[7] B. Mertens and E. F. Lambin, "Spatial modelling of deforestation in southern Cameroon: spatial disaggregation of diverse deforestation processes," *Applied Geography*, vol. 17, no. 2, pp. 143–162, 1997.

[8] D. Kaimowitz and A. Angelsen, *Economic Models of Tropical Deforestation: A Review*, Center for International Forestry Research, Bogor, Indonesia, 1998.

[9] M. G. Turner, V. H. Dale, and R. H. Gardner, "Predicting across scales: theory development and testing," *Landscape Ecology*, vol. 3, no. 3-4, pp. 245–252, 1989.

[10] C. Agarwal, G. M. Green, J. M. Grove, T. P. Evans, and C. M. Schweik, "Review and assessment of land-use change models: dynamics of space, time, and human choice," GTR NE-297, Newton Square, Pa, USA, 2002.

[11] D. G. Brown, R. Walker, S. Manson, and K. Seto, "Modeling land use and land cover change," in *Land Change Science: Observing, Monitoring, and Understanding Trajectories of Change on the Earth's Surface*, G. Gutman, A. Janetos, and C. Justice, Eds., pp. 395–409, Kluwer Academic Publishers, Dordrecht, Netherlands, 2004.

[12] E. F. Lambin, H. J. Geist, and E. Lepers, "Dynamics of land-use and land-cover change in tropical regions," *Annual Review of Environment and Resources*, vol. 28, pp. 205–241, 2003.

[13] P. H. Verburg, "Simulating feedbacks in land use and land cover change models," *Landscape Ecology*, vol. 21, no. 8, pp. 1171–1183, 2006.

[14] S. R. Carpenter and M. G. Turner, "Hares and tortoises: interactions of fast and slow variables in ecosystems," *Ecosystems*, vol. 3, no. 6, pp. 495–497, 2000.

[15] L. Gunderson and C. Holling, *Panarchy: Understanding Transformations in Human and Natural Systems*, Island Press, Washington, DC, USA, 2002.

[16] D. P. C. Peters, R. A. Pielke Sr., B. T. Bestelmeyer, C. D. Allen, S. Munson-McGee, and K. M. Havstad, "Cross-scale interactions, nonlinearities, and forecasting catastrophic events," *Proceedings of the National Academy of Sciences of the United States of America*, vol. 101, no. 42, pp. 15130–15135, 2004.

[17] J. A. Ludwig, J. A. Wiens, and D. J. Tongway, "A scaling rule for landscape patches and how it applies to conserving soil resources in Savannas," *Ecosystems*, vol. 3, no. 1, pp. 84–97, 2000.

[18] N. S. Diffenbaugh, J. S. Pal, R. J. Trapp, and F. Giorgi, "Fine-scale processes regulate the response of extreme events to global climate change," *Proceedings of the National Academy of Sciences of the United States of America*, vol. 102, no. 44, pp. 15774–15778, 2005.

[19] D. P. C. Peters, B. T. Bestelmeyer, and M. G. Turner, "Cross-scale interactions and changing pattern-process relationships: consequences for system dynamics," *Ecosystems*, vol. 10, no. 5, pp. 790–796, 2007.

[20] D. C. Parker, S. M. Manson, M. A. Janssen, M. J. Hoffmann, and P. Deadman, "Multi-agent systems for the simulation of land-use and land-cover change: a review," *Annals of the Association of American Geographers*, vol. 93, no. 2, pp. 314–337, 2003.

[21] M. Bürgi, A. M. Hersperger, and N. Schneeberger, "Driving forces of landscape change—current and new directions," *Landscape Ecology*, vol. 19, no. 8, pp. 857–868, 2005.

[22] C. Gascon, G. B. Williamson, and G. A. B. Da Fonseca, "Receding forest edges and vanishing reserves," *Science*, vol. 288, no. 5470, pp. 1356–1358, 2000.

[23] W. F. Laurance, "Do edge effects occur over large spatial scales?" *Trends in Ecology and Evolution*, vol. 15, no. 4, pp. 134–135, 2000.

[24] E. N. Broadbent, G. P. Asner, M. Keller, D. E. Knapp, P. J. C. Oliveira, and J. N. Silva, "Forest fragmentation and edge effects from deforestation and selective logging in the Brazilian Amazon," *Biological Conservation*, vol. 141, no. 7, pp. 1745–1757, 2008.

[25] D. C. Nepstad, A. Veríssimo, A. Alencar et al., "Large-scale impoverishment of amazonian forests by logging and fire," *Nature*, vol. 398, no. 6727, pp. 505–508, 1999.

[26] G. Asner, M. Keller, R. Pereira, and J. Zweede, "Remote sensing of selective logging in Amazonia: Assessing limitations based on detailed field observations, Landsat ETM+, and textural analysis," *Remote Sensing of Environment*, vol. 80, pp. 483–496, 2002.

[27] G. P. Asner, M. Keller, R. Pereira, J. C. Zweede, and J. N. M. Silva, "Canopy damage and recovery after selective logging in Amazonia: field and satellite studies," *Ecological Applications*, vol. 14, no. 4, pp. S280–S298, 2004.

[28] Millennium Ecosystem Assessment, *Ecosystems and Human Well-Being: A Framework for Assessment*, Island Press, Washington, DC, USA, 2003.

[29] J. A. Foley, G. P. Asner, M. H. Costa et al., "Amazonia revealed: forest degradation and loss of ecosystem goods and services in the Amazon Basin," *Frontiers in Ecology and the Environment*, vol. 5, no. 1, pp. 25–32, 2007.

[30] D. C. Nepstad, C. M. Stickler, B. Soares-Filho, and F. Merry, "Interactions among Amazon land use, forests and climate: prospects for a near-term forest tipping point," *Philosophical Transactions of the Royal Society B*, vol. 363, no. 1498, pp. 1737–1746, 2008.

[31] S. R. Carpenter, "Regime shifts in lake ecosystems: pattern and variation," *Excellence in Ecology*, vol. 15, 2003.

[32] J. A. Foley, M. T. Coe, M. Scheffer, and G. Wang, "Regime shifts in the Sahara and Sahel: interactions between ecological and climatic systems in Northern Africa," *Ecosystems*, vol. 6, no. 6, pp. 524–539, 2003.

[33] C. Folke, S. Carpenter, B. Walker et al., "Regime shifts, resilience, and biodiversity in ecosystem management," *Annual Review of Ecology, Evolution, and Systematics*, vol. 35, pp. 557–581, 2004.

[34] L. K. Snook, *Stand dynamics of mahogany (Swietenia macrophylla King) and associated species after fire and hurricane in the tropical forests of the Yucatan peninsula, Mexico [PhD. dissertation]*, Yale School of Forestry and Environmental Studies, New Haven, Conn, USA, 1993.

[35] R. E. Gullison, S. N. Panfil, J. J. Strouse, and S. P. Hubbell, "Ecology and management of mahogany (*Swietenia macrophylla* King) in the Chimanes Forest, Beni, Bolivia," *Botanical Journal of the Linnean Society*, vol. 122, no. 1, pp. 9–34, 1996.

[36] I. Lombardi and P. Huerta, "Monitoring mahogany," *ITTO Tropical Forest Update*, vol. 17, pp. 5–9, 2007.

[37] I. Lombardi, V. Barrena, C. Vargas et al., "Evaluación de las existencias comerciales y estrategia para el manejo sostenible de la caoba (*Swietenia macrophylla*) en el Perú," Informe técnico, Universidad nacional Agraria La Molina, 2006.

[38] G. P. Asner, D. E. Knapp, E. N. Broadbent, P. J. C. Oliveira, M. Keller, and J. N. Silva, "Ecology: selective logging in the Brazilian Amazon," *Science*, vol. 310, no. 5747, pp. 480–482, 2005.

[39] R. Pereira Jr., J. Zweede, G. P. Asner, and M. Keller, "Forest canopy damage and recovery in reduced-impact and conventional selective logging in eastern Para, Brazil," *Forest Ecology and Management*, vol. 168, no. 1-3, pp. 77–89, 2002.

[40] G. P. Asner, E. N. Broadbent, P. J. C. Oliveira, M. Keller, D. E. Knapp, and J. M. M. Silva, "Condition and fate of logged forests in the Brazilian Amazon," *Proceedings of the National Academy of Sciences of the United States of America*, vol. 103, no. 34, pp. 12947–12950, 2006.

[41] C. M. Pringle and J. P. Benstead, "The effects of logging on tropical river ecosystems," in *The Cutting Edge: Conserving Wildlife in Logged Tropical forests*, R. A. Fimbel, A. Grajal, and J. G. Robinson, Eds., pp. 305–325, Columbia University Press, New York, NY, USA, 2001.

[42] M. A. Cochrane, "Synergistic interactions between habitat fragmentation and fire in evergreen tropical forests," *Conservation Biology*, vol. 15, no. 6, pp. 1515–1521, 2001.

[43] D. C. Nepstad, I. F. Brown, L. Luz, A. Alechandre, and V. Viana, "Biotic impoverishment of Amazonian forests by rubber tappers, loggers, and cattle ranchers," *Advances in Economic Botany*, pp. 1–14, 1992.

[44] R. E. Cossío, *Capacity for timber management among private small-medium forest enterprises in Madre de Dios [Ph.D. dissertation]*, University of Florida, Gainesville, Fla, USA, 2009.

[45] K. N. Suding, K. L. Gross, and G. R. Houseman, "Alternative states and positive sustainable development in imperfect economies," *Environmental and Resource Economics*, vol. 26, pp. 647–685, 2004.

[46] R. J. W. Brienen, P. A. Zuidema, and H. J. During, "Autocorrelated growth of tropical forest trees: unraveling patterns and quantifying consequences," *Forest Ecology and Management*, vol. 237, no. 1–3, pp. 179–190, 2006.

[47] S. N. Panfil and R. E. Gullison, "Short term impacts of experimental timber harvest intensity on forest structure and composition in the Chimanes Forest, Bolivia," *Forest Ecology and Management*, vol. 102, no. 2-3, pp. 235–243, 1998.

[48] A. Verissimo, P. Barreto, R. Tarifa, and C. Uhl, "Extraction of a high-value natural resource in Amazonia: the case of mahogany," *Forest Ecology and Management*, vol. 72, no. 1, pp. 39–60, 1995.

[49] J. Grogan and M. Schulze, "Estimating the number of trees and forest area necessary to supply internationally traded volumes of big-leaf mahogany (Swietenia macrophylla) in Amazonia," *Environmental Conservation*, vol. 35, no. 1, pp. 26–35, 2008.

[50] A. J. Lotka, *Elements of Physical Biology*, Williams Sc Wilkins, Baltimore, Md, USA, 1925.

[51] V. Volterra, "Fluctuations in the abundance of a species considered mathematically," *Nature*, vol. 118, no. 2972, pp. 558–560, 1926.

[52] A. E. Lugo, "Point-counterpoints on the conservation of big-leaf mahogany," Tech. Rep. WO-64, USDA Forest Service, International Institute of Tropical Forestry, Rio Piedras, Puerto Rico, 1999.

[53] C. Uhl, A. Veríssimo, M. M. Mattos, Z. Brandino, and I. C. Guimarães Vieira, "Social, economic, and ecological consequences of selective logging in an Amazon frontier: the case of Tailândia," *Forest Ecology and Management*, vol. 46, no. 3-4, pp. 243–273, 1991.

[54] J. von Neumann, *Theory of self-reproducing automata*, University of Illinois Press, Urbana, Ill, USA, 1966.

[55] D. Tilman and P. Kareiva, *Spatial Ecology*, Princeton University Press, Princeton, NJ, USA, 1997.

[56] H. Caswell, "Community structure: a neutral model analysis," *Ecological Monographs*, vol. 46, pp. 327–354, 1976.

[57] R. H. Gardner, B. T. Milne, M. G. Turnei, and R. V. O'Neill, "Neutral models for the analysis of broad-scale landscape pattern," *Landscape Ecology*, vol. 1, no. 1, pp. 19–28, 1987.

[58] W. A. Duerr, *Fundamentals of Forestry Economics*, McGraw Hill Book Company, New York, NY, USA, 1960.

[59] K. McGarigal, S. A. Cushman, M. C. Neel, and E.. Ene, "FRAGSTATS: Spatial Pattern Analysis Program for Categorical Maps," University of Massachusetts, Amherst, Mass, USA, http://www.tcshelp.com.

[60] R. K. Didham and J. H. Lawton, "Edge structure determines the magnitude of changes in microclimate and vegetation structure in tropical forest fragments," *Biotropica*, vol. 31, no. 1, pp. 17–30, 1999.

[61] J. F. Franklin and R. T. T. Forman, "Creating landscape patterns by forest cutting: ecological consequences and principles," *Landscape Ecology*, vol. 1, no. 1, pp. 5–18, 1987.

[62] H. Li, J. F. Franklin, F. J. Swanson, and T. A. Spies, "Developing alternative forest cutting patterns: a simulation approach," *Landscape Ecology*, vol. 8, no. 1, pp. 63–75, 1993.

[63] E. J. Gustafson and T. R. Crow, "Simulating spatial and temporal context of forest management using hypothetical landscapes," *Environmental Management*, vol. 22, no. 5, pp. 777–787, 1998.

[64] P. Sist, T. Nolan, J. G. Bertault, and D. Dykstra, "Harvesting intensity versus sustainability in Indonesia," *Forest Ecology and Management*, vol. 108, no. 3, pp. 251–260, 1998.

[65] F. E. Putz, G. M. Blate, K. H. Redford, R. Fimbel, and J. Robinson, "Tropical forest management and conservation of biodiversity: an overview," *Conservation Biology*, vol. 15, no. 1, pp. 7–20, 2001.

[66] J. J. Gerwing, "Degradation of forests through logging and fire in the eastern Brazilian Amazon," *Forest Ecology and Management*, vol. 157, no. 1–3, pp. 131–141, 2002.

[67] A. M. Z. Martini, A. N. De Rosa, and C. Uhl, "An attempt to predict which Amazonian tree species may be threatened by logging activities," *Environmental Conservation*, vol. 21, no. 2, pp. 152–162, 1994.

[68] A. C. Barros and C. Uhl, "Logging along the Amazon River and estuary: patterns, problems and potential," *Forest Ecology and Management*, vol. 77, no. 1-3, pp. 87–105, 1995.

[69] P. Turchin, "Does population ecology have general laws?" *Oikos*, vol. 94, no. 1, pp. 17–26, 2001.

[70] M. L. Rosenzweig, "Paradox of enrichment: destabilization of exploitation ecosystems in ecological time," *Science*, vol. 171, no. 3969, pp. 385–387, 1971.

[71] C. Barber, S. D'Angelo, and T. Fernandes, "The future of the Brazilian Amazon," *Science*, vol. 291, pp. 438–439, 2001.

[72] J. A. Brander and M. S. Taylor, "The simple economics of Easter Island: a Ricardo-Malthus model of renewable resource

use," *American Economic Review*, vol. 88, no. 1, pp. 119–138, 1998.

[73] J. M. Anderies, "On modeling human behavior and institutions in simple ecological economic systems," *Ecological Economics*, vol. 35, no. 3, pp. 393–412, 2000.

[74] C. S. Fletcher and D. W. Hilbert, "Resilience in landscape exploitation systems," *Ecological Modelling*, vol. 201, no. 3-4, pp. 440–452, 2007.

[75] M. Pinedo-Vasquez, D. Zarin, and P. Jipp, "Economic returns from forest conversion in the Peruvian Amazon," *Ecological Economics*, vol. 6, no. 2, pp. 163–173, 1992.

[76] R. T. Walker, A. Homma, A. Conto et al., "Farming systems and economic performance in the Brazilian Amazon," in *Proceedings of the Congresso Brasileiro sobre Sistemas Agroflorestais*, pp. 415–429, Annais, EMBRAPA, Colombo, Brazil, 1994.

[77] C. D. Allen, "Interactions across spatial scales among forest dieback, fire, and erosion in northern New Mexico landscapes," *Ecosystems*, vol. 10, no. 5, pp. 797–808, 2007.

[78] D. A. Falk, C. Miller, D. McKenzie, and A. E. Black, "Cross-scale analysis of fire regimes," *Ecosystems*, vol. 10, no. 5, pp. 809–823, 2007.

[79] R. L. Schooley and L. C. Branch, "Spatial heterogeneity in habitat quality and cross-scale interactions in metapopulations," *Ecosystems*, vol. 10, no. 5, pp. 846–853, 2007.

[80] M. R. Willig, C. P. Bloch, N. Brokaw, C. Higgins, J. Thompson, and C. R. Zimmermann, "Cross-scale responses of biodiversity to hurricane and anthropogenic disturbance in a tropical forest," *Ecosystems*, vol. 10, no. 5, pp. 824–838, 2007.

[81] E. Y. Arima, R. T. Walker, S. G. Perz, and M. Caldas, "Loggers and forest fragmentation: behavioral models of road building in the Amazon basin," *Annals of the Association of American Geographers*, vol. 95, no. 3, pp. 525–541, 2005.

[82] G. S. Amacher, F. D. Merry, and M. S. Bowman, "Smallholder timber sale decisions on the Amazon frontier," *Ecological Economics*, vol. 68, no. 6, pp. 1787–1796, 2009.

[83] M. A. Cochrane, "Fire science for rainforests," *Nature*, vol. 421, no. 6926, pp. 913–919, 2003.

[84] R. Walker, S. A. Drzyzga, Y. Li et al., "A behavioral model of landscape change in the Amazon basin: the colonist case," *Ecological Applications*, vol. 14, no. 4, pp. S299–S312, 2004.

# 7

# Leaf Area and Structural Changes after Thinning in Even-Aged *Picea rubens* and *Abies balsamea* Stands in Maine, USA

**R. Justin DeRose[1] and Robert S. Seymour[2]**

[1] *Forest Inventory and Analysis, Rocky Mountain Research Station, 507 25th Street, Ogden, UT 84401, USA*
[2] *School of Forest Resources, University of Maine 5755 Nutting Hall, Orono, ME 04469, USA*

Correspondence should be addressed to R. Justin DeRose, rjustinderose@gmail.com

Academic Editor: Scott D. Roberts

We tested the hypothesis that changes in leaf area index (LAI $m^2\,m^{-2}$) and mean stand diameter following thinning are due to thinning type and residual density. The ratios of pre- to postthinning diameter and LAI were used to assess structural changes between replicated crown, dominant, and low thinning treatments to 33% and 50% residual density in even-aged *Picea rubens* and *Abies balsamea* stands with and without a precommercial thinning history in Maine, USA. Diameter ratios varied predictably by thinning type: low thinnings were <0.7, crown thinnings were >0.7 but <1.0, and dominant thinnings were >1.0 . LAI change was affected by type and intensity of thinning. On average, 33% density reduction removed <50% of LAI, whereas 50% density reduction removed >50% of LAI. Overall reduction of LAI was generally greatest in dominant thinnings (54%), intermediate in crown thinnings (46%), and lowest in low thinnings (35%). Upon closer examination by crown classes, the postthinning distribution of LAI between upper and lower crown classes varied by thinning history, thinning method, and amount of density reduction.

## 1. Introduction

Leaf area index (LAI $m^2\,m^{-2}$) is an important factor determining both tree and stand level production [1]. While LAI is biologically important due to the fact that foliage is where photosynthesis occurs [2], it is also arguably the most appropriate descriptor of growing space occupancy whether one is characterizing simple-structured, single-species forests [3, 4] or complex-structured, multispecies forests [5–7]. This importance is due to the strong well-established relationship between LAI and volume increment for both individual trees [8–11] and stands [12–14]. Accordingly, leaf area-based stocking models that incorporate these relationships have been developed as tools [6, 7, 15] giving managers the ability to base silvicultural retention on individual tree growth capability as opposed to simply diameter (DBH), although implementation is aided by taking advantage of the observed relationships between individual tree DBH and leaf area [9]. Given that LAI describes growing space occupancy and productivity, quantification of its relationship to silvicultural activities, for example, intermediate thinning, should be of primary importance.

Thinning as a means to increase stand-level production, or more specifically the growth of some predetermined individuals, is an important part of silviculture [16]. Though we understand thinning does not increase overall growth *per se* , it can redirect the efficient use of growing space (i.e., LAI) to desired individuals [17, 18]. Traditionally, thinning is implemented in even-aged stands by reducing stand density through thoughtful removal of undesired individuals. Although the selection of trees for removal in thinning treatments can be characterized using the DBH distribution and DBH ratios [19, 20], actual implementation usually focuses on individual tree crown class, height, and position relative to neighbors. Because the amount of individual tree leaf area varies substantially, and predictably, between crown classes in even-aged stands [8, 9] and between cohorts in multiaged stands [3, 10, 13, 21], silviculturists are actually redistributing the LAI in a given stand. For example, when using thinning methods such as low thinning, crown thinning, or selection thinning (henceforth called dominant thinning) removal is focused on trees from particular crown classes [19, 22]. Low thinning, or thinning from below, focuses removal on intermediate or suppressed trees with small crowns containing relatively little leaf area. Crown thinning, or thinning from above, focuses removal on codominant or dominant trees which have more leaf area than trees from the lower crown classes. Dominant thinning focuses removal on the largest crowned trees, which have substantial leaf area.

Given that thinning methods specify removing trees from different crown classes, we would expect the postthinning reduction in LAI to vary accordingly. The potentially large variation in postthinning LAI could have profound implications not only on subsequent stand volume increment [17], but also on ecological processes that have been linked to the redistribution of LAI through thinning [1]. For example, in a single-species stand, we might expect a low thinning to maintain a relatively large amount of LAI because primarily the intermediate and overtopped trees are removed [19]. In contrast, we might expect a crown or dominant thinning to have more drastic reductions in total LAI as a result of the removal of primarily dominant and codominant trees [19]. Holding thinning intensity constant, reoccupation of the growing space and higher postthinning volume growth would be achieved sooner after a low thinning than a crown or dominant thinning [17, 18, 23, 24] because primarily the upper crown classes are retained. Results of LAI changes due to thinning in mixed-species stands would likely vary proportionally with the species-specific [25, 26], and crown class differences in LAI, especially when species preference is part of the marking prescription. Previous work describing crown class changes as a result of intermediate thinning [27, 28] and natural stand development [29] provide a solid background for further understanding dynamics of LAI distribution between crown classes due to thinning methods.

In this study, we used a network of replicated commercial thinning treatments in Maine, the Cooperative Forestry Research Unit's Commercial Thinning Research Network (CTRN), to examine changes in structure and LAI after thinning activities in stands dominated by *Abies balsamea* (L.) Mill. and *Picea rubens* Sarg. [30]. Specifically, we question whether traditional commercial thinning methods that base selection for removal on crown classes result in predictable changes in postthinning structure and leaf area. We further question whether thinning type or intensity has an effect on the relationship between crown classes and leaf area.

## 2. Methods

Study sites, field methods, and leaf area sampling for this study were part of a larger research project on the production ecology of northeastern forests and were detailed in previously published papers [11, 31]. Methodological considerations specific to this study were reiterated for completeness.

*2.1. Study Sites.* The 12 CTRN sites span the Acadian forest region [32] in the state of Maine in the northeastern United States and are the result of even-aged regeneration methods (often one-cut shelterwoods [19], still common in the region [30, 33]). Combinations of replicated low thinning, crown thinning, and dominant thinning treatments, to 33% and 50% relative density (RD) [34], were performed on the CTRN sites, which were divided evenly between a history of precommercial thinning (PCT), and no history of precommercial thinning (NOPCT) [30]. Six sites received PCT 15–20 years prior to the commercial thinning treatments (exact dates were undocumented), which occurred in 2000 and 2001. The PCT thinning focused on leaving evenly spaced trees regardless of diameter. The PCT sites were dominated by *A. balsamea*, ranged in age from 23 to 40 years and ranged in site index from 18.4 m to 24.3 m. The other six sites did not receive PCT (NOPCT) and were codominated by *P. rubens* and *A. balsamea*. The NOPCT sites ranged in age from 33 to 73 years and ranged in site index from 13.1 m to 21.5 m. Each thinning history (PCT versus NOPCT) independently consisted of a randomized-complete-block experimental design with replication of 6 plots per treatment. On the NOPCT sites, two intensities of thinning reduction (33% and 50% RD) and three commercial thinning methods (crown, dominant, and low) were performed. On the PCT sites, commercial thinnings to 33% and 50% of RD were performed on a time schedule. For this study, only one thinning entry had been conducted in the PCT ($n = 6$ per thinning intensity, Table 1), so that the other plots were treated as controls ($n = 30$). Each of the 12 CTRN sites had one designated control plot. Seven 0.02 ha measurement plots, nested within 0.08 ha treatment plots, were located at each site.

Prethinning data were used to calculate RD [34] for each site to guide marking. For the PCT sites, each of the seven plots within a site were ranked based on prethinning RD and assigned a commercial thinning treatment that did not result in stands of similar RD. Plots with median RDs were assigned as the control. Plots with first and the fifth highest ranking RDs were assigned to 33% RD and 50% RD reductions. Plots within the NOPCT sites were randomly assigned thinning treatments. Each plot was marked to the RD targets of 33% and 50%. Thinning goals were to remove all hardwoods, all *A. balsamea* greater than 21.6 cm DBH and to favor *P. rubens*. No consideration to RD reduction was given for harvester trails created during thinning operations. Common harvesting techniques included hand felling and four types of single-grip processors.

*2.2. Field Measurements.* Prior to thinning in each plot, every tree was measured for DBH and species was noted. In addition, crown class was noted for each tree, that is, codominant and dominant trees were considered upper crown classes, and intermediate and overtopped trees considered lower crown classes [19, 22].

Measurements taken after the thinning (~1-2 years later) had two goals, to characterize stand structure, and to predict individual tree leaf area from sapwood area. These data are referred to as stand structure data and sapwood area-leaf area data, respectively. Following harvest, every tree in each treatment unit (including control plots) was tagged with an individual tree number and measurements taken for DBH, height, height to the base of the live crown, crown class,

TABLE 1: Mean (standard deviation) and sample size by treatment type and thinning history for pre- and postthinning leaf area index (LAI) and relative density[a] (RD).

| Thinning treatment | Prethinning LAI ($m^2 m^{-2}$) | RD | Postthinning LAI ($m^2 m^{-2}$) | RD | Number of plots |
|---|---|---|---|---|---|
| | | | NOPCT | | |
| Crown 33% | 10.7 (2.3) | 0.59 (0.06) | 6.4 (1.9) | 0.35 (0.02) | 6 |
| Crown 50% | 10.9 (2.8) | 0.61 (0.09) | 5.1 (1.2) | 0.28 (0.02) | 6 |
| Dominant 33% | 10.7 (1.8) | 0.59 (0.05) | 5.9 (1.9) | 0.34 (0.06) | 6 |
| Dominant 50% | 11.0 (3.3) | 0.58 (0.09) | 4.2 (1.4) | 0.26 (0.05) | 6 |
| Low 33% | 9.8 (1.7) | 0.59 (0.09) | 7.2 (1.9) | 0.36 (0.05) | 6 |
| Low 50% | 10.2 (2.2) | 0.59 (0.04) | 5.8 (1.7) | 0.28 (0.05) | 6 |
| Control | 9.7 (1.7) | 0.60 (0.09) | 10.9 (2.1) | 0.69 (0.12) | 6 |
| | | | PCT | | |
| 33% Reduction | 12.2 (0.9) | 0.36 (0.03) | 7.8 (0.7) | 0.24 (0.02) | 6 |
| 50% Reduction | 10.1 (1.3) | 0.31 (0.05) | 4.9 (0.6) | 0.15 (0.02) | 6 |
| Control | 10.4 (1.7) | 0.31 (0.05) | 11.6 (1.9) | 0.36 (0.05) | 30 |

[a]Relative density calculated according to Wilson et al. [34]. Note: PCT—precommercial thinning history, NOPCT—no history of precommercial thinning.

and species. A systematic sample was developed from the postthinning tally of tagged trees by arraying them as three live crown ratio classes (crown length/height) by 5.1 cm DBH classes. From each cell in this matrix, every tenth tree was selected for sampling. For each of these trees, measurements included DBH; bark thickness, measured at breast height on the north and east sides of the tree; height; height to the lowest live branch. Crown class designations were given to each tree [19, 22, 35]. Two increment cores per tree were taken 90° apart at breast height using a cordless power drill and increment borer [36]. The sapwood boundary was delineated immediately in the field by holding each core to light and marking the boundary between the translucent sapwood and opaque heartwood. Cores were glued into routed boards where both bark and sapwood boundaries were marked with an indelible pencil for subsequent laboratory analysis. Trees were measured one or two growing seasons after thinning, prior to detectable DBH or leaf area growth responses. Sampling for sapwood area-leaf area was conducted in the dominant thinning, low thinning, and control plots at each installation of the NOPCT history. For the PCT sites, two treated plots (33% and 50% reduction) and two untreated plots with the greatest RD were measured.

*2.3. Leaf Area Sampling.* Relationships for indirectly estimating individual tree leaf area were developed by destructively sampling seven or eight trees per site during summer 2001. A total of 88 trees covering a range of DBHs and crown classes were chosen from plot buffers at each of the 12 sites. *A. balsamea* was sampled from PCT plots, while *P. rubens* was sampled from NOPCT plots. After felling, total tree height and height to base of the live crown were measured directly. Sample branches were taken from three random locations within the crown to build predictive models of branch leaf area. The sampling locations were one from the upper half of the crown; one each from the two lower quartiles, divided this way to account for foliage recession on older branches and for potential differences in specific leaf area with crown depth [37–39]. Branch basal diameters and branch length were measured. Approximately 100 needles were removed from across the range of needle ages for each branch for subsequent analysis.

*2.4. Sapwood Area Measurement.* Increment cores were measured for tree-ring increment (0.001 mm) and to determine the radial length of sapwood. Before making any calculations, ring widths were adjusted for shrinkage using the following factors: *A. balsamea*, 2.9%; *P. rubens*, 3.8% [40]. Estimates of sapwood area were then determined for each tree by subtracting the north and east core sapwood radii from the inside bark diameter, then these estimates of heartwood area were averaged before subtracting from the averaged inside bark basal areas. Final estimates of sapwood area were used in combination with other tree metrics for the prediction of leaf area.

*2.5. Branch and Tree Leaf Area Prediction.* Projected needle areas were measured using WinSeedle scanning software (Regent Instruments). Needle area measurements were combined with needle weights (g) to determine specific leaf area for each randomly sampled branch. Specific leaf area was then multiplied by the weight of all needles to determine branch leaf area. Branch-level equations from [11] were applied to every branch in each tree crown before summing for individual tree leaf area.

For *A. balsamea*,

$$\sqrt{(\mathrm{BRLA})} = 129.79 * \mathrm{BRD}^{1.049} * \mathrm{RDINC}^{(1.923-1)} \exp\left[-(1.959 * \mathrm{RDINC}^{1.923})\right],\ R^2 = 0.91,$$

for *P. rubens* codominant, dominant, and intermediate trees,

$$\ln \mathrm{BRLA} = 8.375 + 2.084 * \ln \mathrm{BRD} + 0.705 * \ln \mathrm{RDINC} - 2.652 * \mathrm{RDINC},\ R^2 = 0.83,$$

log bias ratio correction factor = 1.1657 [41], and

for *P. rubens* overtopped trees,

$$\ln \mathrm{BRLA} = 2.754 + 2.872 * \mathrm{BRD},\ R^2 = 0.79,$$

log bias ratio correction factor = 1.1826,

where BRLA is predicted branch leaf area ($cm^2$); BRD is branch basal diameter (mm); RDINC is relative depth into the crown, measured from the leader to the lowest live branch.

Equations for predicting individual tree leaf area (PLA) from sapwood area and crown length reported in [11, 31] were applied to the sapwood area-leaf area data in this study. For *A. balsamea*,

$$\ln \text{PLA} = -1.692+0.863*\ln \text{SA}+0.797*\ln \text{CL},\ R^2 = 0.96,\ n = 43,\ \text{log bias ratio correction factor} = 1.0322,$$

for *P. rubens*,

$$\ln \text{PLA} = -2.722+0.966*\ln \text{SA}+ 0.844*\ln \text{CL},\ R^2 = 0.91,\ n = 45,\ \text{log bias ratio correction factor} = 0.9753,$$

where PLA is projected individual tree leaf area in $m^2$ per $m^{-2}$ of ground area, SA is sapwood area ($cm^2$), and CL is crown length (m).

The sapwood area equations were then used to estimate PLA for every tree measured in the stand structure data. From these results, site-specific [42] PLA models were developed using stepwise regression techniques for the purpose of predicting leaf area of noncored trees. Combinations of the model form $\ln \text{PLA} = \text{b0}+\text{b1} \ln \text{DBH}+\text{b2} \ln \text{DBH}^2+\text{SITE}+\varepsilon$ were used where the SITE is a site-specific parameter and $\varepsilon$ represents random error. Where application of a site-specific model was not possible, due to limited representation of a species, a model fit from the destructively sampled trees (which span all sites) was used. Resultant models performed well (84%–95% variation accounted for), and residual analysis for each model ensured unbiased predictions across independent variables and validated linear model assumptions.

*2.6. LAI Estimation.* PLA for individual trees on each plot were converted to all-sided leaf areas using published ratios of 2.3 for *A. balsamea* [43] and 2.9 for *P. rubens* [44] in order to take into account differences in needle morphology between species [2]. All-sided leaf area likely better integrates physiological processes such as light interception and respiration because it takes into account the entire photosynthesizing surface of needles in comparison to the more common one-sided PLA. More importantly, standardizing leaf area estimates between *A. balsamea* and *P. rubens* before combining into plot level LAI should reduce or eliminate any species-specific affect on LAI. LAI was calculated by summing individual tree all-sided leaf area for all *A. balsamea* and *P. rubens* on each plot during the prethinning analysis, and again for the postthinning stand structure data (Table 1). LAI estimates reported henceforth in this study are all-sided leaf areas. The change in LAI for each plot was calculated by dividing the post-thinning LAI by the prethinning LAI and subtracting 1.0 to reflect the decrease in LAI that occurred in all plots except the untreated controls.

*2.7. Analysis.* Given the differences in study design and implementation, no statistical tests between attributes associated with PCT and NOPCT histories were made. To assess the effect of thinning method on stand structure, a ratio ($d/D$) of mean plot DBH before thinning to mean plot DBH after treatment was calculated and analyzed between treatments using one-way ANOVAs. This ratio is similar to the "$d/D$" ratio mentioned in [19] and was used for yield projections in relation to thinning methods [45]; however, DBH of trees removed during thinning was not measured in this study, instead mean posttreatment DBH was used. To assess the effect of thinning method and species composition change (Table 1) on LAI, the proportional change in LAI from pre- to postthinning for each treatment was analyzed using one-way ANOVAs. Potential LAI differences associated with species composition effects were assessed by including a variable which described the net change in percent *P. rubens* for each plot from pre-to postthinning for the NOPCT histories (range −0.02%–0.05%). PCT stands were not tested for species composition effects because the net change in composition pre-to postthinning was <1%. Preliminary results suggested the percent *P. rubens* effect ($P = 0.556$) and the percent *P. rubens* by treatment effect ($P = 0.338$) were insignificant; therefore, subsequent analyses were interpreted solely as treatment effects. We also examined the possible postthinning shift in the distribution of LAI between crown classes within thinning types. This was done by calculating the postthinning proportional change in LAI by upper and lower crown classes relative to the prethinning LAI and plotting them for all but the crown thinning treatment. Due to time constraints, canopy information was not collected for the crown thinning treatment during posttreatment measurements, so they were not included in this analysis. As a result of limited pretreatment measurements, posttreatment crown classes were grouped into upper (dominant and codominant trees) and lower (overtopped and intermediate trees) crown classes in order to investigate the shift in LAI by treatment. We confirmed that independence, normality, and constancy of error variance assumptions for ANOVA were met. Pair-wise means tests were performed using Tukey's test. All statistical tests were considered significant at the $P < 0.05$ level.

## 3. Results

*3.1. NOPCT.* The $d/D$ ratio was <1.0 in the NOPCT low thinning treatment plots, which indicated that indeed smaller DBH trees were removed during the thinning. Similarly, the dominant thinning in the NOPCT resulted in $d/D > 1.0$, but only for the 50% reduction. The 33% reduction had mixed results and was not significantly different from 1.0. Both crown thinning reduction levels had $d/D$s well below 1.0, which indicated that, although dominant and codominant trees may have been selected during thinning, proportionally more small DBH trees were removed (Figure 1). NOPCT control plots had $d/D\text{s} < 1.0$ which indicated diameter increment between measurement periods (∼1 to 2 years).

Although strictly marked to postthinning relative densities, the LAI reductions for NOPCT plots were proportional to thinning intensity (33% or 50%) and to a lesser extent thinning type (low, crown, or dominant). For NOPCT

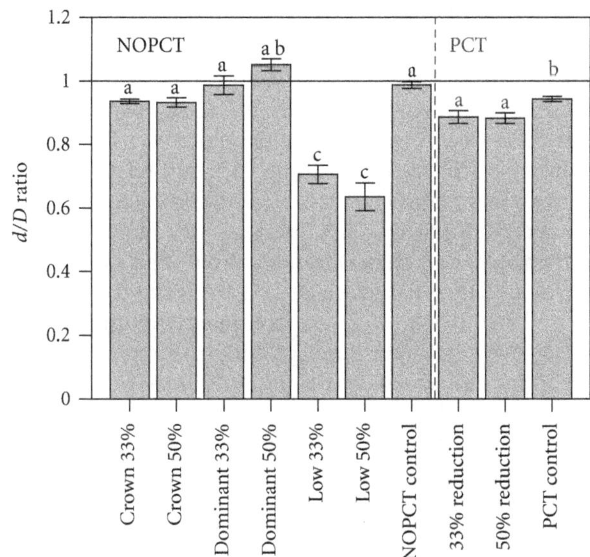


FIGURE 1: Mean ± one standard error pre- to postthinning average diameter ratios ($d/D$) for the commercial thinning types (crown, dominant, low) and reduction levels (33% and 50%) implemented by the Cooperative Forestry Research Unit Commercial Thinning Research Network. Statistical analysis was conducted independently for PCT (precommercial thinning history) and NOPCT (no history of precommercial thinning) groups. Letters represent significant differences determined using Tukey's test.

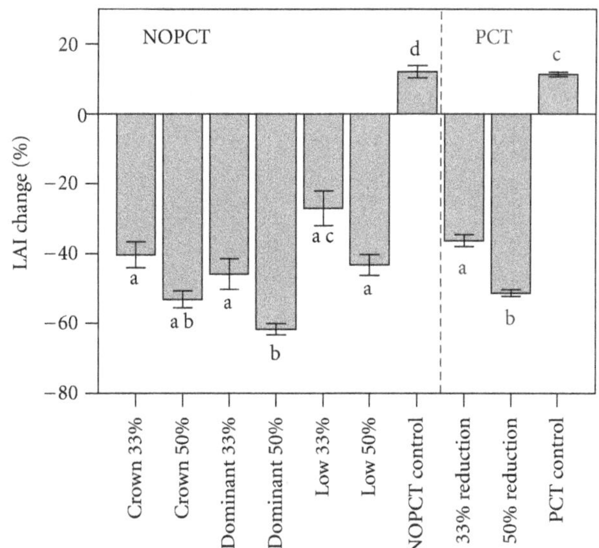


FIGURE 2: Direction and magnitude of net leaf area index (LAI) changes for the commercial thinning types (crown, dominant, low) and reduction levels (33% and 50%) implemented by the Cooperative Forestry Research Unit Commercial Thinning Research Network. Statistical analysis was conducted independently for PCT (precommercial thinning history) and NOPCT (no history of precommercial thinning) groups. Letters represent significant differences determined using Tukey's test.

thinning types, absolute LAI reduction averaged 35% for low thinning, 46% for crown thinning, and 54% for dominant thinning (Figure 2). Similarly, for each thinning type, the difference in the reduction of LAI between the 33% and 50% RD treatments averaged 13% for crown thinning, 16% for dominant thinning, and 14% for low thinning (Figure 2). The NOPCT control plots increased LAI by ~12% due to growth between measurement periods.

The reduction in LAI due to thinning type resulted from drastically different postthinning distributions of LAI between crown classes. The NOPCT low thinning 33% reduction treatment resulted in 10% more leaf area in the upper crown classes and 60% less in the lower crown classes relative to prethinning (Figure 3). Here, trees marked for removal were generally smaller in DBH and from the lower crown classes, and resulted in a 27% reduction in LAI for a 33% reduction in RD. Dominant thinning to 50% resulted in relatively more postthinning LAI in the lower crown classes and relatively less postthinning LAI in the upper canopy classes, as was expected when removing primarily dominants (large DBH), and which resulted in a 62% reduction in LAI for a 50% reduction in RD. In contrast, dominant thinning to 33% resulted in a 55% decrease in LAI for a 33% reduction in RD, even though the relative amount of postthinning LAI in the upper crown classes increased (Figure 3). NOPCT controls had a net negative change in LAI in the lower crown classes and only a small positive LAI change in the upper crown classes, which likely indicated plots were undergoing strong size class differentiation. With the possible exception of low thinning, patterns of postthinning LAI redistribution by crown class were not readily predictable by thinning type or intensity.

*3.2. PCT.* The $d/D$ ratio was <1.0 for PCT plots with 33% and 50% RD reductions, which indicated primarily smaller DBH trees were removed during thinning (Figure 1). The $d/D$ ratio was also <1.0 for PCT control plots, the result of diameter growth between measurement periods.

LAI change in the PCT plots was more severe with a 50% RD reduction relative to the 33% RD reduction. Absolute LAI reduction in PCT plots averaged 36% for the 33% RD reduction and 51% for the 50% RD reduction (Figure 2). In addition, the difference in total LAI reduction between the 33% and 50% RD treatments was only 8% on average (Figure 2). Like the NOPCT controls, the PCT control plots increased in LAI between measurement periods by ~11% as a result of growth between measurement periods.

Changes in PCT plot LAIs due to RD reduction resulted from strikingly different postthinning distributions of LAI between crown classes. While the 50% RD treatment resulted in a 49% reduction in total LAI and a net increase in lower crown class LAI (Figure 3), the 33% RD treatment resulted in a 36% reduction in total LAI through a net decrease in lower crown class LAI (Figure 3). In contrast to the NOPCT controls, PCT control plots had an unexpected net positive change in lower crown class LAI, suggesting an influence of PCT stand history on subsequent stand dynamics. Similar to the NOPCT results, patterns of postthinning LAI between crown classes were not predictable from RD treatments.

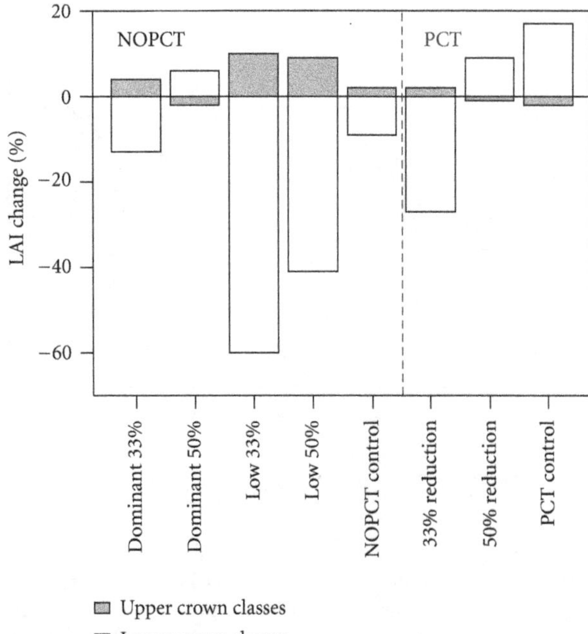


FIGURE 3: Direction and magnitude of leaf area change by crown classes (upper and lower) for the commercial thinning types (dominant, low) and reduction levels (33% and 50%) implemented by the Cooperative Forestry Research Unit Commercial Thinning Research Network. Note. PCT—precommercial thinning history, NOPCT—no history of precommercial thinning. Crown thinning canopy level information was not collected.

## 4. Discussion

Based on the thinning types, one would expect low thinning to have a $d/D$ of <1.0, dominant thinning >1.0 and crown thinning approximately equal to 1.0. These expectations are similar to the guidelines proposed by Vezina [46] (low thinning <0.7, crown thinning 0.7 to 1.0 and dominant >1.0). However, the relationships for dominant and crown thinning in our data are more vague, likely a result of two factors: (1) the use of posttreatment DBH in our calculations of $d/D$ that are not directly comparable with the guidelines suggested by Vezina [46]; (2) strict adherence to percent RD reductions (33% and 50%) during marking the CTRN. For example, in one case codominant and dominant trees were iteratively marked and remarked until the RD target for crown thinning was met. That the dominant 50% treatment has a higher $d/D$ in every case suggests larger trees had to be removed to meet the RD target, which has a measurable effect on the shift in proportion of LAI between upper and lower crown classes postthinning (Figure 3). Although the NOPCT low thinning treatment $d/D$s are substantially lower than 1.0, both PCT RD reductions (33% and 50%) also averaged just under 1.0 (~0.89), resulting in stand structure that suggests lower crown classes were more likely marked for thinning.

While overall LAI reduction was predictable from stand history (NOPCT versus PCT), RD reduction, and thinning type, differences in the postthinning LAI distribution between crown classes requires further scrutiny. For example, the PCT 50% RD reduction and the dominant thinning 50% RD reduction both show a shift in the distribution of LAI from the upper to the lower crown classes. We suggest the similarity in crown class results for these two treatments stems from the need to mark enough trees for removal to meet the RD target, resulting in some relatively large-crowned trees being removed. In addition, the $d/D$ for the dominant thinning to 50% RD suggests that more large DBH trees (upper crown classes) were removed. This explains why there is a relative reduction of LAI from the upper crown classes. Whereas the $d/D$ for dominant thinning to 33% RD suggests trees were removed uniformly from the DBH distribution, as opposed to large DBH trees as prescribed, so that the postthinning distribution of LAI between crown classes is similar to the NOPCT control (Figure 3) [27].

Another possible reason postthinning crown class LAI distribution was variable might be the differences in stand history (i.e., NOPCT and PCT). The similarity in $d/D$ between the NOPCT and PCT controls does not explain their postthinning difference in LAI. In the PCT control plots, the proportion of postthinning LAI in the lower crown classes increased, whereas the proportion of LAI in upper crown classes increased on the NOPCT control plots (Figure 3). One possible explanation is that the older NOPCT plots are undergoing strong self-thinning so mortality of intermediate and overtopped trees is likely due to competition related, density-dependent mortality [47] which has been found to maintain dominant and codominant trees [27, 29]. Conversely, the younger PCT plots have wider spacing than NOPCT plots and homogenized diameters a result of the PCT. This likely explains why the $d/D$s are not substantially lower than 1.0. While the PCT plots are likely undergoing differentiation into size classes [28] because of widening crowns [48], they are probably not self-thinning because of the precommercial thinning, so that any intermediate and overtopped trees that were not cut have persisted and subsequently enlarged their crowns. Furthermore, when conducting a precommercial thinning, multiple criteria are taken into consideration [49, 50]. Typically, workers simultaneously consider intertree spacing, species preference, and size. The PCT treatment history in this study area prioritized size and spacing, but not species. As a result, primarily the largest diameter *A. balsamea*, which tend to grow more rapidly than *P. rubens* early in stand development [51], were retained during thinning and smaller individuals were removed, which would further narrow the size distribution. Another possible explanation is that the subjectivity of crown class determination [35] is exacerbated when plots that have been precommercially thinned are scrutinized during field measurements.

Differences in the distribution of LAI between the upper and lower crown classes could have implications for the ability of the postthinning stands to reoccupy the growing space. In general, posttreatment LAI in the PCT plots should reoccupy the growing space faster than the NOPCT plots due to their higher potential productivity [11, 52], which is in part a result of the fact that higher quality sites are typically chosen to receive PCT [49]. However, in order to positively affect the potential increase of individual tree

LA and growth typically sought by reducing stand density [53], careful attention must be paid to whether the type of thinning employed removes trees from the appropriate crown classes. For example, the PCT 33% RD reduction resulted in a net loss of LAI from the lower crown classes and a net gain in LAI from the upper crown classes. In general, these dominant and codominant trees have more individual tree leaf area [8, 9], and their higher growth potential ought to translate into an increase in postthinning individual tree growth increment, compared to trees of similar size on control plots [54]. Juodvalkis et al. [48] note that the effect of thinning on individual tree growth (DBH and crown reaction) is most pronounced when thinning occurs early in stand age, as was the case for the CTRN precommercial thinning. However, as a result of this historic spacing in the PCT stands, higher individual tree growth [55] might come at the expense of stand volume growth if the trees do not fully reoccupy the growing space, which may be especially problematic in the 50% RD reduction treatments.

Because stand volume production is strongly linked to LAI [13, 14], and thinning has been demonstrated to result in subsequent rapid LAI increase relative to controls [56], it is realistic to expect changes in volume increment proportional to the quantity of LAI as a result of thinning type and intensity [57]. A thorough review of long-term stand volume response to thinning to various intensities reveals mixed results regarding comparisons between thinning intensities, thinning types, and their controls over fixed time periods [58]. Furthermore, while a lengthy literature discusses potential stand volume changes due to thinning [17], there is likely a point beyond which thinning intensity reduces potential subsequent stand yield. This idea of an optimum stand density for growth is not new [17, 24, 59] but the empirical evidence is mixed. For example, long-term increases in stand volume production were correlated with thinning intensity, where maximum volume increment for *P. abies* was ~15% greater than controls for 30-to-40-year-old stands, and the volume increase was most pronounced for the youngest thinned stands [48]. In another example, low thinning was found to result in the highest long-term production of *P. abies*, followed by crown thinning, and dominant thinning [60]. In contrast, mid-term (15 year) stand volume of all thinning intensities were found to be lower than controls in mixed stands of *P. abies-Betula pubescens* Erhr [61]. Similarly, Slodicak et al. [62] noted long-term (29 years) stand volume reductions ranging from 14 to 23%, with larger reductions in crown thinning, which they attributed to site or "other" conditions. Based on our results, it is possible those "other" conditions might be explained as the size (i.e., DBH) of trees that were selected for removal.

One possible limitation of our study is the prediction of leaf area in mixed species stands. For example, although stands of *Pinus contorta* var. *latifolia* (Engelm.) Critchfield and *Abies lasiocarpa* (Hook.) Nutt have relatively similar stand growth efficiencies (stemwood increment per unit leaf area) [25], this was likely the result of the relatively efficient individual tree use of lower absolute LAI in *P. contorta* than in *A. lasiocarpa*, and this was likely mediated by stand structure [63]. Conversely, both LAI and growth efficiency differ greatly in stands of mixed *Picea abies* (L.) Karst-*Pinus sylvestris* L. in Finland where the difference was related to species composition [26]. In this study, we account for the potential differences in leaf area between *A. balsamea* and *P. rubens*, which are both shade tolerant and had very similar individual tree leaf areas to begin with [9, 10], in two ways: (1) we used all-sided leaf area conversion factors before estimating LAI in our methodology; (2) we subsequently conducted a preliminary analysis for species effect using net change in percent *P. rubens* as a measure of composition change from pre- to postthinning for the NOPCT data, which had nonsignificant results. All-sided conversions minimize potential differences in absolute LAIs due to species, which likely explains the insignificance of net change in percent *P. rubens* composition when included in the analysis. In addition, although it is possible to have similar LAIs at drastically different structures, all CTRN stands are even-aged with strong unimodal DBH distributions, so that DBH distributions probably do not explain LAI differences *sensu* O'Hara [12]. Therefore, our results and interpretation reflect the influence of treatment type and RD reduction on LAI changes and not species-specific effects.

## 5. Management Considerations

Although not necessarily desirable in practice, the relatively severe reductions of LAI imposed as part of the CTRN provide useful benchmarks for future work. In the CTRN study, strict adherence to marking guidelines ensured that contrasts between treatments were enhanced [30]; however, relaxing these assumptions in practice might have a less drastic effect on LAI and its potentially unintended redistribution between crown classes. Of course, the practical trade-off for low thinning is the need to remove a larger number of small DBH stems in order to meet the RD target. This is often not practical from an economical harvesting perspective without removing some larger DBH stems. In contrast, when marking a dominant thinning perhaps the most difficult procedure is balancing the removal of poorly formed dominants with the desired residual density while not succumbing to high grading [19]. Stands with a history of PCT, on the other hand, will have more merchantable stems earlier in the rotation [64], potentially making removal of the smaller portion of stems financially feasible.

Based on our results, commercial thinning tends to redistribute LAI predictably; however, there is some variability in the observed relationships. This suggests that, although a particular type of thinning is prescribed (e.g., dominant thinning), actual RD reduction will not necessarily meet the goals associated with the treatment. In a region where arguably little silviculture is conducted, silviculturists designing marking guidelines for even-aged *P. rubens-A. balsamea* forests might consider relaxing strict adherence to density reduction targets, which may result in unanticipated shifts in LAI from one crown class to another and result in unanticipated effects on subsequent growth. Instead, we recommend focusing on leaving trees of appropriate species and growth capability (i.e., *P. rubens*, or small-crowned codominants) so that growing stock is not overly reduced and growing space can be rapidly reoccupied after treatment.

## Acknowledgment

Financial support came from the Cooperative Forestry Research Unit and the Maine Agricultural and Forest Experiment Station (no. 3258).

## References

[1] J. N. Long, T. J. Dean, and S. D. Roberts, "Linkages between silviculture and ecology: examination of several important conceptual models," *Forest Ecology and Management*, vol. 200, no. 1–3, pp. 249–261, 2004.

[2] H. Lambers, F. S. I. Chapin, and T. L. Pons, *Plant Physiological Ecology*, Springer, New York:, NY, USA, 1998.

[3] C. L. Kollenberg and K. L. O'Hara, "Leaf area and tree increment dynamics of even-aged and multiaged lodgepole pine stands in Montana," *Canadian Journal of Forest Research*, vol. 29, no. 6, pp. 687–695, 1999.

[4] D. E. B. Reid, V. J. Lieffers, and U. Silins, "Growth and crown efficiency of height repressed lodgepole pine; are suppressed trees more efficient?" *Trees*, vol. 18, no. 4, pp. 390–398, 2004.

[5] K. L. O'Hara, E. Lähde, O. Laiho, Y. Norokorpi, and T. Saksa, "Leaf area allocation as a guide to stocking control in multiaged, mixed-conifer forests in southern Finland," *Forestry*, vol. 74, no. 2, pp. 171–185, 2001.

[6] K. L. O'Hara, N. I. Valappil, and L. M. Nagel, "Stocking control procedures for multiaged ponderosa pine stands in the Inland Northwest," *Western Journal of Applied Forestry*, vol. 18, no. 1, pp. 5–14, 2003.

[7] K. L. O'hara and C. L. Kollenberg, "Stocking control procedures for multiaged lodgepole pine stands in the Northern Rocky Mountains," *Western Journal of Applied Forestry*, vol. 18, no. 1, pp. 15–21, 2003.

[8] K. L. O'Hara, "Stand structure and growing space efficiency following thinning in an even-aged Douglas-fir stand," *Canadian Journal of Forest Research*, vol. 18, no. 7, pp. 859–866, 1988.

[9] D. W. Gilmore, R. S. Seymour, and D. A. Maguire, "Foliage—sapwood area relationships for Abies balsamea in central Maine, USA," *Canadian Journal of Forest Research*, vol. 26, no. 12, pp. 2071–2079, 1996.

[10] D. A. Maguire, J. C. Brissette, and L. Gu, "Crown structure and growth efficiency of red spruce in uneven-aged, mixed-species stands in Maine," *Canadian Journal of Forest Research*, vol. 28, no. 8, pp. 1233–1240, 1998.

[11] R. J. DeRose and R. S. Seymour, "The effect of site quality on growth efficiency of upper crown class *Picea rubens* and *Abies balsamea* in Maine, USA," *Canadian Journal of Forest Research*, vol. 39, no. 4, pp. 777–784, 2009.

[12] K. L. O'hara, "Stand growth efficiency in a douglas fir thinning trial," *Forestry*, vol. 62, no. 4, pp. 409–418, 1989.

[13] K. L. O'Hara, "Dynamics and stocking-level relationships of multi-aged ponderosa pine stands," *Forest Science*, vol. 42, no. 4, pp. 1–34, 1996.

[14] J. N. Long and F. W. Smith, "Determinants of stemwood production in *Pinus contorta* var. latifolia forests: the influence of site quality and stand structure," *Journal of Applied Ecology*, vol. 27, no. 3, pp. 847–856, 1990.

[15] K. L. O'Hara and N. I. Valappil N., "MASAM—a flexible stand density management model for meeting diverse structural objectives in multiaged stands," *Forest Ecology and Management*, vol. 118, no. 1–3, pp. 57–71, 1999.

[16] B. Zeide, "Self-thinning and stand density," *Forest Science*, vol. 37, no. 2, pp. 517–523, 1991.

[17] E. Assmann, *The Principles of Forest Yield Study: Studies in the Organic Production, Structure, Increment and Yield of Forest Stands*, Pergamon Press, 1970.

[18] C. M. Moller, "The effect of thinning, age, and site on foliage, increment, and loss of dry matter," *Journal of Forestry*, vol. 45, no. 6, pp. 393–404, 1947.

[19] D. M. Smith, B. C. Larson, M. J. Kelty, and P. M. S. Ashton, *The Practice of Silviculture: Applied Forest Ecology*, John Wiley and Sons, New York, NY, USA, 9th edition, 1997.

[20] H. Mäkinen and A. Isomäki, "Thinning intensity and long-term changes in increment and stem form of Norway spruce trees," *Forest Ecology and Management*, vol. 201, no. 2-3, pp. 295–309, 2004.

[21] J. P. Berrill and K. L. O'Hara, "Patterns of leaf area and growing space efficiency in young even-aged and multiaged coast redwood stands," *Canadian Journal of Forest Research*, vol. 37, no. 3, pp. 617–626, 2007.

[22] R. D. Nyland, *Silviculture: Concepts and Applications*, New York, NY, USA, 2nd edition, 2002.

[23] B. Zeide, "Thinning and growth: a full turnaround," *Journal of Forestry*, vol. 99, no. 1, pp. 20–25, 2001.

[24] G. J. Hamilton, "The effects of high intensity thinning on yield," *Forestry*, vol. 54, no. 1, pp. 1–15, 1981.

[25] F. W. Smith and J. N. Long, "A comparison of stemwood production in monocultures and mixtures of *Pinus contorta* var. latifolia and *Abies lasiocarpa*," in *The Ecology of Mixed-Species Stands of Trees*, pp. 87–98, Blackwell Scientific, Cambridge, Mass, USA, 1992.

[26] K. L. O'Hara, E. Lähde, O. Laiho, Y. Norokorpi, and T. Saksa, "Leaf area and tree increment dynamics on a fertile mixed-conifer site in southern Finland," *Annals of Forest Science*, vol. 56, no. 3, pp. 237–247, 1999.

[27] G. Warrack, "Comparative observations of the changes in classes in a thinned and natural stand of immature Douglas-fir," *Forestry Chronicle*, vol. 28, no. 2, pp. 46–56, 1952.

[28] H. Mäkinen, P. Nöjd, and A. Isomäki, "Radial, height and volume increment variation in *Picea abies* (L.) Karst. stands with varying thinning intensities," *Scandinavian Journal of Forest Research*, vol. 17, no. 4, pp. 304–316, 2002.

[29] J. S. Ward and G. R. Stephens, "Crown class transition rates of maturing northern red oak (*Quercus rubra* L.)," *Forest Science*, vol. 40, no. 2, pp. 221–237, 1994.

[30] D. J. McConville, R. G. Wagner, and R. S. Seymour, "Maine's commercial thinning research network: a long-term research installation designed to improve our understanding about how forests respond to thinning," in *Proceedings of the the New England Society of American Foresters 85th Winter Meeting: Changing Forests—Challenging Times*, L. S. Kenefic and M. J. Twery, Eds., USDA FS GTR-NE-235, 2005.

[31] R. J. DeRose and R. S. Seymour, "Patterns of leaf area index during stand development in even-aged balsam fir—red spruce stands," *Canadian Journal of Forest Research*, vol. 40, no. 4, pp. 629–637, 2010.

[32] J. S. Rowe, "Forest regions of Canada," *Canadian Forestry Service*, no. 1300, 1972.

[33] R. S. Seymour, P. R. Hannah, J. R. Grace, and D. A. Marquis, "Silviculture: the next 30 years, the past 30 years. Part IV. The Northeast," *Journal of Forestry*, vol. 84, no. 7, pp. 31–38, 1986.

[34] D. S. Wilson, R. S. Seymour, and D. A. Maguire, "Density management diagram for northeastern red spruce and balsam

fir forests," *Northern Journal of Applied Forestry*, vol. 16, no. 1, pp. 48–56, 1999.

[35] N. S. Nicholas, T. G. Gregoire, and S. M. Zedakcr, "The reliability of tree crown position classification," *Canadian Journal of Forest Research*, vol. 21, no. 5, pp. 698–701, 1991.

[36] F. A. Baker and J. N. Long, "How to use a cordless drill to extract increment cores," *Journal of Forestry*, vol. 101, no. 5, p. 4, 2003.

[37] M. G. Keane and G. F. Weetman, "Leaf area—sapwood cross-sectional area relationships in repressed stands of lodgepole pine," *Canadian Journal of Forest Research*, vol. 17, no. 3, pp. 205–209, 1987.

[38] D. A. Maguire and W. S. Bennett, "Patterns in vertical distribution of foliage in young coastal Douglas- fir," *Canadian Journal of Forest Research*, vol. 26, no. 11, pp. 1991–2005, 1996.

[39] J. D. Marshall and R. A. Monserud, "Foliage height influences specific leaf area of three conifer species," *Canadian Journal of Forest Research*, vol. 33, no. 1, pp. 164–170, 2003.

[40] Forest Products Laboratory, "Wood handbook: wood as an engineering material," *USDA Forest Service GTR-113*, 1999.

[41] P. Snowdon, "A ratio estimator for bias correction in logarithmic regressions," *Canadian Journal of Forest Research*, vol. 21, no. 5, pp. 720–724, 1991.

[42] D. W. Gilmore and R. S. Seymour, "Foliage-sapwood area equations for balsam fir require local validation," *Forest Science*, vol. 50, no. 4, pp. 566–570, 2004.

[43] E. R. J. Hunt, M. B. Lavigne, and S. E. Franklin, "Factors controlling the decline of net primary production with stand age for balsam fir in Newfoundland assessed using an ecosystem simulation model," *Ecological Modelling*, vol. 122, no. 3, pp. 151–164, 1999.

[44] M. E. Day, "Influence of temperature and leaf-to-air vapor pressure deficit on net photosynthesis and stomatal conductance in red spruce (*Picea rubens*)," *Tree Physiology*, vol. 20, no. 1, pp. 57–63, 2000.

[45] G. Warrack, *Forecast of Yield in Relation to Thinning Regimes in Douglas Fir*, vol. 51 of *Forest Service technical publication*, 1959.

[46] P. E. Vezina, "Objective measures of thinning grades and methods," *S. L. Report*, vol. 62, no. 12, p. 15, 1963.

[47] C. D. Oliver and B. C. Larson, *Forest Stand Dynamics*, John Wiley and Sons, New York, NY, USA, 1996.

[48] A. Juodvalkis, L. Kairiukstis, and R. Vasiliauskas, "Effects of thinning on growth of six tree species in north-temperate forests of Lithuania," *European Journal of Forest Research*, vol. 124, no. 3, pp. 187–192, 2005.

[49] R. D. Briggs and R. C. Lemin, "Soil drainage class effects on early response of balsam fir to precommercial thinning," *Soil Science Society of America Journal*, vol. 58, no. 4, pp. 1231–1239, 1994.

[50] R. S. Seymour and C. J. Gadzik, "A nomogram for predicting precommercial thinning costs in overstocked spruce-fir stands," *Northern Journal of Applied Forestry*, vol. 2, no. 2, pp. 37–40, 1985.

[51] P. J. Strauch, *The Establishment and Early Growth of Red Spruce and Balsam Fir Regeneration*, Resource Conservation Services Inc, Bangor, ME, USA, 1991.

[52] R. A. Williams, B. F. Hoffman, and R. S. Seymour, "Comparison of site index and biomass production of spruce-fir stands by soil drainage class in Maine," *Forest Ecology and Management*, vol. 41, no. 3-4, pp. 279–290, 1991.

[53] J. N. Long, "A practical approach to density management," *Forestry Chronicle*, vol. 61, no. 1, pp. 23–27, 1985.

[54] M. B. Lavigne, "Effects of thinning on the allocation of growth and respiration in young stands of balsam fir," *Canadian Journal of Forest Research*, vol. 21, no. 2, pp. 186–192, 1991.

[55] A. L. Bernardo, M. G. F. Reis, G. G. Reis, R. B. Harrison, and D. J. Firme, "Effect of spacing on growth and biomass distribution in *Eucalyptus camaldulensis*, *E. pellita* and *E. urophylla* plantations in southeastern Brazil," *Forest Ecology and Management*, vol. 104, no. 1–3, pp. 1–13, 1998.

[56] I. Markova, R. Pokorny, and M. V. Marek, "Transformation of solar radiation in Norway spruce stands into produced biomass—the effect of stand density," *Journal of Forest Science*, vol. 57, no. 6, pp. 233–241, 2011.

[57] D. Binkley and P. Reid, "Long-term responses of stem growth and leaf area to thinning and fertilization in a Douglas-fir plantation," *Canadian Journal of Forest Research*, vol. 14, no. 5, pp. 656–660, 1984.

[58] C. Wallentin, *Thinning of Norway spruce*, Ph.D. thesis, Swedish University of Agricultural Sciences, Alnarp, Sweden, 2007.

[59] D. W. Gilmore, T. C. O'Brien, and H. M. Hoganson, "Thinning red pine plantations and the langsaeter hypothesis: a Northern Minnesota case study," *Northern Journal of Applied Forestry*, vol. 22, no. 1, pp. 19–26, 2005.

[60] G. J. Hamilton, "The bowmont Norway spruce thinning experiment 1930–1974," *Forestry*, vol. 49, no. 2, pp. 109–121, 1976.

[61] J. Repola, H. Hökkä, and T. Penttilä, "Thinning intensity and growth of mixed spruce-birch stands on drained peatlands in Finland," *Silva Fennica*, vol. 40, no. 1, pp. 83–99, 2006.

[62] M. Slodicak, J. Novak, and J. P. Skovsgaard, "Wood production, litter fall and humus accumulation in a Czech thinning experiment in Norway spruce (*Picea abies* (L.) Karst.)," *Forest Ecology and Management*, vol. 209, no. 1-2, pp. 157–166, 2005.

[63] S. D. Roberts, J. N. Long, and F. W. Smith, "Canopy stratification and leaf area efficiency: a conceptualization," *Forest Ecology and Management*, vol. 60, no. 1-2, pp. 143–156, 1993.

[64] R. J. Barbour, R. E. Bailey, and J. A. Cook, "Evaluation of relative density, diameter growth, and stem form in a red spruce (*Picea rubens*) stand 15 years after precommercial thinning," *Canadian Journal of Forest Research*, vol. 22, no. 2, pp. 229–238, 1992.

# Biology of the Wild Silkmoth *Anaphe panda* (Boisduval) in the Kakamega Forest of Western Kenya

**N. Mbahin,[1,2] S. K. Raina,[1] E. N. Kioko,[3] and J. M. Mueke[2]**

[1] *Environmemtal Health Division, International Centre of Insect Physiology and Ecology (icipe), P.O. Box 30772-00100, Nairobi, Kenya*
[2] *Department of Biological Sciences, Kenyatta University, P.O. Box 43844-00100, Nairobi, Kenya*
[3] *Biodiversity Conservation Division, National Museums of Kenya, P.O. Box 40658-00100, Nairobi, Kenya*

Correspondence should be addressed to N. Mbahin, mnorber@icipe.org

Academic Editor: Piermaria Corona

A study on the life cycle of the silkmoth *Anaphe panda* (Boisduval) was conducted in two different habitats of the Kakamega Forest in western Kenya: Ikuywa, an indigenous forest, and Isecheno, a mixed indigenous forest. Eggs were laid in clusters, and the incubation period ranged from 40 to 45 days. Larvae fed on *Bridelia micrantha* (Hochst) and passed through seven instars. The developmental period took between 83 to 86 days in the dry season and 112 to118 days in the rainy season. The pupal period ranged between 158 and 178 days in the rainy season and, on the other hand, between 107 and 138 days in the dry season. But the later caught up in development with those that formed earlier. Moths emerged from mid-October until mid-May. Longevity of adult *Anaphe panda* moths took between 4 and 6 days, but generally females seemed to live longer than males. The moth also seems to have higher lifespan in the indigenous forest compared to the mixed indigenous forest.

## 1. Introduction

Wild silk farming is a supplementary activity for income generation for rural communities that mainly depend on subsistence farming and assist in conserving the wild silkmoth and its habitat. *Anaphe panda* Boisduval (Lepidoptera: Thaumetopoeidae) is one of the indigenous silkmoth species in the Kakamega Forest that is currently used in silk production by the community surrounding the Kakamega Forest. The huge nest of *Anaphe* pupal cocoons can be degummed to produce brown silk that is of high quality. Although aspects on the conservation of this wild silkmoth for economic incentives to rural communities [1, 2], its spatial distribution [3, 4], and the use of sleeve nets to improve survival of the Boisduval silkworm [5] are known, their biology and life cycle have not been studied. Biological information is essential to formulate suitable management strategies for the sustainable utilization of these wild silkmoth resources [6]. At this crucial time of climate change, the objective of the present study was to gather more information on the development and annual life cycle and seasonal occurrence of *A. panda* silkmoth which is one of the treasures of people living around the Kakamega Forest of western Kenya.

## 2. Materials and Methods

*2.1. Study Area.* Four different types of forests are found in the Kakamega Forest: forest with only indigenous species (indigenous forest), mixed indigenous forest (forest with indigenous and exotic species), hardwood (forest with exotic hard species only), and softwood plantations (forest with exotic soft species only). The study was conducted in two different habitats of the Kakamega Forest, Isecheno and Ikuywa at Lunyu and Ikuywa sublocations, Ileho Division, in the Kakamega Forest, western Kenya. The Kakamega Forest is located between latitudes 0° 10′ and 0° 21′ North and longitudes 34° 47′ and 34° 58′ East (Figure 1). The entire Kakamega Forest covers a total area of approximately 265 km$^2$. It comprises several separate blocks of forest of which Isecheno (415 ha) belongs to Lunyu sublocation and Ikuywa (380 ha) belongs to Ikuywa sublocation. Sampling was carried out during a period of three years (2005–2007) in two project sites: Chirobani (mixed indigenous forest) and Musembe village (indigenous forest) (Figure 1). The natural vegetation of Kakamega Forest is tropical rainforest. Apart from the indigenous forest and mixed indigenous forest, the area has a variety of other habitats including hardwood

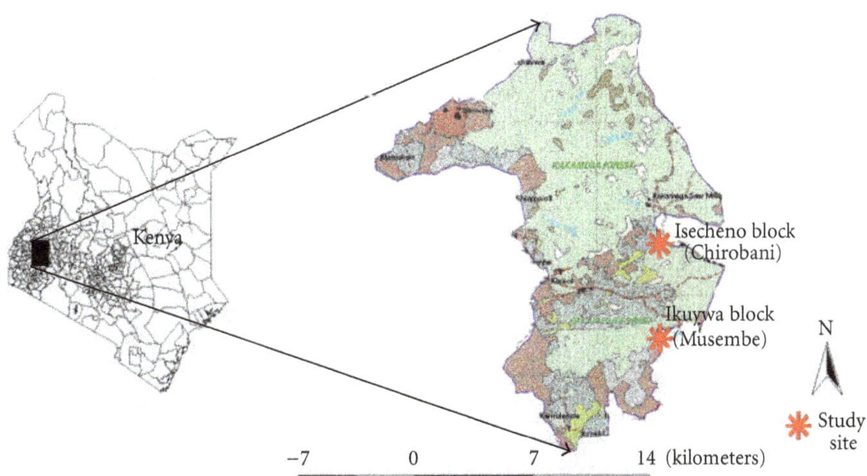


FIGURE 1: Study sites in the Kakamega Forest of Western Kenya.

and softwood plantations. There are about 150 species of woody trees, 90 species of dicotyledonous herbs, 80 species of monocotyledonous herbs of which about 60 are orchids, and a further 62 species of ferns totalling to about 380 identified species of vascular plants [7].

*2.2. Egg Cluster Characteristics and Incubation Period.* To know the egg size, one hundred eggs obtained from twenty-five *A. panda* female moth were measured using a stereomicroscope with an ocular micrometer. A Vernier calliper was used to determine the average length of eighty-three freshly laid egg clusters. Their weight with golden brown hairs was determined by electronic balance Mettler PJ 360 DeltaRange. One hundred and ninety-five egg clusters which were laid daily were marked to determine the exact incubation period. Daily observations were made on hatching for thirty-five days.

*2.3. Larval Period.* The larval period was determined by observing from the 1st to 7th instars 180 cohorts of *A. panda* larvae on *B. micrantha* until they span. The observed number was 120 and 78 in the dry season and rainy season, respectively. The rearing was done by protecting the entire larval stage of eighty-six (86) cohorts with net sleeves, and another ninety-four (94) cohorts were unprotected. The number of instars was determined by visual observation of exuviae [8, 9]. Earlier field observations showed that younger larvae rest about two days on the leaf, and older ones (from three to four days) attach themselves on the bark of *B. micrantha* before moulting. Hence, observations were made daily at 10:00 am at that period to observe if the larvae cast their skin (exuviae). The total number of the observed moults was used to determine the number of instars.

*2.4. Pupal Period.* To determine pupal period, spinning date of 100 cocoon nests was marked and kept in the enclosed net sleeve (60 × 40 × 30 cm) in the field until moths emerged.

*2.5. Adults.* Time of emergence was determined by direct observation and also by the marked pupae which were kept on *B. micrantha* in the net sleeve in the field. The life span of male moths was also recorded for comparison with that of the female moths. Wing expanse was determined by measuring with Vernier calliper, the maximum distance between the tips of the forewing after spreading the moths. This was the measurement from the apex of the forewing and its attachment to the thorax.

*2.6. Environmental Data.* A digital hygrothermometer (Zheda Electric Apparatus Inc., http://www.zjlab.com/) was used for recording daily temperature (maximum and minimum), and measurements of relative humidity were recorded four times daily (6 am, 12 am, 3 pm, and 9 pm) at both sites throughout the period of the study. A rain gauge was used for recording rainfall data.

*2.7. Data Analysis.* A *t*-test and chi-square [10] was used for all the means comparisons. The means of the analysis were compared between the two blocks of forest using the Stata7 software [11]. The degree of significance was indicated conventionally as follows:

*: significant ($P < 0.05$),

**: highly significant ($P < 0.001$),

ns: nonsignificant ($P > 0.05$).

## 3. Results

*3.1. Egg Cluster.* The eggs of *A. panda* are usually laid in clusters (Figure 2(d)) underside of leaves of the host plant *B. micrantha* (Figure 6). The mean number of freshly laid eggs in a cluster was 395.167 ± 40.736 and 485.313 ± 89.9 in the mixed indigenous ($n = 54$) and in the indigenous forest ($n = 49$), respectively. Highly significant difference was found to exist between the mean numbers of eggs in the two habitats. *B. micrantha* was the only observed host plant for *A. panda*

FIGURE 2: *Anaphe panda* silkmoth life stages: (a) male; (b) female moths; (c) mating; (d) egg cluster; (e) hatching (1st instar silkworms); (f) 7th instar silkworm on the host plant *Bridelia micrantha*; (g) Cocoon nest; (h) Pupae in the cocoon nest.

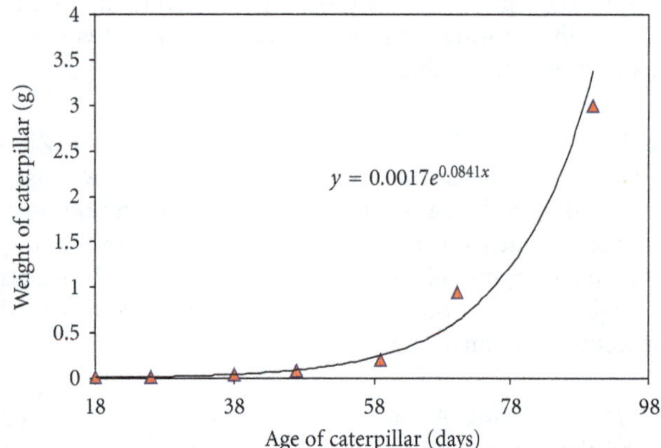


FIGURE 3: Developmental velocity of *Anaphe panda* silkworm in the Kakamega Forest.

oviposition in the Kakamega Forest. The eggs are small, and they are usually discoidal and isodiametric measuring 0.963 ± 0.008 mm as a major axis and 0.61 ± 0.004 mm as a minor axis. The mean length and weight of the egg cluster ($n = 40$) was 1.64 ± 0.16 cm and 0.22 ± 0.04 g in the mixed indigenous forest, whereas in the indigenous forest ($n = 43$) it was 1.66 ± 0.2 cm and 0.25 ± 0.07 g for the length and weight, respectively. No significant difference was observed between the length and the weight of the freshly laid egg cluster of the two habitats.

The incubation period ranged from 40 to 45 days in the dry season and from 45 to 55 days in the rainy season. Mean incubation periods at the two habitats are summarized in Table 1.

In the same habitat, high significant difference was found to exist between the dry season and rainy season. No significant difference was found between the two habitats.

*3.2. Larvae.* The larvae are gregarious (Figure 2(e)) from the first to the seventh instars. The direct observation of cast exuviae by the larvae reared through the entire larval stage showed that six moults occurred from egg hatch to the pupation. The larval stage was between 83 and 86 days in the dry season and 112 and 118 days in the rainy season.

From the egg hatches, young caterpillars follow prodigious growth from 3 mg to more than 3,000 mg (Figure 2(f)). The developmental velocity in the whole larval stage is shown in Figure 3.

A highly significant difference was found to exist between the dry season and rainy season of the larval period in the same habitat. There was no significant difference in larval period between the two habitats.

*3.3. Pupa.* When larvae were fully grown (Figure 2(f)), they started to spin silk. The pupa (Figure 2(h)) was enclosed in a tough silk cocoon (Figure 2(g)). In rainy season, the duration of the pupal period ranged between 158 and 178 days, and between 107 and 138 days for those which spun in dry season. Mean pupal stage duration by habitats is summarized in Table 2.

In the same habitat, highly significant difference was found to exist between the dry season and the rainy season. No significant difference was found between the two habitats. Furthermore, no significant difference was found to exist between the three years of observation of this study.

Table 1: Means (±SD) incubation period in days of *Anaphe panda* in the Indigenous and mixed indigenous forest in Kakamega.

| Years | Isecheno block (mixed indigenous forest) | | | | Ikuywa block (indigenous forest) | | | |
|---|---|---|---|---|---|---|---|---|
| | 1st brood | | 2nd brood | | 1st brood | | 2nd brood | |
| | $n$ | Incubation | $n$ | Incubation | $n$ | Incubation | $n$ | Incubation |
| 2005 | 12 | 41.75 ± 1.82 | 8 | 49.88 ± 3.91** | 13 | 41.69 ± 1.65 | 10 | 49.9 ± 3.57** |
| 2006 | 17 | 42.82 ± 1.91 | 18 | 50.78 ± 3.69** | 16 | 42.19 ± 1.97 | 15 | 49.87 ± 3.58** |
| 2007 | 16 | 42.13 ± 1.86 | 18 | 50.39 ± 3.15** | 18 | 41.83 ± 1.79 | 17 | 50.77 ± 3.54** |

$n$: number of egg clusters; **: highly significant difference between the 1st and 2nd broods.

Table 2: Mean (± SD) pupal period (days) of *A. panda* in the Kakamega Forest.

| Years | Brood | Isecheno (mixed indigenous forest) | | Ikuywa (indigenous forest) | |
|---|---|---|---|---|---|
| | | $n$ | Pupa period (days) | $n$ | Pupa period (days) |
| 2005 | 1 | 6 | 167.5 ± 7.52** | 9 | 167 ± 7.7 |
| | 2 | 7 | 122.14 ± 10.49 | 8 | 126.25 ± 11.88 |
| 2006 | 1 | 10 | 168 ± 6.06** | 8 | 168 ± 7.86 |
| | 2 | 9 | 123.33 ± 12.94 | 6 | 121.5 ± 5.24 |
| 2007 | 1 | 9 | 166.67 ± 5.43** | 8 | 169 ± 8.33 |
| | 2 | 9 | 122.33 ± 10.69 | 11 | 124.64 ± 11.94 |

$n$: number of cocoon nests; **: difference highly significant between the 1st and the 2nd broods.

The length and width of male ranged from 1.8 to 2.7 cm (mean 2.11 ± 0.17) and 0.6 to 1 cm (mean 0.73 ± 0.07), respectively, while the length and width of female pupae were from 2.1 to 3.1 (mean 2.4 ± 0.15) and 0.6 to 1.3 cm (mean 0.82 ± 0.08), respectively. High significant differences ($P < 0.001$) were found to exist between the lengths and also the widths of the two sexes. The weight of the male pupal ranged from 0.23 to 0.9 g (mean 0.5 ± 0.14), whereas female pupae weight ranged from 0.27 to 1.16 g (mean 0.73 ± 0.19). There were also significant differences between the weights of the pupae of the two sexes.

*3.4. Adults.* Moths (Figures 2(a) and 2(b)) were found in the forest from October to April. The peak numbers were observed in late November to late February (Figure 4). The mean wingspan of forewing was 2.347 ± 0.216 cm and 3.019 ± 0.251 cm for males ($n = 2701$) and for the females ($n = 3088$), respectively, whereas the mean lengths of forewing was 1.137 ± 0.106 cm and 1.448 ± 0.138 cm for males and females, respectively.

There was a high significant difference between the wingspan and the mean length of forewing of the male and the female moths. *A. panda* mean life span in Kakamega Forest is summarized in Table 3.

Time of occurrence and developmental periods of various stages of *A. panda* in the Kakamega Forest is summarized in Figures 2 and 4, respectively.

*3.5. Environmental Data.* The environmental data on mean monthly average temperature, humidity, and rainfall for the two locations were reported in Figures 5(a) and 5(b), respectively. Note that rainfall was bimodal in the Kakamega Forest, with a period of long rains from April through June, and short rains in August through November. During the study, the annual rainfall ranged from 180.9 to 265.9 cm at Isecheno and from 188.6 to 224.7 cm at Ikuywa. The annual number of rainy days ranged from 196 to 219 and from 207 to 209 in Isecheno and Ikuywa, respectively. Mean monthly maximum temperature ranged from 15.5°C to 36.8°C at Isecheno, and from 16.5 to 35.6°C at Ikuywa. Mean monthly maximum humidity ranged from 45.4 to 86.2% in the mixed indigenous forest and from 35.6 to 80.9% in the indigenous forest.

## 4. Discussion

Observations in this study showed that one egg cluster had 350 to 539 eggs, contrary to Jolly et al. [12] and Kioko et al. [1] who noted one cluster of 250–350 and 250–300 eggs, respectively. Depending on the health of the female, the number of eggs laid may vary from just a few to one hundred [13]. The number of eggs per cluster does not vary significantly ($P > 0.05$) for different generations in the same habitat (block of forest). Nevertheless, a highly significant difference ($P < 0.001$) was observed between the indigenous and the mixed indigenous forest. The incubation period depends on climatic conditions. Results obtained from this study confirm an earlier study by Jolly et al. [12] who recorded that the embryonic period may last a month or more (45 days) depending on climatic conditions. The fact that there was no significant difference between the two habitats means that temperature may have strongly influenced the incubation period.

The time between hatching from the egg and the completion of larval stages varied. This depends upon the availability of food, a favourable climate, and other factors [13]. The total larval period of *A. panda* is 120 days. This differed from the 140 days reported by Jolly et al.

Table 3: Adult lifespan (in days) of *A. panda* from indigenous and mixed indigenous forest in Kakamega (years 2005–2007).

| Years | Isecheno block (mixed indigenous forest) | | | | | Ikuywa block (indigenous forest) | | | | |
|---|---|---|---|---|---|---|---|---|---|---|
| | $n$ | Male | $n$ | Female | $t$-test | $n$ | Male | $n$ | Female | $t$-test |
| 2005 | 51 | $4.545 \pm 1.036$ | 51 | $5.182 \pm 0.982$ | ns | 50 | $5 \pm 0.943$ | 60 | $6.25 \pm 0.639$ | ** |
| 2006 | 65 | $3.88 \pm 1.013$ | 51 | $5.091 \pm 0.701$ | ** | 60 | $5.25 \pm 1.209$ | 64 | $6.25 \pm 1.032$ | * |
| 2007 | 71 | $4.323 \pm 1.043$ | 56 | $5 \pm 0.632$ | * | 57 | $5.235 \pm 1.251$ | 79 | $6.333 \pm 0.701$ | ** |

ns: non significant difference; *: significant difference between male and female; **: highly significant difference between male and female; $n$: Number of males or females moths.

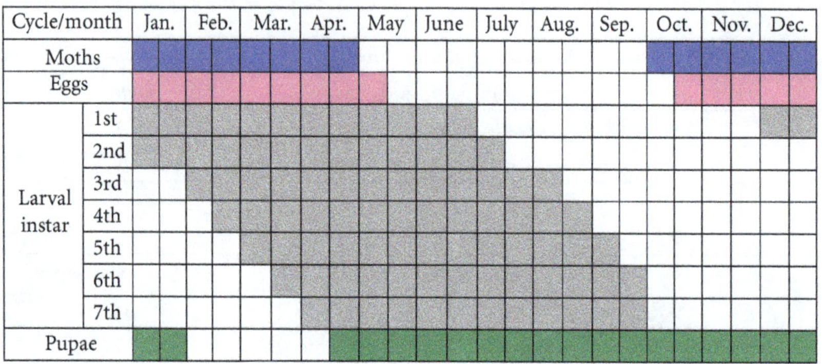


Figure 4: Time of occurrence of the various stages of *Anaphe panda* in the Kakamega Forest.

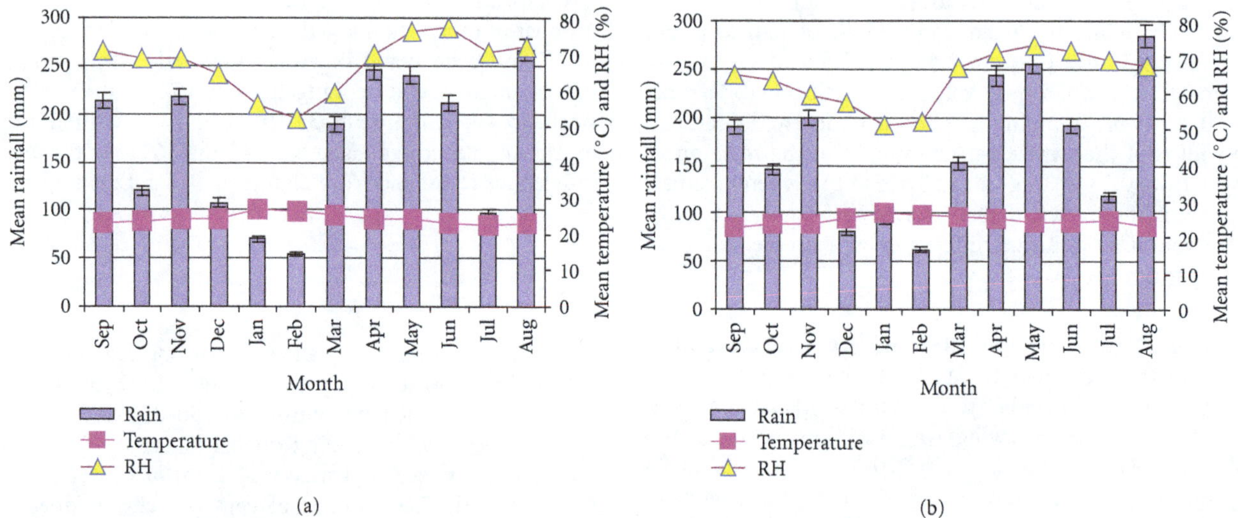


Figure 5: (a) Rainfall, temperature, and relative humidity in the mixed indigenous forest (Isecheno). (b) Rainfall, temperature, and relative humidity in the indigenous forest (Ikuywa).

[12] for *A. venata*. The general relationship between the developmental velocity, expressed in terms of weight and the number of days tends to be nonlinear (Figure 3); it appears to be exponential. This finding is similar to that of Wigglesworth [14]. From the results obtained in these experiments, it appears that the duration of the larval instars is dependent upon temperature. High temperatures during dry season cause a shorter duration, while low temperatures during rainy season prolongs the development. This agrees with observations made in studies both by Geertsema [15] and Chapman [16]. According to Wigglesworth [14], temperatures and nutrition may also have an effect on the number of larval instars.

In this study, a delay of emergence of moths in the larvae was observed in the dry season. Pupal diapause which extended over five months was also recorded in this study. This probably explains the significant difference in the records between the two seasons. Adult moths of the brood of the dry season emerge after 168 days, and the brood of rainy season emergence was after 122 days. This indicates that the rate of development of the pupae differs between the two seasons and is most probably influenced by intrinsic

Figure 6: *A. panda* host plant *Bridelia micrantha* with egg cluster in the Kakamega Forest.

factors. Van den Berg's [17] study on the *N. cytherea clarki* had similar conclusion that pupae formed during the later (warmer) season were able to catch up in development with those formed earlier in the colder season.

*A. panda* longevity in days of male and female does not differ significantly ($P = 0.324 > 0.05$) for different generations. But difference between the two blocks of the forest was highly significant ($P < 0.001$). Generally, moths seem to have higher life span in the indigenous forest compared to the mixed indigenous forest. There was also a highly significant difference between the two sexes in both blocks of the forest ($P < 0.001$). Females live longer because they perhaps must lay the maximum number of eggs before dying. Observations made in this study confirm earlier findings by Jolly et al. [12] who reported that an adult *Anaphe* sp. lives for nearly a week.

Highly significant differences ($P < 0.001$) were observed between the means of the forewing expanse and the length of females and males. Mean forewing expanse and length of females were larger than those of the males. The size of male and female seems to be one of the differences in this insect.

According to this study, February and March are the best months to harvest dry cocoon nests with minimal damage to fresh cocoons when moths are not emerged. To develop wild silkmoth farming, life cycle information is very important. By understanding the life cycle, the species and its food plants can be conserved and the community helped to utilize it for income generation.

The climatic conditions observed in this study (Figures 5(a) and 5(b)) were consistent with reports by Muriuki and Tsingalia [18] and Kokwaro [19]. As poikilothermic organisms, the life cycle, activity, distribution, and abundance of Lepidoptera are influenced by temperature [20]. Pollard and Yates [21] found that temperature and rainfall were likely to influence the survival of butterflies directly and indirectly through the effects on plant growth, disease, predation, or other factors. In light of the present study, further work is warranted to understand why a forest insect like *A. panda* periodically develops high populations in certain well-defined types of forest habitat, but not in all habitats where it occurs.

## Acknowledgments

The authors express their sincere thanks to DAAD (Deutscher Akademischer Austauschdienst) for providing the scholarship and to IFAD (International Fund for Agricultural Development) and GEF (Global Environmental Facility) for funding the research. Thanks are also due to icipe for providing logistical support and research facilities.

## References

[1] E. N. Kioko, S. K. Raina, and J. M. Mueke, "Conservation of the African wild silkmoths for economic incentives to rural communities of the kakamega forest in Kenya," *International Journal of Wild Silkmoth and Silk*, vol. 4, pp. 1–5, 1999.

[2] S. K. Raina, "On developing incentives for community participation in forest conservation through the use of commercial insects in Kenya," First training course, ICIPE, pp. 213, November-December 2004.

[3] N. Mbahin, E. N. Kioko, S. K. Raina, and J. M. Mueke, "Spatial distribution of cocoon nests and egg clusters of the silkmoth *Anaphe panda* (Lepidoptera: Thaumetopoeidae) and its host plant *Bridelia micrantha* (Euphorbiaceae) in the Kakamega Forest of western Kenya," *International Journal of Tropical Insect Science*, vol. 27, no. 3-4, pp. 138–144, 2007.

[4] N. Mbahin, *The ecology and economic potential of wild silkmoth Anaphe panda (Boisduval) (lepidoptera: Thaumetopoeidae) in the Kakamega forest*, Ph.D. thesis, Kenyatta University, Nairobi, Kenya, 2008.

[5] N. Mbahin, S. K. Raina, E. N. Kioko, and J. M. Mueke, "Use of sleeve nets to improve survival of the Boisduval silkworm, *Anaphe panda*, in the Kakamega forest of western Kenya," *Journal of Insect Science*, vol. 10, p. 6, 2010.

[6] E. N. Kioko, S. K. Raina, and J. M. Mueke, "Survey on diversity of wild silk moths species in East Africa," *East African Journal of Sciences*, vol. 2, no. 1, pp. 1–6, 2000.

[7] KIFCON, "Kenya indigenous forest conservation programme," Phase 1 Report Karura Forest Station, Center for Biodiversity, Nairobi, Kenya, 1994.

[8] M. O. Ashiru, "Determination of the number of instars of the silkworm *Anaphe veneta* Butler (*Lepidoptera: Notonidae*)," *Insect Science and its Application*, vol. 9, no. 3, pp. 405–410, 1988.

[9] F. H. Schmidt, R. K. Campbell, and J. R. Trotter, "Errors in determining instar number through head capsule measurements of Lepidoptera. A laboratory study and critique," *Annals of the Entomological Society of America*, vol. 24, pp. 451–466, 1977.

[10] SAS Institute, *SAS/STAT Users' Guid, Version 8*, vol. 2, SAS Institute, Cary, NC, USA, 6th edition, 2003.

[11] Statacorp, *Stata Statistical Software, Release 8.0*, Stata Corporation, College Station, Tex, USA, 2004.

[12] M. S. Jolly, S. K. Sen, T. N. Sonwalker, and G. K. Prasad, "Non-mulberry silks, F.A.O," *Agricultural Services Bulletin*, vol. 29, 164 pages, 1979.

[13] C. V. Covell, *Eastern Moths*, Houghton Mifflin Company, Boston, Mass, USA, 1984.

[14] V. B. Wiggleworth, *The Principles of Insect Physiology*, Methuen, London, UK, 1965.

[15] H. Geertsema, "Studies on the biology, ecology and control of the pine tree emperor moth, *Nudaurelia Cytherea Cytherea* (Fabr.) (Lepidoptera: *Saturniidae*)," *Annale Universiteit Van Stellenbosch*, vol. 50, no. 1, p. 170, 1975.

[16] R. F. Chapman, *The Insects: Structure and Function*, English University Press, London, UK, 1969.

[17] M. A. Van den Berg, *The bio-ecology, economical importance and control of three Saturniidae injurious to forest trees*, Ph.D. thesis, University of Pretoria, South Africa, 1971.

[18] J. W. Muriuki and M. H. Tsingalia, "A new population of de Brazza's monkey in Kenya," *Oryx*, vol. 24, no. 3, pp. 157–162, 1990.

[19] J. O. Kokwaro, "Conservation status of the Kakamega forest in Kenya. The Eastern most relic of the equatorial rain forest of Africa," *Monographs of Systematics and Botanical Gardens*, vol. 25, pp. 471–489, 1988.

[20] J. K. Hill, C. D. Thomas, and B. Huntley, "Climate and habitat availability determine 20th century changes in a butterfly's range margin," *Proceedings of the Royal Society B*, vol. 266, no. 1425, pp. 1197–1206, 1999.

[21] E. Pollard and T. J. Yates, *Monitoring Butterflies for Ecology and Conservation*, Chapman and Hall, London, UK, 1985.

# 9

# Using Different Approaches to Approximate a Pareto Front for a Multiobjective Evolutionary Algorithm: Optimal Thinning Regimes for *Eucalyptus fastigata*

**Oliver Chikumbo**

*Scion, Sustainable Design, Private Bag 3020, Rotorua 3046, New Zealand*

Correspondence should be addressed to Oliver Chikumbo, ochikumb@mac.com

Academic Editor: Andrew Gray

A stand-level, multiobjective evolutionary algorithm (MOEA) for determining a set of efficient thinning regimes satisfying two objectives, that is, value production for sawlog harvesting and volume production for a pulpwood market, was successfully demonstrated for a *Eucalyptus fastigata* trial in Kaingaroa Forest, New Zealand. The MOEA approximated the set of efficient thinning regimes (with a discontinuous Pareto front) by employing a ranking scheme developed by Fonseca and Fleming (1993), which was a Pareto-based ranking (a.k.a Multiobjective Genetic Algorithm—MOGA). In this paper we solve the same problem using an improved version of a fitness sharing Pareto ranking algorithm (a.k.a Nondominated Sorting Genetic Algorithm—NSGA II) originally developed by Srinivas and Deb (1994) and examine the results. Our findings indicate that NSGA II approximates the entire Pareto front whereas MOGA only determines a subdomain of the Pareto points.

## 1. Introduction

Forest estate planning requires the ability to have a suite of possible thinning regimes for each management unit where the estate is an amalgam of the many management units. Satisfying any kinds of goals for the forest estate is determined by choosing a mosaic of thinning regimes for all the management units. The preferred choices are normally achieved through some form of optimisation [1]. Chikumbo and Nicholas [2] demonstrated how a suite of thinning regimes at a management unit level might be approximated.

Chikumbo and Nicholas [2] formulated a stand-level optimisation problem to determine a set of efficient thinning regimes. Each regime defined an initial planting stocking, frequency of thinning, timing of thinning, intensity of thinning, final crop number prior to clearfelling, and rotation length. The problem was solved as a multiobjective optimisation to simultaneously maximise sawlog (or value) production and pulpwood (or volume) production, which are conflicting objectives. Traditionally, thinning regimes for most tree crops are determined either for value or volume production but not for both. This normally means an initially high tree stocking density, a short rotation with no thinning operations for volume production, and a more complex regime with thinning operations, pruning, and longer rotations for value production. The emphasis for value production is to maximise volume of clear wood per tree, rather than maximising biomass growth on as many trees as possible. Such an approach to forest management does not give the forester the flexibility to deal with volatile markets where the demands for sawlog and pulpwood tend to change rapidly. An ability to provide all products at the markets with the flexibility of meeting an increased supply of the product most in demand requires a different tact to forest management. Chikumbo and Nicholas [2] demonstrated how to determine a set of efficient thinning regimes that are suitable for both volume and value production in different ratios or "tradeoff," where the choice of a single regime was made on the basis of the product most in demand.

As such, the critical part of approximating the set of efficient thinning regimes was to find trade-off solutions (i.e., nondominated solutions) where for each solution an improvement in one objective did not lead to worsening in the other [3]. For the mathematically savvy, nondominated solutions can be defined as follows.

Assume two solutions $(\mathbf{x}, \mathbf{u})$, $(\mathbf{x}', \mathbf{y}') \in \Omega$, where $\Omega$ is denoted as the solution space. Then $(\mathbf{x}, \mathbf{u})$ is said to dominate $(\mathbf{x}', \mathbf{y}')$ (also written $(\mathbf{x}, \mathbf{u}) \succ (\mathbf{x}', \mathbf{y}')$) if and only if

$$\begin{aligned} &\forall i \in \{1, 2, \ldots, n\} : J_i(\mathbf{x}, \mathbf{u}) \geq J_i(\mathbf{x}', \mathbf{u}') \wedge, \\ &\exists m \in \{1, 2, \ldots, n\} : J_m(\mathbf{x}, \mathbf{u}) > J_m(\mathbf{x}', \mathbf{u}'). \end{aligned} \tag{1}$$

However, $(\mathbf{x}, \mathbf{u})$ is said to cover $(\mathbf{x}', \mathbf{u}')$ $((\mathbf{x}, \mathbf{u}) \succeq (\mathbf{x}', \mathbf{u}'))$ if and only if

$$\begin{aligned} &\forall i \in \{1, 2, \ldots, n\} : J_i(\mathbf{x}, \mathbf{u}) \leq J_i(\mathbf{x}', \mathbf{u}') \wedge, \\ &\exists m \in \{1, 2, \ldots, n\} : J_m(\mathbf{x}, \mathbf{u}) < J_m(\mathbf{x}', \mathbf{u}'). \end{aligned} \tag{2}$$

In this case $(\mathbf{x}, \mathbf{u})$ is nondominated by $(\mathbf{x}', \mathbf{y}')$.

Therefore, the set of nondominated solutions, $\Omega^N$, within the entire search space, $\Omega$, is called the Pareto-optimal set.

The set of solutions to the multiobjective problem was determined using a competitive co-evolutionary genetic algorithm [4] with five subpopulations of 100 individuals each, computed over 1000 generations. This is an island model characterised by multiple sub-populations instead of a single large population. These subpopulations evolve independently for a certain number of generations (isolation time). After the isolation time a number of individuals are distributed between subpopulations (a process called migration). Each subpopulation exerts selective pressure on the other, thereby maintaining diversity a lot longer than each subpopulation would do solitarily, thereby guarding against premature convergence. When competition is superimposed between the subpopulations, the ones with higher mean fitness values will maintain larger subpopulation sizes and receive more capable individuals, because they have more chances of finding the global optimum [5].

The Fonseca and Fleming ranking scheme [6] was used to determine the nondominated solutions. Only six nondominated thinning regimes were obtained, which was far less than expected because there should be as many nondominated solutions as possible, up to the size of the population [7]. A close inspection revealed that the Fonseca and Fleming ranking scheme quickly lost diversity resulting in "genetic drift" where convergence only occurs in one "niche" of the search space. The island model formulation is not sufficient to offset this genetic drift or avoid loss of diversity. Chikumbo and Nicholas [2] had no notion of the inability of MOGA to prevent "crowding" (i.e., inability to maintain diversity and encourage a good spread of solutions), in their first attempt at estimating efficient thinning regimes for *E. fastigata*. The more generations the algorithm was run, the more the Pareto front points exhibited crowding resulting in points superimposed on each other on certain subregions of the Pareto front. The workings of MOGA are explained later. This paper explores an alternative ranking scheme that is better at maintaining diversity such that not only a few points on a limited number of sub-regions of the Pareto front are estimated, rather that all the points on the Pareto front are determined.

The concept of applying evolutionary algorithms (or population-based search algorithms that mimic the evolutionary process) to multiobjective optimisation first appeared in the literature in the late 1960s, and it was not until 1985 when Schaffer [8] published the first algorithm. Figure 1 shows a diagrammatic representation of an evolutionary algorithm and how it works. Evolutionary algorithms work on populations of individuals instead of single solutions. In this way the search is performed in a parallel manner [9].

At the beginning of the computation, a number of individuals (the population) determined by the user are randomly initialised. The objective function(s), based on the "fitness value(s)," is/are then evaluated for these individuals; thus, the first/initial generation is produced. If the optimisation criteria are not met, the creation of a new generation starts, which is guided by a process of "selection," "crossover/recombination," and "mutation." Individuals are selected according to their fitness for the production of offspring. Parents are recombined to produce offspring. All offspring will be mutated with a certain probability. The fitness of the offspring is then computed. The offspring are inserted into the population replacing the parents, producing a new generation. This cycle is performed until the optimisation criteria are reached [9].

Today there exist many variants of evolutionary algorithms, namely, genetic algorithms [10], evolutionary strategies [11, 12], and evolutionary programming [13].

The subject of multiobjective optimisation problems is extremely complex and mathematically difficult, with many underresearched areas and outstanding problems [14]. Evolutionary algorithms have become established as the method at hand for exploring the search space in multi-objective optimisation problems [15].

In the absence of preference information among the objectives, there is no single solution for a multi-objective optimisation problem [16]. This is because under these circumstances, optimality takes a whole new meaning. This was first proposed by Edgeworth [17] and later generalised by Pareto [18]. Determining a set of nondominated solutions is what is now commonly known as Pareto optimality. Evolutionary algorithms have become the default method for approximating the Pareto front for multiobjective optimisation problems. This is because there are hardly any alternative methods. Also, evolutionary algorithms have the capability to search many Pareto-optimal solutions in parallel and exploit similarities of solutions by the recombination/crossover operation [15]. For multiobjective evolutionary algorithms (MOEAs), the Pareto-optimal solutions are approximated in a single optimisation run, for efficiency.

Various implementations of MOEA have appeared in the literature each with a different approach and algorithm for estimating the Pareto front. Some of the more advanced algorithms currently in use are as follows:

(a) Schaeffer's Vector Evaluated Genetic Algorithm (VEGA)—switching objective type [8];

(b) Hajela and Lin's Weighting-Based Genetic Algorithm (HLGA)—parameter variation type [19];

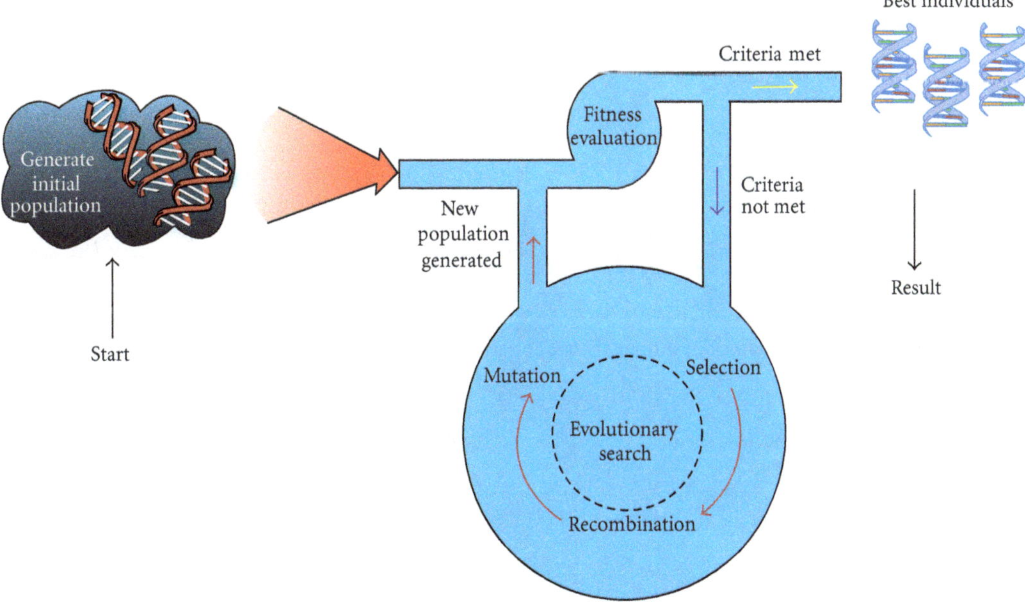


FIGURE 1: Evolutionary algorithm.

(c) Fonseca and Fleming's Multiobjective Genetic Algorithm (MOGA)—Pareto-based goal interaction [6];

(d) Horn, Nafpliotis, and Goldberg's Niched Pareto Genetic Algorithm (NPGA)—cooperative sharing [20];

(e) Srinivas and Deb's Nondominated Sorting Genetic Algorithm (NSGA)—fitness sharing Pareto [16];

(f) Coello's Min-Max Optimisation (CMMO)—ideal non-Pareto feasibility vetting [21];

(g) Zitzler and Thiele's Strength Pareto Evolutionary Algorithm (SPEA)—elitist external niching type [22];

(h) Corne, Knowles, and Oates' Pareto Envelope-Based selection Algorithm (PESA)—integrated selection and diversity maintenance via a hypergrid crowding strategy [23];

(i) Knowles and Corne's Pareto Archived Evolution Strategy (PAES)—single parent, local search [24].

In this paper the optimal thinning regime problem for *Eucalyptus fastigata* [2] is revisited and solved using single population MOGA and NSGA II algorithms. The GEATbx software that runs in MATLAB was used in the analysis, and it employs MOGA, which became our default. The reason NSGA II was chosen as a comparison is because of its wide use and reputation in the field of evolutionary computation. The results are compared, not to find which is better of the two but rather to determine a greater depth of the Pareto front in order to provide more thinning regimes to choose from. A brief description of the data and model formulation are given in the next section followed by a concise account of the MOGA and NSGA II algorithms. Sections 4 and 5 present in some detail the findings, and a conclusion on the extent of the Pareto front is presented.

## 2. Data, Objective Functions, and State Equations

The data for approximating the Pareto front using the MOGA and NSGA II algorithms came from a Nelder trial [25] of *E. fastigata* established in 1979 in Kaingaroa Forest, New Zealand. The details can be found in Chikumbo and Nicholas [2].

The emphasis of this paper is to approximate the solutions, or vectors, of decision variables of MOEA that satisfy constraints and optimise an objective vector function whose elements represent two objective functions. The decision variables, $\vec{x} = [x_1, x_2, \ldots, x_9]$, form the 9 genes of each chromosome or individual, which represent a potential solution with their bound constraints, for both lower bound (lb) and upper bound (ub) as shown in Figure 2. Therefore, an evolutionary algorithm is initiated with a set of randomly generated potential solutions that are driven towards optimality over generations, where each generation is defined by selection, recombination/crossover, and mutation for a set of potential solutions or population.

These objective functions or fitness functions form a mathematical description of performance, which are in conflict with each other:

$$\text{maximise } [f_1(\vec{x}),\ f_2(\vec{x})] \tag{3}$$

subject to $m$ inequality constraints,

$$g_i(\vec{x}) \leq 0, \quad i = 1, 2, \ldots, m, \tag{4}$$

and $r$ equality constraints,

$$h_i(\vec{x}) = 0, \quad i = 1, 2, \ldots, r, \tag{5}$$

where the objective functions, $f_1(\vec{x})$ or $f_2(\vec{x})$: $\mathbb{R}^n \rightarrow \mathbb{R}$ and the vector of decision variables, $\vec{x} = [x_1, x_2, \ldots, x_9]$.

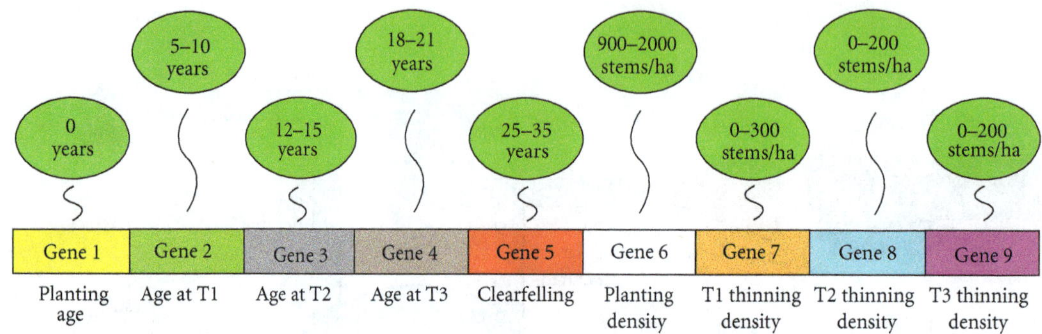


FIGURE 2: Decision variables with their ub and lb for each chromosome in the MOEA formulation.

The objective functions were as follows:

$$f_1(\vec{x}) = \max_{u(t)} \sum_{t=T-(n-1)}^{T} \frac{u(t)}{\text{sph}(t)} * \text{sba}(t) * \text{mht}(t) * \text{fhv}_{\text{value}}(t) * \frac{\text{vol}(t)}{\text{sph}(t)} * p_{\vec{x}}, \quad (6)$$

for sawlog production (NZ\$m$^6$ stems$^{-1}$ ha$^{-2}$), and

$$f_2(\vec{x}) = \max_{u(t)} \sum_{t=T-(n-1)}^{T} \frac{u(t)}{\text{sph}(t)} * \text{sba}(t) * \text{mht}(t) * \text{fhv}_{\text{vol}}(t) * p_{\vec{x}}, \quad (7)$$

for pulpwood production (NZ\$m$^3$ ha$^{-2}$), where $t$ is time in years, $\text{fhv}_{\text{value}}$ and $\text{fhv}_{\text{vol}}$ are forest holding values [26] for sawlog production and pulpwood production, respectively,

$$\text{sph}(t) = \text{sph}(t-1) - u(t), \quad (8)$$

$$\text{sba}(t+1) = a_1(\text{sph}(t))\text{sba}(t) + a_2(\text{sph}(t))\text{sba}(t-1) + b_1(\text{sph}(t)), \quad (9)$$

where

$a_1(t) = a\_1 * \exp(b\_1 * \text{sph}(t)) + c\_1 * \exp(d\_1 * \text{sph}(t))$;

$a_2(t) = a\_2 * \exp(b\_2 * \text{sph}(t)) + c\_2 * \exp(d\_2 * \text{sph}(t))$; and

$b_1(t) = a\_3 * \text{sph}(t)^{b\_3} + c\_3$.

$$\text{mht}(t+1) = a_3(\text{sph}(t))\text{mht}(t) + a_4(\text{sph}(t))\text{mht}(t-1) + b_2, \quad (10)$$

where

$a_3(t) = -1.055 * a_4(t) + 0.9548$;

$a_4(t) = -7510/(\text{sph}(t) + 1.13e + 04)$; and

$b_2(t) = 0.6579$.

$$\text{vol}(t+1) = a_5(\text{sph}(t))\ \text{vol}(t) + a_6(\text{sph}(t))\text{vol}(t-1) + b_3(\text{sph}(t)), \quad (11)$$

where,

$a_5(t) = a\_6 * \exp(b\_4 * \text{sph}(t)) + c\_6 * \exp(d\_3 * \text{sph}(t))$;

$a_6(t) = a\_7 * \exp(b\_5 * \text{sph}(t)) + c\_7 * \exp(d\_4 * \text{sph}(t))$; and

$b_3(t) = a\_8 * \text{sph}(t)^{b\_6} + c\_8$.

$$p_{\vec{x}} = \begin{cases} 1, & \text{if } x_6 + (x_7 + x_8 + x_9) \geq 0, \\ 0, & \text{otherwise.} \end{cases} \quad (12)$$

The relationship vol($t$)/sph($t$) in objective function (6) concentrates volume production per tree, which is the goal for sawlog production. State equation (8) tracks the number of trees per hectare (sph) harvested (i.e., $u$ or control) subject to the two objectives, and growth dynamics in state equations (9)–(11), which are stand basal area in square metres per hectare (sba), mean stand height of the 100 largest trees per hectare (mht), and volume in cubic metres per hectare (vol), respectively. The state equations (9)–(11) are discrete-time dynamical equations [27], which have been demonstrated to accurately determine growth dynamics influenced by changes in tree spacing, initial planting stocking, and thinning treatment [28]. The parameters of these state equations are shown in Table 1.

The constraint

$$x_6 - (x_7 + x_8 + x_9) \geq 0 \quad (13)$$

ensures that the total number of trees thinned over the complete rotation will not exceed the initial number of trees planted and is handled by a penalty function, $p_{\vec{x}}$, in the objective functions (3) and (4), which is equal to one when constraint (13) is not violated and zero otherwise.

The thinning regime problem for *E. fastigata* is a nonlinear optimal control problem of dynamical equations (9)–(11). These dynamical equations experience state changes in time under some rules, that is, the growth dynamics of *E. fastigata*. These equations are mathematical "black box" representations of the underlying physical growth processes of *E. fastigata*. The state is described by a collection of variables that fully characterise growth. These parameters are time

TABLE 1: Parameters of the state equations (7)–(9) [2].

| Parameter | Value | 95% confidence bounds | |
|---|---|---|---|
| | | Lower bound | Upper bound |
| $a\_1$ | 0.1467 | 0.1088 | 0.1847 |
| $b\_1$ | −000337 | −0.005411 | −0.001328 |
| $c\_1$ | 1.871 | 1.844 | 1.898 |
| $d\_1$ | $-4.817e-06$ | $-1.013e-05$ | $4.922e-07$ |
| $a\_2$ | −0.1428 | −0.1788 | −0.1069 |
| $b\_2$ | −0.00358 | −0.005614 | −0.001547 |
| $c\_2$ | −0.8829 | −0.9068 | −0.859 |
| $d\_2$ | $-8.491e-06$ | $-1.852e-05$ | $1.536e-06$ |
| $a\_3$ | 0.001906 | 0.0001262 | 0.003685 |
| $b\_3$ | 0.8226 | 0.7127 | 0.9326 |
| $c\_3$ | 0.1109 | 0.03341 | 0.1885 |
| $a\_4$ | −1.055 | −1.135 | −0.9745 |
| $c\_4$ | 0.9548 | 0.9056 | 1.004 |
| $a\_5$ | −7510 | $-1.051e+04$ | −4508 |
| $c\_5$ | $1.13e+04$ | 6447 | $1.615e+04$ |
| $a\_6$ | 0.1103 | 0.07602 | 0.1446 |
| $b\_4$ | −0.001873 | −0.003263 | −0.000483 |
| $c\_6$ | 1.973 | 1.933 | 2.013 |
| $d\_3$ | $-7.167e-06$ | $-1.331e-05$ | $-1.02e-06$ |
| $a\_7$ | −0.1119 | −0.1447 | −0.07914 |
| $b\_5$ | −0.001911 | −0.00325 | −0.0005712 |
| $c\_7$ | −0.9798 | −1.018 | −0.9417 |
| $d\_4$ | $-1.365e-05$ | $-2.558e-05$ | $-1.725e-06$ |
| $a\_8$ | 0.004516 | −0.002179 | 0.01121 |
| $b\_6$ | 0.8976 | 0.722 | 1.073 |
| $c\_8$ | 0.5459 | 0.07415 | 1.018 |

TABLE 2: MOGA and NSGA II specifications for the stand-level optimal thinning regime problem for *E. fastigata*.

| | MOGA | NSGA II |
|---|---|---|
| Population size | 500 | 500 |
| Maximum generations | 500 | 500 |
| Number of objectives | 2 | 2 |
| Number of variables | 9 | 9 |
| Number of constraints | 1 | 1 |
| Mutation | Random | Gaussian |
| Recombination/crossover | Discrete | Intermediate |
| Selection | Stochastic universal sampling | Tournament |

variant as they vary with the changing number of stems per hectare. The state is a vector of values in a Euclidean state space and evolves continuously with time along a continuous trajectory through the state space. The value of the state at any time is governed by its past values, as well as by the past values of all external influences on the system. The external influences that are manipulated to regulate the state constitute the control inputs, which are the decision variables.

The aim in the *E. fastigata* control problem is to determine the control inputs that produce a desired outcome, represented by the two objective functions, (4) and (5). Finding analytical solutions to nonlinear optimal control problems is a nontrivial task. It has only been achieved satisfactorily for limited classes of systems. Even more difficult a task is to analytically solve the optimal control problem. However, we employ evolutionary algorithms for solving the *E. fastigata* control problem as first demonstrated by Chikumbo [29] for a single objective control problem. A summary of the genetic algorithms for the *E. fastigata* control problem is shown in Table 2 for MOGA and NSGA II algorithms that are explained in more detail in the following section.

## 3. Ranking Schemes for Estimating the Pareto Front

This section explains the basis of the algorithms used to estimate the Pareto front, particularly MOGA and NSGA II. The explanation here is in sink with the findings in this paper.

Many approaches are now in use for MOEA (with different ranking schemes that alter the "fitness landscape" of the search space by adding more "ordering information"), and the key to their success lies in how they accomplish fitness assignment and selection and how they maintain diversity to achieve a well-distributed Pareto (trade-off) front [15]. For fitness assignment and selection, there exist two groups of ranking schemes, that is, non-Pareto [8, 19] and Pareto-based [30] schemes. The MOGA and NSGA II considered here are both Pareto-based approaches.

Goldberg [30] made a key observation that, for a multimodal search space with multiple peaks, different subdomains (which he called niches) exist with unique stable subgroups of individuals (analogous to species). A conventional evolutionary algorithm will tend to converge its individuals to only one niche (a phenomenon known as genetic drift). However, important information that could have been determined from the other peaks or sub-domains of importance is missed. This is the reason why Chikumbo and Nicholas [2] could only find 6 nondominated solutions for the *E. fastigata* problem, using the MOGA algorithm. The use of ranking schemes encourages the formation of artificial niches, thereby preventing genetic drift and predicting nondominated solutions from most of these important subdomains of a search space.

To maintain diversity in the population, fitness sharing may be used [31]. The idea of fitness sharing is to subdivide the population into several subpopulations based on the similarity among individuals. Note that "similarity" in the context of MOEA can be measured in the space of the decision variables or in the space of the objective functions. Fitness sharing is defined in the following way:

$$\phi\left(d_{ij}\right) = \begin{cases} 1 - \left(\dfrac{d_{ij}}{\sigma_{\text{share}}}\right)^{\alpha}, & d_{ij} < \sigma_{\text{share}} \\ 0, & \text{otherwise.} \end{cases} \tag{14}$$

In expression (14), $\alpha = 1$, $d_{ij}$ indicates the distance between solutions/individuals $i$ and $j$ (in any space defined), and $\sigma_{\text{share}}$ is a parameter (or threshold) that defines the size of a niche or neighbourhood. Any solutions within this distance will be considered as part of the same niche. The fitness, $f_s$, of an individual $i$ is modified as follows:

$$f_{s_i}^{\text{modified}} = \frac{f_{s_i}}{\sum_{j=1}^{M} \phi\left(d_{ij}\right)}, \tag{15}$$

where $M$ is the number of individuals located in the neighborhood (or niche) of the $i$th individual.

Several other schemes are possible to maintain diversity and to encourage a good spread of solutions, such as crowding. In a crowding model, crowding of solutions anywhere in the search space is discouraged, thereby providing the diversity needed to maintain multiple nondominated solutions. The crowding distance in the model is determined from the sum of distances between solution's neighbours in Euclidean space that is defined by objective function values.

*3.1. Fonseca and Fleming's Multiobjective Genetic Algorithm (MOGA).* The ranking scheme proposed by Fonseca and Fleming [6] involved assigning an individual's rank (in the objective function space) equal to the number of population individuals that dominated that individual; that is, the ranking was done according to the degree of domination; the more members of the current population that dominate a particular individual, the lower its rank. It, therefore, uses fitness sharing in the objective function space and mating is also restricted. Reproduction probabilities were determined by means of exponential ranking. Afterwards the fitness values were averaged and shared among individuals having identical ranks [15]. Finally, stochastic universal sampling, which provides zero bias and minimum spread (i.e., the range of possible values for the number of offspring of an individual) was used to fill the sampling pool.

The main strength of MOGA is that it is efficient and relatively easy to implement. It has also been successful in solving optimal control problems where it has exhibited very good overall performance [32].

*3.2. Srinivas and Deb's Nondominated Sorting Genetic Algorithm (NSGA II).* NSGA [16] is a nondomination-based genetic algorithm for multi-objective optimisation. It is an effective algorithm that was initially popular. However, it was later criticised for its computational complexity, lack of elitism (i.e., a mechanism for retaining in the population the best individual hitherto), and for choosing the optimal value for the sharing parameter [33]. Consequently Deb et al. [34] developed a second-generation algorithm, NSGA II. This improved version is a better sorting algorithm that incorporates elitism and does not choose the sharing parameter *a priori*. NSGA II is one of the most successful evolutionary multiobjective optimisation algorithms.

Convergence to the Pareto front is made possible by nondominated sorting. This sorting ranks individuals by iteratively determining the nondominated solutions in the population, assigning those individuals the next best rank and removing them from the population. Favouring individuals with large crowding distances maintain diversity within one rank. NSGA II, being elitist, keeps as many nondominated solutions as possible, up to the size of the population [7]. The elitist selection is as follows. After evaluating the offspring's fitness (nondominated rank and crowding distance), parents and offspring fight for survival as Pareto dominance is applied to the combined population of parents and offspring. Then, the least dominated $N$ solutions survive to make the population of the next generation. The sampling pool is filled using the tournament selection process, where a number of individuals (ranging from 2 to $N$) are chosen randomly and the "best" individual from that group is selected as parent for the next generation.

## 4. Results

The MOGA for the optimal thinning regime optimisation problem ran for 25 minutes using the MATLAB [35] "SPMD" parallel computing for the objective function evaluations on a MacBook Pro, with a 2.93 GHz Intel Core 2 Duo processor. The SPMD is a block of MATLAB statements that denote a *s*ingle *p*rogram to be executed by the available processors, and *m*ultiple *d*ata are assigned to the processors for computation in parallel. In this case it was more of "MPSD," that is, *m*ultiple *p*rograms *s*ingle *d*ata, since each

Table 3: Summary of the approaches for the *E. fastigata* optimal thinning regime problem using MOGA, NSGA II, and ccGA as computed on a MacBook Pro with a 2.93 GHz Intel Core 2 Duo processor.

| Model | Turnaround time (mins) | MATLAB parallel computing | Population size | Max generations | No. of nondominated solutions |
|---|---|---|---|---|---|
| MOGA | 25 | SPMD | 500 | 500 | 153 |
| NSGA II | 82 | PARFOR | 500 | 500 | 500 |
| ccGA-MOGA [2] | 39 | n/a | 500 | 1000 | 6 |

Table 4: Incomplete set of the Pareto front estimated using MOGA.

| Regime | Age at 1st thinning (years) | Age at 2nd thinning (years) | Age at 3rd thinning (years) | Clear felling age (years) | Initial planting density (stems $ha^{-1}$) | Stems thinned at 1st thinning (stems $ha^{-1}$) | Stems thinned at 2nd thinning (stems $ha^{-1}$) | Stems thinned at 3rd thinning (stems $ha^{-1}$) | Obj1-value prodn | Obj2-volume prodn |
|---|---|---|---|---|---|---|---|---|---|---|
| 1 | 6 | 13 | 21 | 35 | 930 | 298 | 101 | 183 | $4.79E+24$ | $2.79E+26$ |
| 2 | 6 | 13 | 21 | 35 | 930 | 298 | 101 | 183 | $4.79E+24$ | $2.79E+26$ |
| 3 | 6 | 13 | 21 | 35 | 930 | 298 | 101 | 183 | $4.79E+24$ | $2.79E+26$ |
| 4 | 6 | 13 | 21 | 35 | 930 | 298 | 101 | 183 | $4.79E+24$ | $2.79E+26$ |
| 5 | 6 | 13 | 21 | 35 | 930 | 298 | 101 | 183 | $4.79E+24$ | $2.79E+26$ |
| 6 | 6 | 13 | 21 | 35 | 930 | 298 | 101 | 183 | $4.79E+24$ | $2.79E+26$ |
| ... | ... | ... | ... | ... | ... | ... | ... | ... | ... | ... |
| 153 | 5 | 13 | 21 | 34 | 902 | 275 | 135 | 190 | $1.01E+24$ | $1.18E+26$ |

processor computed a different objective function each from the same decision variable vector. The NSGA II had a longer turnaround time of 82 minutes, and it also utilised parallel computing using the MATLAB "PARFOR" for the objective function evaluations. PARFOR is an extension of the normal FOR loop statement, where the iterations of statements in the loop are executed in parallel on separate processors.

The MOGA had a total of 153 nondominated solutions whereas NSGA II had 500. Chikumbo and Nicholas [2] ran their competitive coevolutionary genetic algorithm (ccGA) with 5 subpopulations, using MOGA for Pareto front estimation and the same hardware but with no parallel computing. The computation ran over twice as many generations and in 39 minutes, but approximated only 6 nondominated solutions, because of genetic drift. Table 3 shows a summary of these results.

Many of the MOGA nondominated solutions were replicates and superimposed on each other when plotted on graph as shown in Figure 3. It is also clear from the graph that the Pareto front was disjointed, a reminder of the difficulty of solving such kinds of problems using conventional techniques as experienced by Chikumbo and Mareels [36]. Table 4 shows initial nondominated solutions and the complete solutions are found in Table 6.

NSGA II nondominated solutions were more widely spread. This revealed a greater extent of the Pareto front that was not obvious with MOGA as shown in Figure 4. The Pareto front was even more disjointed being composed of a set of disjoint continuous sets in the objective function space. The disjointed parts gave rise to isolated nondominated solution subsets, while the continuous parts (or topologies) were associated with the actual isolated solutions. Each topology was associated with a specific network of decision

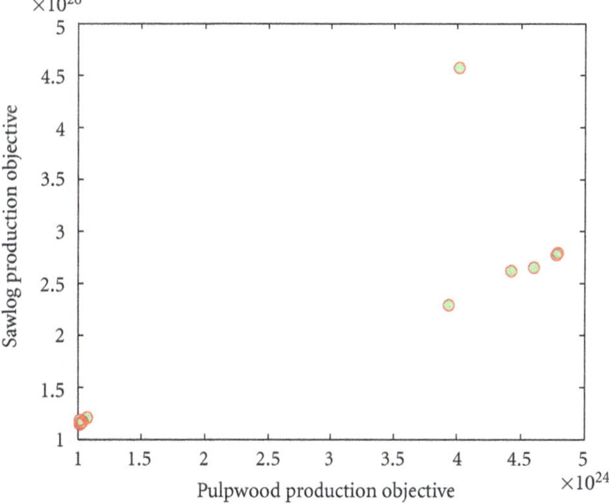


Figure 3: Disjointed Pareto front as predicted by the MOGA.

variables or thinning regimes. A snippet of the isolated nondominated solutions is shown in Table 5, and the complete results are found in Table 7.

When the MOGA solutions are superimposed on the NSGA II ones it becomes clear that these solutions are originating from the same niches in the search space as shown in Figure 5.

The 6 nondominated solutions predicted by Chikumbo and Nicholas [2] occupied the niche close to the origin but were not included in the graphical plot in Figure 5 because they were not possible to see clearly.

TABLE 5: Incomplete set of the Pareto front estimated using NSGA II.

| Regime | Age at 1st thinning (years) | Age at 2nd thinning (years) | Age at 3rd thinning (years) | Clear felling age (years) | Initial planting density (stems $ha^{-1}$) | Stems thinned at 1st thinning (stems $ha^{-1}$) | Stems thinned at 2nd thinning (stems $ha^{-1}$) | Stems thinned at 3rd thinning (stems $ha^{-1}$) | Obj1-value prodn | Obj2-volume prodn |
|---|---|---|---|---|---|---|---|---|---|---|
| 1 | 10 | 15 | 20 | 25 | 2000 | 0 | 0 | 0 | 0 | 0 |
| 2 | 10 | 15 | 20 | 25 | 2000 | 0 | 0 | 200 | $-2.8E+08$ | $2.18E+09$ |
| 3 | 5 | 12 | 18 | 35 | 900 | 300 | 200 | 200 | $6.34E+24$ | $7.16E+26$ |
| 4 | 5 | 13 | 18 | 35 | 928 | 268 | 187 | 187 | $5.51E+24$ | $6.30E+26$ |
| 5 | 5 | 15 | 18 | 35 | 900 | 161 | 191 | 187 | $3.39E+24$ | $3.84E+26$ |
| 6 | 6 | 14 | 19 | 34 | 900 | 226 | 179 | 193 | $6.52E+23$ | $3.76E+25$ |
| ... | ... | ... | ... | ... | ... | ... | ... | ... | ... | ... |
| 500 | 5 | 13 | 18 | 35 | 902 | 226 | 200 | 193 | $4.76E+24$ | $5.38E+26$ |

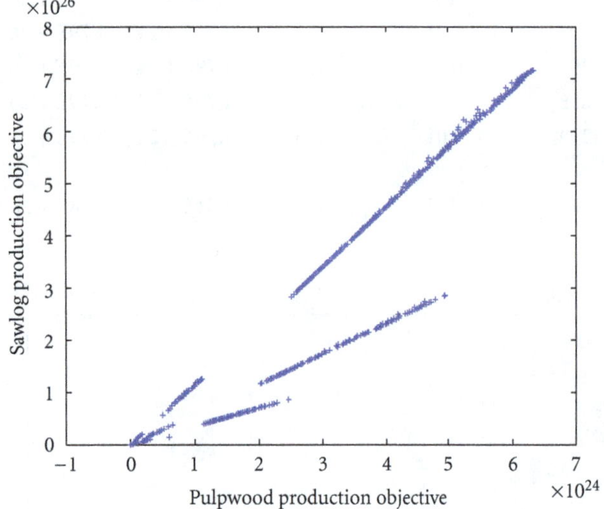


FIGURE 4: Disjointed Pareto front as predicted by the NSGA II.

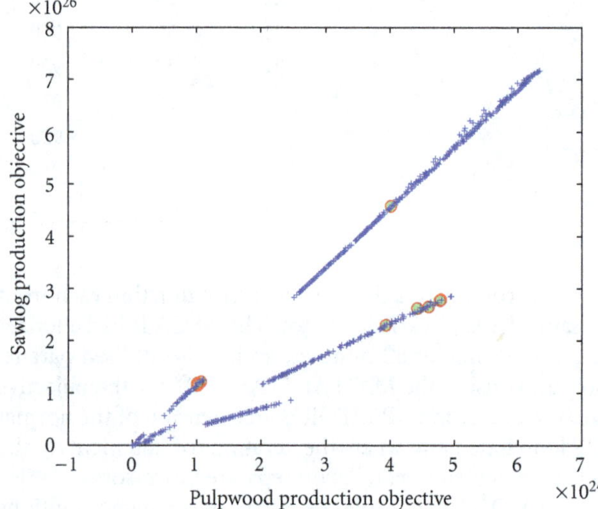


FIGURE 5: Superimposed Pareto fronts from MOGA and NSGA II.

Investigating the implications of the overlaps, or lack thereof, of the Pareto sets in the objective function domain involved the comparison in the decision variables domain (or phenotype) using the Kruskal-Willis test [37]. This is a nonparametric version of the classical one-way analysis of variance and an extension of the Wilcoxon rank sum test [38], specifically for testing equality of population medians among groups (which in this case were the three Pareto sets, ccGA-MOGA, MOGA, and NSGA II). Ordering the data from smallest to largest across all groups and taking the numeric index of this ordering determined the ranks.

*4.1. Initial Planting Stocking.* A $P$ value of $3.8626e-17$ which is practically zero means that the null hypothesis is rejected, suggesting that at least one sample median of the initial planting stockings from the different Pareto sets is significantly different. The box plots in Figure 6 show that the NSGA II Pareto set is different from the other two, which are significantly similar within the 25th and 75th percentiles. The NSGA II shows other initial planting densities (shown as "+" signs in Figure 6) that are higher than the range that overlaps those from the ccGA-MOGA and MOGA Pareto sets. The Wilcoxon rank sum test, plotted in Figure 7, shows more clearly the difference between the population medians of these Pareto sets for the initial planting stockings.

*4.2. Final Crop Number.* A $P$ value of 0.1542 and the box plots in Figure 8 suggest some overlap between the final crop numbers from the three Pareto sets, where the MOGA and NSGA II sets are lower than and higher than the common overlapping range of final crop numbers, respectively (as shown by the "+" signs in Figure 8). The mean ranking in Figure 9 emphasizes the degree of overlap of the final crop numbers from the three Pareto sets.

*4.3. Rotation Length.* There was a difference in the rotation length medians amongst the Pareto sets because of the $P$ value of $1.0275e-04$. The NSGA II Pareto set showed a much greater range of 25–35 years, whereas ccGA-MOGA was consistently 35 years and MOGA only ranged between 34 and 35 years. The box plots are shown in Figure 10. The summary of the mean ranks of the rotation lengths in Figure 11 clarifies

Table 6: A Pareto set estimated using MOGA.

| Regime | Age at 1st thinning (years) | Age at 2nd thinning (years) | Age at 3rd thinning (years) | Clear felling age (years) | Initial planting density (stems ha$^{-1}$) | Stems thinned at 1st thinning (stems ha$^{-1}$) | Stems thinned at 2nd thinning (stems ha$^{-1}$) | Stems thinned at 3rd thinning (stems ha$^{-1}$) | Obj1-value prodn | Obj2 volume prodn |
|---|---|---|---|---|---|---|---|---|---|---|
| 1 | 6 | 13 | 21 | 35 | 930 | 298 | 101 | 183 | $4.7873E+24$ | $2.793E+26$ |
| 2 | 6 | 13 | 21 | 35 | 930 | 298 | 101 | 183 | $4.7873E+24$ | $2.793E+26$ |
| 3 | 6 | 13 | 21 | 35 | 930 | 298 | 101 | 183 | $4.7873E+24$ | $2.793E+26$ |
| 4 | 6 | 13 | 21 | 35 | 930 | 298 | 101 | 183 | $4.7873E+24$ | $2.793E+26$ |
| 5 | 6 | 13 | 21 | 35 | 930 | 298 | 101 | 183 | $4.7873E+24$ | $2.793E+26$ |
| 6 | 6 | 13 | 21 | 35 | 930 | 298 | 101 | 183 | $4.7873E+24$ | $2.793E+26$ |
| 7 | 6 | 13 | 21 | 35 | 930 | 298 | 101 | 183 | $4.7873E+24$ | $2.793E+26$ |
| 8 | 6 | 13 | 21 | 35 | 930 | 298 | 101 | 183 | $4.7873E+24$ | $2.793E+26$ |
| 9 | 6 | 13 | 21 | 35 | 930 | 298 | 101 | 183 | $4.7873E+24$ | $2.793E+26$ |
| 10 | 6 | 13 | 21 | 35 | 930 | 298 | 101 | 183 | $4.7873E+24$ | $2.793E+26$ |
| 11 | 6 | 13 | 21 | 35 | 930 | 298 | 101 | 183 | $4.7873E+24$ | $2.793E+26$ |
| 12 | 6 | 13 | 21 | 35 | 930 | 298 | 101 | 183 | $4.7873E+24$ | $2.793E+26$ |
| 13 | 6 | 13 | 21 | 35 | 930 | 298 | 101 | 183 | $4.7873E+24$ | $2.793E+26$ |
| 14 | 6 | 13 | 21 | 35 | 930 | 298 | 101 | 183 | $4.7873E+24$ | $2.793E+26$ |
| 15 | 6 | 13 | 21 | 35 | 930 | 298 | 101 | 183 | $4.7873E+24$ | $2.793E+26$ |
| 16 | 6 | 13 | 21 | 35 | 930 | 298 | 101 | 183 | $4.7873E+24$ | $2.793E+26$ |
| 17 | 6 | 13 | 21 | 35 | 930 | 298 | 101 | 183 | $4.7873E+24$ | $2.793E+26$ |
| 18 | 6 | 13 | 21 | 35 | 930 | 298 | 101 | 183 | $4.7873E+24$ | $2.793E+26$ |
| 19 | 6 | 13 | 21 | 35 | 930 | 298 | 101 | 183 | $4.7873E+24$ | $2.793E+26$ |
| 20 | 6 | 13 | 21 | 35 | 930 | 298 | 101 | 183 | $4.7873E+24$ | $2.793E+26$ |
| 21 | 6 | 13 | 21 | 35 | 930 | 298 | 101 | 183 | $4.7873E+24$ | $2.793E+26$ |
| 22 | 6 | 13 | 21 | 35 | 930 | 298 | 101 | 183 | $4.7873E+24$ | $2.793E+26$ |
| 23 | 6 | 13 | 21 | 35 | 930 | 298 | 101 | 183 | $4.7873E+24$ | $2.793E+26$ |
| 24 | 6 | 13 | 21 | 35 | 930 | 298 | 101 | 183 | $4.7873E+24$ | $2.793E+26$ |
| 25 | 6 | 13 | 21 | 35 | 930 | 298 | 101 | 183 | $4.7873E+24$ | $2.793E+26$ |
| 26 | 6 | 13 | 21 | 35 | 930 | 298 | 101 | 183 | $4.7873E+24$ | $2.793E+26$ |
| 27 | 6 | 13 | 21 | 35 | 930 | 298 | 101 | 183 | $4.7873E+24$ | $2.793E+26$ |
| 28 | 6 | 13 | 21 | 35 | 930 | 298 | 101 | 183 | $4.7873E+24$ | $2.793E+26$ |
| 29 | 6 | 13 | 21 | 35 | 930 | 298 | 101 | 183 | $4.7873E+24$ | $2.793E+26$ |
| 30 | 6 | 13 | 21 | 35 | 930 | 298 | 101 | 183 | $4.7873E+24$ | $2.793E+26$ |
| 31 | 6 | 13 | 21 | 35 | 930 | 298 | 101 | 183 | $4.7873E+24$ | $2.793E+26$ |
| 32 | 6 | 13 | 21 | 35 | 930 | 298 | 101 | 183 | $4.7873E+24$ | $2.793E+26$ |
| 33 | 6 | 13 | 21 | 35 | 930 | 298 | 101 | 183 | $4.7873E+24$ | $2.793E+26$ |
| 34 | 6 | 13 | 21 | 35 | 930 | 298 | 101 | 183 | $4.7873E+24$ | $2.793E+26$ |
| 35 | 6 | 13 | 21 | 35 | 930 | 298 | 101 | 183 | $4.7873E+24$ | $2.793E+26$ |
| 36 | 6 | 13 | 21 | 35 | 930 | 298 | 101 | 183 | $4.7873E+24$ | $2.793E+26$ |
| 37 | 6 | 13 | 21 | 35 | 930 | 298 | 101 | 183 | $4.7873E+24$ | $2.793E+26$ |
| 38 | 6 | 13 | 21 | 35 | 930 | 298 | 101 | 183 | $4.7873E+24$ | $2.793E+26$ |
| 39 | 6 | 13 | 21 | 35 | 930 | 298 | 101 | 183 | $4.7873E+24$ | $2.793E+26$ |
| 40 | 6 | 13 | 21 | 35 | 930 | 298 | 101 | 183 | $4.7873E+24$ | $2.793E+26$ |
| 41 | 6 | 13 | 21 | 35 | 930 | 298 | 101 | 183 | $4.7873E+24$ | $2.793E+26$ |
| 42 | 6 | 13 | 21 | 35 | 930 | 298 | 101 | 183 | $4.7873E+24$ | $2.793E+26$ |
| 43 | 6 | 13 | 21 | 35 | 930 | 298 | 101 | 183 | $4.7873E+24$ | $2.793E+26$ |
| 44 | 6 | 13 | 21 | 35 | 930 | 298 | 101 | 183 | $4.7873E+24$ | $2.793E+26$ |
| 45 | 6 | 13 | 21 | 35 | 930 | 298 | 101 | 183 | $4.7873E+24$ | $2.793E+26$ |

Table 6: Continued.

| Regime | Age at 1st thinning (years) | Age at 2nd thinning (years) | Age at 3rd thinning (years) | Clear felling age (years) | Initial planting density (stems $ha^{-1}$) | Stems thinned at 1st thinning (stems $ha^{-1}$) | Stems thinned at 2nd thinning (stems $ha^{-1}$) | Stems thinned at 3rd thinning (stems $ha^{-1}$) | Obj1-value prodn | Obj2-volume prodn |
|---|---|---|---|---|---|---|---|---|---|---|
| 46 | 6 | 13 | 21 | 35 | 930 | 298 | 101 | 183 | $4.7873E+24$ | $2.793E+26$ |
| 47 | 6 | 13 | 21 | 35 | 930 | 298 | 101 | 183 | $4.7873E+24$ | $2.793E+26$ |
| 48 | 6 | 13 | 21 | 35 | 930 | 298 | 101 | 183 | $4.7873E+24$ | $2.793E+26$ |
| 49 | 6 | 13 | 21 | 35 | 930 | 298 | 101 | 183 | $4.7873E+24$ | $2.793E+26$ |
| 50 | 6 | 13 | 21 | 35 | 930 | 298 | 101 | 183 | $4.7873E+24$ | $2.793E+26$ |
| 51 | 6 | 13 | 21 | 35 | 930 | 298 | 101 | 183 | $4.7873E+24$ | $2.793E+26$ |
| 52 | 6 | 13 | 21 | 35 | 930 | 298 | 101 | 183 | $4.7873E+24$ | $2.793E+26$ |
| 53 | 6 | 13 | 21 | 35 | 930 | 298 | 101 | 183 | $4.7873E+24$ | $2.793E+26$ |
| 54 | 6 | 13 | 21 | 35 | 930 | 298 | 101 | 183 | $4.7873E+24$ | $2.793E+26$ |
| 55 | 6 | 13 | 21 | 35 | 930 | 298 | 101 | 183 | $4.7873E+24$ | $2.793E+26$ |
| 56 | 6 | 13 | 21 | 35 | 930 | 298 | 101 | 183 | $4.7873E+24$ | $2.793E+26$ |
| 57 | 6 | 13 | 21 | 35 | 930 | 298 | 101 | 183 | $4.7873E+24$ | $2.793E+26$ |
| 58 | 6 | 13 | 21 | 35 | 930 | 298 | 101 | 183 | $4.7873E+24$ | $2.793E+26$ |
| 59 | 6 | 13 | 21 | 35 | 930 | 298 | 101 | 183 | $4.7873E+24$ | $2.793E+26$ |
| 60 | 6 | 13 | 21 | 35 | 930 | 298 | 101 | 183 | $4.7873E+24$ | $2.793E+26$ |
| 61 | 6 | 13 | 21 | 35 | 930 | 298 | 101 | 183 | $4.7873E+24$ | $2.793E+26$ |
| 62 | 6 | 13 | 21 | 35 | 930 | 298 | 101 | 183 | $4.7873E+24$ | $2.793E+26$ |
| 63 | 6 | 13 | 21 | 35 | 930 | 298 | 101 | 183 | $4.7873E+24$ | $2.793E+26$ |
| 64 | 6 | 13 | 21 | 35 | 930 | 298 | 101 | 183 | $4.7873E+24$ | $2.793E+26$ |
| 65 | 6 | 13 | 21 | 35 | 930 | 298 | 101 | 183 | $4.7873E+24$ | $2.793E+26$ |
| 66 | 6 | 13 | 21 | 35 | 930 | 298 | 101 | 183 | $4.7873E+24$ | $2.793E+26$ |
| 67 | 6 | 13 | 21 | 35 | 930 | 298 | 101 | 183 | $4.7873E+24$ | $2.793E+26$ |
| 68 | 6 | 13 | 21 | 35 | 930 | 298 | 101 | 183 | $4.7873E+24$ | $2.793E+26$ |
| 69 | 6 | 13 | 21 | 35 | 930 | 298 | 101 | 183 | $4.7873E+24$ | $2.793E+26$ |
| 70 | 6 | 13 | 21 | 35 | 930 | 298 | 101 | 183 | $4.7873E+24$ | $2.793E+26$ |
| 71 | 6 | 15 | 20 | 35 | 921 | 295 | 175 | 196 | $4.7768E+24$ | $2.778E+26$ |
| 72 | 6 | 14 | 18 | 35 | 901 | 279 | 197 | 122 | $4.601E+24$ | $2.6556E+26$ |
| 73 | 6 | 12 | 18 | 35 | 979 | 287 | 171 | 187 | $4.4224E+24$ | $2.6228E+26$ |
| 74 | 6 | 12 | 18 | 35 | 979 | 287 | 171 | 187 | $4.4224E+24$ | $2.6228E+26$ |
| 75 | 6 | 12 | 18 | 35 | 979 | 287 | 171 | 187 | $4.4224E+24$ | $2.6228E+26$ |
| 76 | 6 | 12 | 18 | 35 | 979 | 287 | 171 | 187 | $4.4224E+24$ | $2.6228E+26$ |
| 77 | 6 | 12 | 18 | 35 | 979 | 287 | 171 | 187 | $4.4224E+24$ | $2.6228E+26$ |
| 78 | 6 | 12 | 18 | 35 | 979 | 287 | 171 | 187 | $4.4224E+24$ | $2.6228E+26$ |
| 79 | 6 | 12 | 18 | 35 | 979 | 287 | 171 | 187 | $4.4224E+24$ | $2.6228E+26$ |
| 80 | 6 | 12 | 18 | 35 | 979 | 287 | 171 | 187 | $4.4224E+24$ | $2.6228E+26$ |
| 81 | 6 | 12 | 18 | 35 | 979 | 287 | 171 | 187 | $4.4224E+24$ | $2.6228E+26$ |
| 82 | 6 | 12 | 18 | 35 | 979 | 287 | 171 | 187 | $4.4224E+24$ | $2.6228E+26$ |
| 83 | 6 | 12 | 18 | 35 | 979 | 287 | 171 | 187 | $4.4224E+24$ | $2.6228E+26$ |
| 84 | 6 | 12 | 18 | 35 | 979 | 287 | 171 | 187 | $4.4224E+24$ | $2.6228E+26$ |
| 85 | 6 | 12 | 18 | 35 | 979 | 287 | 171 | 187 | $4.4224E+24$ | $2.6228E+26$ |
| 86 | 6 | 12 | 18 | 35 | 979 | 287 | 171 | 187 | $4.4224E+24$ | $2.6228E+26$ |
| 87 | 6 | 12 | 18 | 35 | 979 | 287 | 171 | 187 | $4.4224E+24$ | $2.6228E+26$ |
| 88 | 6 | 12 | 18 | 35 | 979 | 287 | 171 | 187 | $4.4224E+24$ | $2.6228E+26$ |
| 89 | 6 | 12 | 18 | 35 | 979 | 287 | 171 | 187 | $4.4224E+24$ | $2.6228E+26$ |
| 90 | 6 | 12 | 18 | 35 | 979 | 287 | 171 | 187 | $4.4224E+24$ | $2.6228E+26$ |

Table 6: Continued.

| Regime | Age at 1st thinning (years) | Age at 2nd thinning (years) | Age at 3rd thinning (years) | Clear felling age (years) | Initial planting density (stems ha$^{-1}$) | Stems thinned at 1st thinning (stems ha$^{-1}$) | Stems thinned at 2nd thinning (stems ha$^{-1}$) | Stems thinned at 3rd thinning (stems ha$^{-1}$) | Obj1-value prodn | Obj2-volume prodn |
|---|---|---|---|---|---|---|---|---|---|---|
| 91 | 6 | 12 | 18 | 35 | 979 | 287 | 171 | 187 | $4.4224E+24$ | $2.6228E+26$ |
| 92 | 6 | 12 | 18 | 35 | 979 | 287 | 171 | 187 | $4.4224E+24$ | $2.6228E+26$ |
| 93 | 6 | 12 | 18 | 35 | 979 | 287 | 171 | 187 | $4.4224E+24$ | $2.6228E+26$ |
| 94 | 6 | 12 | 18 | 35 | 979 | 287 | 171 | 187 | $4.4224E+24$ | $2.6228E+26$ |
| 95 | 6 | 12 | 18 | 35 | 979 | 287 | 171 | 187 | $4.4224E+24$ | $2.6228E+26$ |
| 96 | 6 | 12 | 18 | 35 | 979 | 287 | 171 | 187 | $4.4224E+24$ | $2.6228E+26$ |
| 97 | 6 | 12 | 18 | 35 | 979 | 287 | 171 | 187 | $4.4224E+24$ | $2.6228E+26$ |
| 98 | 6 | 12 | 18 | 35 | 979 | 287 | 171 | 187 | $4.4224E+24$ | $2.6228E+26$ |
| 99 | 6 | 12 | 18 | 35 | 979 | 287 | 171 | 187 | $4.4224E+24$ | $2.6228E+26$ |
| 100 | 6 | 12 | 18 | 35 | 979 | 287 | 171 | 187 | $4.4224E+24$ | $2.6228E+26$ |
| 101 | 6 | 12 | 18 | 35 | 979 | 287 | 171 | 187 | $4.4224E+24$ | $2.6228E+26$ |
| 102 | 6 | 12 | 18 | 35 | 979 | 287 | 171 | 187 | $4.4224E+24$ | $2.6228E+26$ |
| 103 | 6 | 12 | 18 | 35 | 979 | 287 | 171 | 187 | $4.4224E+24$ | $2.6228E+26$ |
| 104 | 6 | 12 | 18 | 35 | 979 | 287 | 171 | 187 | $4.4224E+24$ | $2.6228E+26$ |
| 105 | 6 | 12 | 18 | 35 | 979 | 287 | 171 | 187 | $4.4224E+24$ | $2.6228E+26$ |
| 106 | 6 | 12 | 18 | 35 | 979 | 287 | 171 | 187 | $4.4224E+24$ | $2.6228E+26$ |
| 107 | 6 | 12 | 18 | 35 | 979 | 287 | 171 | 187 | $4.4224E+24$ | $2.6228E+26$ |
| 108 | 6 | 12 | 18 | 35 | 979 | 287 | 171 | 187 | $4.4224E+24$ | $2.6228E+26$ |
| 109 | 6 | 12 | 18 | 35 | 979 | 287 | 171 | 187 | $4.4224E+24$ | $2.6228E+26$ |
| 110 | 6 | 12 | 18 | 35 | 979 | 287 | 171 | 187 | $4.4224E+24$ | $2.6228E+26$ |
| 111 | 6 | 12 | 18 | 35 | 979 | 287 | 171 | 187 | $4.4224E+24$ | $2.6228E+26$ |
| 112 | 6 | 12 | 18 | 35 | 979 | 287 | 171 | 187 | $4.4224E+24$ | $2.6228E+26$ |
| 113 | 6 | 12 | 18 | 35 | 979 | 287 | 171 | 187 | $4.4224E+24$ | $2.6228E+26$ |
| 114 | 6 | 12 | 18 | 35 | 979 | 287 | 171 | 187 | $4.4224E+24$ | $2.6228E+26$ |
| 115 | 6 | 12 | 18 | 35 | 979 | 287 | 171 | 187 | $4.4224E+24$ | $2.6228E+26$ |
| 116 | 6 | 12 | 18 | 35 | 979 | 287 | 171 | 187 | $4.4224E+24$ | $2.6228E+26$ |
| 117 | 6 | 12 | 18 | 35 | 979 | 287 | 171 | 187 | $4.4224E+24$ | $2.6228E+26$ |
| 118 | 6 | 12 | 18 | 35 | 979 | 287 | 171 | 187 | $4.4224E+24$ | $2.6228E+26$ |
| 119 | 6 | 12 | 18 | 35 | 979 | 287 | 171 | 187 | $4.4224E+24$ | $2.6228E+26$ |
| 120 | 6 | 12 | 18 | 35 | 979 | 287 | 171 | 187 | $4.4224E+24$ | $2.6228E+26$ |
| 121 | 5 | 14 | 21 | 35 | 920 | 194 | 171 | 200 | $4.0137E+24$ | $4.5726E+26$ |
| 122 | 6 | 13 | 21 | 35 | 930 | 245 | 192 | 158 | $3.9339E+24$ | $2.2947E+26$ |
| 123 | 5 | 14 | 20 | 35 | 901 | 290 | 74 | 190 | $1.0703E+24$ | $1.2109E+26$ |
| 124 | 5 | 13 | 18 | 35 | 900 | 280 | 118 | 160 | $1.0343E+24$ | $1.1696E+26$ |
| 125 | 5 | 14 | 21 | 34 | 908 | 279 | 60 | 175 | $1.0227E+24$ | $1.1601E+26$ |
| 126 | 5 | 15 | 18 | 34 | 998 | 299 | 165 | 171 | $1.0132E+24$ | $1.1855E+26$ |
| 127 | 5 | 13 | 21 | 34 | 902 | 275 | 135 | 190 | $1.0139E+24$ | $1.1473E+26$ |
| 128 | 5 | 13 | 21 | 34 | 902 | 275 | 135 | 190 | $1.0139E+24$ | $1.1473E+26$ |
| 129 | 5 | 13 | 21 | 34 | 902 | 275 | 135 | 190 | $1.0139E+24$ | $1.1473E+26$ |
| 130 | 5 | 13 | 21 | 34 | 902 | 275 | 135 | 190 | $1.0139E+24$ | $1.1473E+26$ |
| 131 | 5 | 13 | 21 | 34 | 902 | 275 | 135 | 190 | $1.0139E+24$ | $1.1473E+26$ |
| 132 | 5 | 13 | 21 | 34 | 902 | 275 | 135 | 190 | $1.0139E+24$ | $1.1473E+26$ |
| 133 | 5 | 13 | 21 | 34 | 902 | 275 | 135 | 190 | $1.0139E+24$ | $1.1473E+26$ |
| 134 | 5 | 13 | 21 | 34 | 902 | 275 | 135 | 190 | $1.0139E+24$ | $1.1473E+26$ |
| 135 | 5 | 13 | 21 | 34 | 902 | 275 | 135 | 190 | $1.0139E+24$ | $1.1473E+26$ |

Table 6: Continued.

| Regime | Age at 1st thinning (years) | Age at 2nd thinning (years) | Age at 3rd thinning (years) | Clear felling age (years) | Initial planting density (stems ha$^{-1}$) | Stems thinned at 1st thinning (stems ha$^{-1}$) | Stems thinned at 2nd thinning (stems ha$^{-1}$) | Stems thinned at 3rd thinning (stems ha$^{-1}$) | Obj1-value prodn | Obj2-volume prodn |
|---|---|---|---|---|---|---|---|---|---|---|
| 136 | 5 | 13 | 21 | 34 | 902 | 275 | 135 | 190 | $1.0139E+24$ | $1.1473E+26$ |
| 137 | 5 | 13 | 21 | 34 | 902 | 275 | 135 | 190 | $1.0139E+24$ | $1.1473E+26$ |
| 138 | 5 | 13 | 21 | 34 | 902 | 275 | 135 | 190 | $1.0139E+24$ | $1.1473E+26$ |
| 139 | 5 | 13 | 21 | 34 | 902 | 275 | 135 | 190 | $1.0139E+24$ | $1.1473E+26$ |
| 140 | 5 | 13 | 21 | 34 | 902 | 275 | 135 | 190 | $1.0139E+24$ | $1.1473E+26$ |
| 141 | 5 | 13 | 21 | 34 | 902 | 275 | 135 | 190 | $1.0139E+24$ | $1.1473E+26$ |
| 142 | 5 | 13 | 21 | 34 | 902 | 275 | 135 | 190 | $1.0139E+24$ | $1.1473E+26$ |
| 143 | 5 | 13 | 21 | 34 | 902 | 275 | 135 | 190 | $1.0139E+24$ | $1.1473E+26$ |
| 144 | 5 | 13 | 21 | 34 | 902 | 275 | 135 | 190 | $1.0139E+24$ | $1.1473E+26$ |
| 145 | 5 | 13 | 21 | 34 | 902 | 275 | 135 | 190 | $1.0139E+24$ | $1.1473E+26$ |
| 146 | 5 | 13 | 21 | 34 | 902 | 275 | 135 | 190 | $1.0139E+24$ | $1.1473E+26$ |
| 147 | 5 | 13 | 21 | 34 | 902 | 275 | 135 | 190 | $1.0139E+24$ | $1.1473E+26$ |
| 148 | 5 | 13 | 21 | 34 | 902 | 275 | 135 | 190 | $1.0139E+24$ | $1.1473E+26$ |
| 149 | 5 | 13 | 21 | 34 | 902 | 275 | 135 | 190 | $1.0139E+24$ | $1.1473E+26$ |
| 150 | 5 | 13 | 21 | 34 | 902 | 275 | 135 | 190 | $1.0139E+24$ | $1.1473E+26$ |
| 151 | 5 | 13 | 21 | 34 | 902 | 275 | 135 | 190 | $1.0139E+24$ | $1.1473E+26$ |
| 152 | 5 | 13 | 21 | 34 | 902 | 275 | 135 | 190 | $1.0139E+24$ | $1.1473E+26$ |
| 153 | 5 | 13 | 21 | 34 | 902 | 275 | 135 | 190 | $1.0139E+24$ | $1.1473E+26$ |

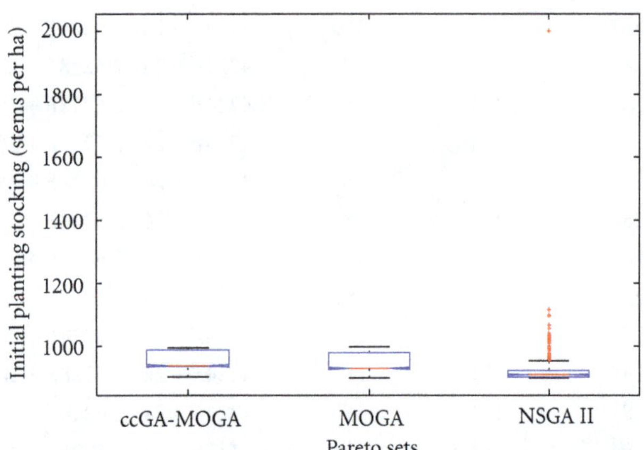


Figure 6: Box plots of the initial planting stockings from the ccGA-MOGA, MOGA, and NSGA II Pareto sets. The edges of each box represent the 25th and 75th percentiles; the ends of the whiskers are the minimum and maximum; outliers are plotted individually with the "+" sign.

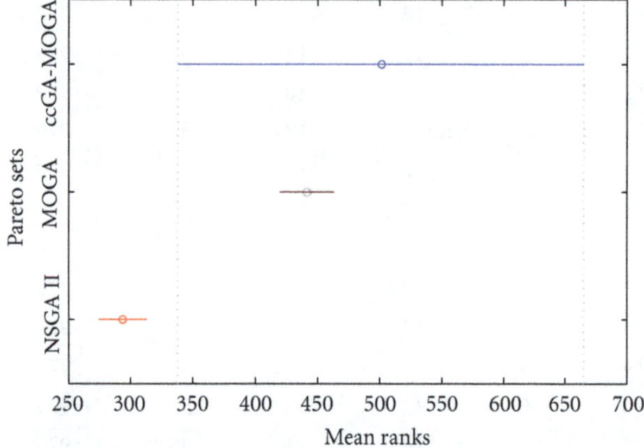


Figure 7: Wilcoxon rank sum test for the initial planting stockings from the ccGA-MOGA, MOGA, and NSGA II Pareto sets.

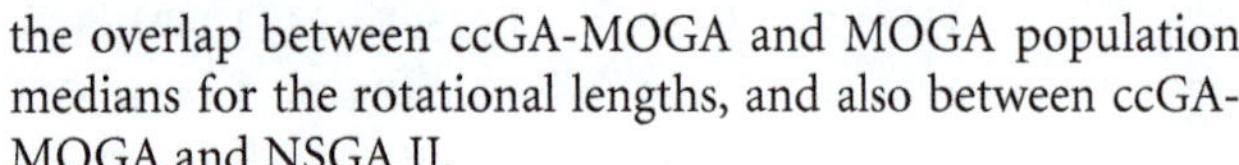

the overlap between ccGA-MOGA and MOGA population medians for the rotational lengths, and also between ccGA-MOGA and NSGA II.

*4.4. Frequency of Thinning.* The NSGA II Pareto set was the only one that had a regime with no thinnings at all where the initial planting stocking was 2000 stems ha$^{-1}$ with a final harvesting at age 25 years. Also in the set was included a regime with a single thinning at age 20 of 200 stems ha$^{-1}$ and final harvesting at age 25 years. As for the rest of the thinning regimes from all the Pareto sets, the frequency of thinning at a total of three in one rotation was the norm with small differences in the timings and intensities.

*4.5. Timing and Intensity of Thinning.* For brevity only the description (without the box plots and the plotted mean ranks) is given here for the timing and intensity of thinnings.

Timings for the first thinnings from the three Pareto sets showed similarities between NSGA II and ccGA-MOGA with overlaps of the timing medians at 5 years. MOGA had a higher median at 6 years. The intensities of the first thinnings

TABLE 7: A Pareto set estimated using NSGA II.

| Regime | Age at 1st thinning (years) | Age at 2nd thinning (years) | Age at 3rd thinning (years) | Clear felling age (years) | Initial planting density (stems ha$^{-1}$) | Stems thinned at 1st thinning (stems ha$^{-1}$) | Stems thinned at 2nd thinning (stems ha$^{-1}$) | Stems thinned at 3rd thinning (stems ha$^{-1}$) | Obj1- value prodn | Obj2-volume prodn |
|---|---|---|---|---|---|---|---|---|---|---|
| 1 | 10 | 15 | 20 | 25 | 2000 | 0 | 0 | 0 | 0 | 0 |
| 2 | 10 | 15 | 20 | 25 | 2000 | 0 | 0 | 200 | $-2.8E+08$ | $2.18E+09$ |
| 3 | 5 | 12 | 18 | 35 | 900 | 300 | 200 | 200 | $6.34E+24$ | $7.16E+26$ |
| 4 | 5 | 13 | 18 | 35 | 928 | 268 | 187 | 187 | $5.51E+24$ | $6.30E+26$ |
| 5 | 5 | 15 | 18 | 35 | 900 | 161 | 191 | 187 | $3.39E+24$ | $3.84E+26$ |
| 6 | 6 | 14 | 19 | 34 | 900 | 226 | 179 | 193 | $6.52E+23$ | $3.76E+25$ |
| 7 | 6 | 15 | 18 | 35 | 903 | 138 | 197 | 196 | $2.27E+24$ | $1.31E+26$ |
| 8 | 5 | 13 | 18 | 35 | 912 | 300 | 189 | 187 | $6.26E+24$ | $7.11E+26$ |
| 9 | 7 | 14 | 19 | 35 | 900 | 214 | 197 | 200 | $1.89E+24$ | $6.64E+25$ |
| 10 | 5 | 14 | 18 | 35 | 900 | 164 | 192 | 190 | $3.46E+24$ | $3.91E+26$ |
| 11 | 7 | 14 | 19 | 35 | 901 | 136 | 199 | 195 | $1.20E+24$ | $4.22E+25$ |
| 12 | 6 | 14 | 19 | 35 | 904 | 123 | 199 | 200 | $2.02E+24$ | $1.17E+26$ |
| 13 | 5 | 15 | 18 | 35 | 901 | 123 | 200 | 199 | $2.59E+24$ | $2.93E+26$ |
| 14 | 5 | 13 | 18 | 35 | 923 | 300 | 191 | 169 | $6.20E+24$ | $7.07E+26$ |
| 15 | 5 | 13 | 18 | 35 | 911 | 263 | 194 | 196 | $5.49E+24$ | $6.24E+26$ |
| 16 | 5 | 13 | 19 | 35 | 923 | 248 | 156 | 198 | $5.12E+24$ | $5.84E+26$ |
| 17 | 5 | 12 | 18 | 35 | 926 | 253 | 187 | 190 | $5.22E+24$ | $5.95E+26$ |
| 18 | 7 | 14 | 19 | 34 | 925 | 152 | 158 | 168 | $4.02E+22$ | $1.42E+24$ |
| 19 | 5 | 13 | 18 | 35 | 906 | 300 | 111 | 188 | $6.30E+24$ | $7.14E+26$ |
| 20 | 5 | 12 | 18 | 35 | 939 | 272 | 195 | 188 | $5.54E+24$ | $6.35E+26$ |
| 21 | 5 | 13 | 18 | 35 | 911 | 209 | 197 | 194 | $4.36E+24$ | $4.95E+26$ |
| 22 | 5 | 14 | 18 | 34 | 945 | 223 | 185 | 185 | $7.90E+23$ | $9.08E+25$ |
| 23 | 6 | 13 | 19 | 35 | 908 | 287 | 177 | 189 | $4.70E+24$ | $2.72E+26$ |
| 24 | 5 | 15 | 18 | 35 | 911 | 186 | 196 | 195 | $3.88E+24$ | $4.41E+26$ |
| 25 | 5 | 13 | 18 | 34 | 907 | 272 | 181 | 174 | $9.98E+23$ | $1.13E+26$ |
| 26 | 6 | 14 | 19 | 35 | 900 | 300 | 174 | 199 | $4.95E+24$ | $2.86E+26$ |
| 27 | 5 | 14 | 18 | 35 | 913 | 190 | 195 | 197 | $3.96E+24$ | $4.50E+26$ |
| 28 | 5 | 15 | 18 | 35 | 908 | 215 | 194 | 194 | $4.50E+24$ | $5.10E+26$ |
| 29 | 5 | 12 | 18 | 35 | 918 | 260 | 189 | 193 | $5.40E+24$ | $6.14E+26$ |
| 30 | 6 | 15 | 18 | 35 | 902 | 168 | 196 | 196 | $2.76E+24$ | $1.60E+26$ |
| 31 | 5 | 14 | 18 | 35 | 915 | 191 | 194 | 191 | $3.97E+24$ | $4.51E+26$ |
| 32 | 6 | 14 | 19 | 35 | 924 | 286 | 176 | 190 | $4.62E+24$ | $2.69E+26$ |
| 33 | 5 | 13 | 18 | 34 | 1009 | 300 | 198 | 184 | $1.01E+24$ | $1.18E+26$ |
| 34 | 5 | 14 | 18 | 35 | 900 | 215 | 198 | 190 | $4.53E+24$ | $5.13E+26$ |
| 35 | 5 | 14 | 18 | 35 | 906 | 166 | 196 | 197 | $3.48E+24$ | $3.94E+26$ |
| 36 | 5 | 13 | 18 | 35 | 911 | 207 | 197 | 194 | $4.32E+24$ | $4.90E+26$ |
| 37 | 5 | 15 | 18 | 35 | 911 | 176 | 196 | 197 | $3.67E+24$ | $4.17E+26$ |
| 38 | 5 | 15 | 18 | 35 | 909 | 160 | 198 | 198 | $3.34E+24$ | $3.79E+26$ |
| 39 | 6 | 15 | 18 | 35 | 1067 | 145 | 189 | 200 | $6.36E+22$ | $3.88E+24$ |
| 40 | 5 | 13 | 18 | 35 | 901 | 252 | 186 | 185 | $5.31E+24$ | $6.01E+26$ |
| 41 | 6 | 15 | 18 | 35 | 905 | 143 | 196 | 197 | $2.35E+24$ | $1.36E+26$ |
| 42 | 7 | 14 | 18 | 35 | 903 | 210 | 194 | 182 | $1.85E+24$ | $6.51E+25$ |
| 43 | 5 | 12 | 18 | 35 | 900 | 285 | 181 | 185 | $6.02E+24$ | $6.80E+26$ |
| 44 | 5 | 15 | 18 | 35 | 908 | 141 | 197 | 197 | $2.95E+24$ | $3.34E+26$ |
| 45 | 5 | 13 | 18 | 35 | 918 | 268 | 175 | 181 | $5.56E+24$ | $6.33E+26$ |
| 46 | 5 | 13 | 18 | 35 | 901 | 202 | 200 | 186 | $4.26E+24$ | $4.81E+26$ |

Table 7: Continued.

| Regime | Age at 1st thinning (years) | Age at 2nd thinning (years) | Age at 3rd thinning (years) | Clear felling age (years) | Initial planting density (stems $ha^{-1}$) | Stems thinned at 1st thinning (stems $ha^{-1}$) | Stems thinned at 2nd thinning (stems $ha^{-1}$) | Stems thinned at 3rd thinning (stems $ha^{-1}$) | Obj1- value prodn | Obj2-volume prodn |
|---|---|---|---|---|---|---|---|---|---|---|
| 47 | 5 | 13 | 18 | 35 | 930 | 222 | 194 | 193 | $4.55E+24$ | $5.21E+26$ |
| 48 | 6 | 13 | 18 | 35 | 901 | 255 | 193 | 178 | $4.21E+24$ | $2.43E+26$ |
| 49 | 5 | 14 | 18 | 35 | 910 | 234 | 191 | 189 | $4.89E+24$ | $5.55E+26$ |
| 50 | 5 | 15 | 18 | 35 | 905 | 131 | 198 | 197 | $2.75E+24$ | $3.11E+26$ |
| 51 | 6 | 14 | 19 | 35 | 976 | 300 | 200 | 199 | $4.63E+24$ | $2.75E+26$ |
| 52 | 5 | 15 | 18 | 35 | 903 | 178 | 196 | 190 | $3.74E+24$ | $4.23E+26$ |
| 53 | 5 | 15 | 18 | 34 | 1019 | 191 | 200 | 182 | $6.35E+23$ | $7.48E+25$ |
| 54 | 5 | 12 | 18 | 35 | 915 | 282 | 184 | 198 | $5.87E+24$ | $6.68E+26$ |
| 55 | 5 | 15 | 18 | 35 | 900 | 157 | 199 | 200 | $3.31E+24$ | $3.74E+26$ |
| 56 | 6 | 15 | 18 | 35 | 902 | 144 | 196 | 197 | $2.37E+24$ | $1.37E+26$ |
| 57 | 5 | 15 | 18 | 35 | 900 | 121 | 200 | 200 | $2.55E+24$ | $2.88E+26$ |
| 58 | 7 | 13 | 18 | 35 | 903 | 217 | 195 | 178 | $1.91E+24$ | $6.73E+25$ |
| 59 | 6 | 13 | 18 | 35 | 937 | 264 | 186 | 189 | $4.21E+24$ | $2.46E+26$ |
| 60 | 5 | 15 | 18 | 35 | 929 | 161 | 199 | 197 | $3.30E+24$ | $3.77E+26$ |
| 61 | 5 | 15 | 18 | 35 | 910 | 177 | 192 | 193 | $3.69E+24$ | $4.19E+26$ |
| 62 | 5 | 14 | 18 | 35 | 918 | 250 | 188 | 186 | $5.19E+24$ | $5.90E+26$ |
| 63 | 5 | 15 | 18 | 35 | 908 | 161 | 197 | 197 | $3.37E+24$ | $3.82E+26$ |
| 64 | 7 | 14 | 19 | 35 | 901 | 129 | 199 | 197 | $1.14E+24$ | $4.00E+25$ |
| 65 | 5 | 13 | 18 | 35 | 900 | 279 | 191 | 182 | $5.89E+24$ | $6.66E+26$ |
| 66 | 6 | 14 | 18 | 35 | 904 | 156 | 196 | 190 | $2.56E+24$ | $1.48E+26$ |
| 67 | 7 | 14 | 19 | 35 | 923 | 202 | 194 | 192 | $1.75E+24$ | $6.19E+25$ |
| 68 | 6 | 14 | 19 | 35 | 906 | 179 | 197 | 194 | $2.93E+24$ | $1.70E+26$ |
| 69 | 5 | 14 | 18 | 35 | 919 | 206 | 195 | 193 | $4.27E+24$ | $4.86E+26$ |
| 70 | 5 | 13 | 18 | 35 | 931 | 299 | 183 | 181 | $6.14E+24$ | $7.02E+26$ |
| 71 | 6 | 14 | 19 | 35 | 927 | 284 | 176 | 189 | $4.57E+24$ | $2.67E+26$ |
| 72 | 6 | 14 | 19 | 35 | 901 | 150 | 191 | 197 | $2.47E+24$ | $1.43E+26$ |
| 73 | 5 | 12 | 18 | 35 | 929 | 250 | 186 | 191 | $5.14E+24$ | $5.87E+26$ |
| 74 | 7 | 13 | 18 | 35 | 902 | 237 | 193 | 171 | $2.09E+24$ | $7.35E+25$ |
| 75 | 7 | 14 | 19 | 35 | 904 | 141 | 198 | 195 | $1.24E+24$ | $4.36E+25$ |
| 76 | 5 | 13 | 18 | 35 | 908 | 283 | 120 | 187 | $5.93E+24$ | $6.72E+26$ |
| 77 | 5 | 13 | 18 | 35 | 903 | 265 | 190 | 195 | $5.58E+24$ | $6.31E+26$ |
| 78 | 5 | 15 | 18 | 35 | 903 | 136 | 199 | 196 | $2.86E+24$ | $3.23E+26$ |
| 79 | 6 | 13 | 18 | 35 | 935 | 255 | 193 | 191 | $4.08E+24$ | $2.38E+26$ |
| 80 | 6 | 14 | 19 | 35 | 929 | 167 | 198 | 195 | $2.68E+24$ | $1.56E+26$ |
| 81 | 5 | 13 | 18 | 35 | 914 | 221 | 194 | 196 | $4.60E+24$ | $5.23E+26$ |
| 82 | 5 | 14 | 18 | 35 | 913 | 223 | 191 | 192 | $4.65E+24$ | $5.28E+26$ |
| 83 | 6 | 13 | 18 | 35 | 900 | 248 | 197 | 170 | $4.09E+24$ | $2.36E+26$ |
| 84 | 5 | 13 | 18 | 35 | 1010 | 285 | 192 | 192 | $5.47E+24$ | $6.42E+26$ |
| 85 | 7 | 14 | 19 | 35 | 919 | 235 | 183 | 200 | $2.04E+24$ | $7.22E+25$ |
| 86 | 5 | 15 | 18 | 35 | 903 | 140 | 200 | 199 | $2.94E+24$ | $3.33E+26$ |
| 87 | 5 | 13 | 18 | 35 | 961 | 300 | 194 | 188 | $6.00E+24$ | $6.93E+26$ |
| 88 | 5 | 13 | 18 | 35 | 900 | 244 | 181 | 182 | $5.15E+24$ | $5.82E+26$ |
| 89 | 5 | 13 | 18 | 35 | 934 | 246 | 200 | 194 | $5.03E+24$ | $5.76E+26$ |
| 90 | 5 | 13 | 18 | 35 | 921 | 228 | 195 | 194 | $4.72E+24$ | $5.37E+26$ |
| 91 | 5 | 15 | 18 | 35 | 903 | 130 | 199 | 199 | $2.73E+24$ | $3.09E+26$ |
| 92 | 5 | 13 | 18 | 35 | 910 | 268 | 191 | 193 | $5.60E+24$ | $6.36E+26$ |

TABLE 7: Continued.

| Regime | Age at 1st thinning (years) | Age at 2nd thinning (years) | Age at 3rd thinning (years) | Clear felling age (years) | Initial planting density (stems $ha^{-1}$) | Stems thinned at 1st thinning (stems $ha^{-1}$) | Stems thinned at 2nd thinning (stems $ha^{-1}$) | Stems thinned at 3rd thinning (stems $ha^{-1}$) | Obj1- value prodn | Obj2-volume prodn |
|---|---|---|---|---|---|---|---|---|---|---|
| 93 | 7 | 13 | 18 | 35 | 902 | 235 | 196 | 168 | $2.07E+24$ | $7.29E+25$ |
| 94 | 5 | 13 | 18 | 35 | 916 | 237 | 190 | 195 | $4.93E+24$ | $5.60E+26$ |
| 95 | 6 | 14 | 19 | 35 | 900 | 148 | 194 | 187 | $2.44E+24$ | $1.41E+26$ |
| 96 | 5 | 13 | 18 | 33 | 900 | 271 | 192 | 191 | $1.75E+23$ | $1.98E+25$ |
| 97 | 5 | 12 | 18 | 35 | 944 | 300 | 171 | 189 | $6.09E+24$ | $6.99E+26$ |
| 98 | 5 | 13 | 18 | 35 | 900 | 275 | 187 | 194 | $5.81E+24$ | $6.56E+26$ |
| 99 | 5 | 15 | 18 | 35 | 903 | 128 | 200 | 200 | $2.69E+24$ | $3.04E+26$ |
| 100 | 6 | 14 | 19 | 35 | 975 | 217 | 178 | 168 | $5.86E+23$ | $3.47E+25$ |
| 101 | 5 | 15 | 18 | 35 | 900 | 165 | 177 | 183 | $6.08E+23$ | $6.88E+25$ |
| 102 | 5 | 15 | 18 | 35 | 913 | 180 | 193 | 192 | $3.75E+24$ | $4.26E+26$ |
| 103 | 5 | 13 | 18 | 35 | 910 | 296 | 177 | 194 | $6.19E+24$ | $7.03E+26$ |
| 104 | 6 | 14 | 19 | 35 | 1096 | 199 | 200 | 177 | $8.56E+22$ | $5.27E+24$ |
| 105 | 5 | 12 | 18 | 34 | 900 | 300 | 187 | 186 | $1.11E+24$ | $1.25E+26$ |
| 106 | 5 | 12 | 18 | 35 | 900 | 282 | 200 | 182 | $5.96E+24$ | $6.73E+26$ |
| 107 | 5 | 13 | 18 | 35 | 961 | 275 | 200 | 200 | $5.49E+24$ | $6.35E+26$ |
| 108 | 5 | 15 | 18 | 35 | 904 | 142 | 195 | 195 | $2.98E+24$ | $3.37E+26$ |
| 109 | 7 | 14 | 19 | 35 | 921 | 230 | 183 | 199 | $1.99E+24$ | $7.06E+25$ |
| 110 | 5 | 13 | 18 | 35 | 922 | 295 | 184 | 181 | $6.10E+24$ | $6.96E+26$ |
| 111 | 5 | 13 | 18 | 35 | 938 | 222 | 192 | 194 | $4.52E+24$ | $5.18E+26$ |
| 112 | 6 | 13 | 19 | 35 | 900 | 208 | 197 | 188 | $3.43E+24$ | $1.98E+26$ |
| 113 | 5 | 14 | 18 | 34 | 958 | 220 | 179 | 182 | $7.70E+23$ | $8.89E+25$ |
| 114 | 5 | 13 | 18 | 35 | 917 | 281 | 188 | 187 | $5.84E+24$ | $6.64E+26$ |
| 115 | 5 | 14 | 18 | 35 | 919 | 200 | 199 | 182 | $4.14E+24$ | $4.72E+26$ |
| 116 | 5 | 12 | 18 | 35 | 908 | 297 | 200 | 200 | $6.23E+24$ | $7.06E+26$ |
| 117 | 7 | 14 | 19 | 35 | 900 | 139 | 195 | 200 | $1.23E+24$ | $4.31E+25$ |
| 118 | 7 | 14 | 19 | 35 | 900 | 155 | 186 | 191 | $1.37E+24$ | $4.81E+25$ |
| 119 | 5 | 12 | 18 | 35 | 902 | 285 | 195 | 193 | $6.01E+24$ | $6.80E+26$ |
| 120 | 5 | 15 | 18 | 35 | 902 | 129 | 200 | 199 | $2.71E+24$ | $3.07E+26$ |
| 121 | 5 | 13 | 18 | 35 | 915 | 277 | 188 | 189 | $5.77E+24$ | $6.56E+26$ |
| 122 | 7 | 14 | 18 | 35 | 967 | 201 | 196 | 193 | $1.67E+24$ | $6.02E+25$ |
| 123 | 5 | 13 | 18 | 35 | 900 | 288 | 199 | 196 | $6.08E+24$ | $6.88E+26$ |
| 124 | 5 | 15 | 18 | 35 | 910 | 182 | 195 | 194 | $3.80E+24$ | $4.31E+26$ |
| 125 | 5 | 12 | 18 | 35 | 916 | 278 | 194 | 198 | $5.79E+24$ | $6.58E+26$ |
| 126 | 5 | 14 | 18 | 35 | 907 | 202 | 193 | 194 | $4.23E+24$ | $4.80E+26$ |
| 127 | 5 | 13 | 18 | 35 | 991 | 265 | 194 | 179 | $5.16E+24$ | $6.02E+26$ |
| 128 | 7 | 14 | 19 | 35 | 922 | 151 | 200 | 200 | $1.31E+24$ | $4.63E+25$ |
| 129 | 5 | 14 | 18 | 35 | 953 | 213 | 199 | 195 | $4.28E+24$ | $4.93E+26$ |
| 130 | 8 | 14 | 19 | 35 | 916 | 161 | 189 | 193 | $5.95E+23$ | $1.42E+25$ |
| 131 | 5 | 12 | 18 | 35 | 934 | 282 | 187 | 193 | $5.77E+24$ | $6.61E+26$ |
| 132 | 5 | 15 | 18 | 35 | 907 | 149 | 196 | 196 | $3.12E+24$ | $3.54E+26$ |
| 133 | 5 | 13 | 18 | 35 | 928 | 252 | 166 | 200 | $5.18E+24$ | $5.92E+26$ |
| 134 | 5 | 12 | 18 | 35 | 922 | 281 | 191 | 190 | $5.82E+24$ | $6.63E+26$ |
| 135 | 6 | 14 | 19 | 35 | 902 | 162 | 191 | 199 | $2.67E+24$ | $1.54E+26$ |
| 136 | 5 | 15 | 18 | 35 | 900 | 144 | 200 | 191 | $3.03E+24$ | $3.43E+26$ |
| 137 | 7 | 14 | 18 | 35 | 906 | 208 | 194 | 195 | $1.83E+24$ | $6.44E+25$ |
| 138 | 6 | 14 | 19 | 35 | 913 | 148 | 196 | 197 | $2.41E+24$ | $1.40E+26$ |

Table 7: Continued.

| Regime | Age at 1st thinning (years) | Age at 2nd thinning (years) | Age at 3rd thinning (years) | Clear felling age (years) | Initial planting density (stems ha$^{-1}$) | Stems thinned at 1st thinning (stems ha$^{-1}$) | Stems thinned at 2nd thinning (stems ha$^{-1}$) | Stems thinned at 3rd thinning (stems ha$^{-1}$) | Obj1- value prodn | Obj2-volume prodn |
|---|---|---|---|---|---|---|---|---|---|---|
| 139 | 5 | 15 | 18 | 35 | 911 | 147 | 195 | 195 | $3.06E + 24$ | $3.48E + 26$ |
| 140 | 5 | 15 | 18 | 35 | 903 | 150 | 200 | 199 | $3.15E + 24$ | $3.57E + 26$ |
| 141 | 6 | 15 | 18 | 35 | 948 | 101 | 199 | 190 | $2.79E + 23$ | $1.64E + 25$ |
| 142 | 5 | 13 | 18 | 35 | 911 | 240 | 200 | 198 | $5.01E + 24$ | $5.69E + 26$ |
| 143 | 6 | 14 | 18 | 35 | 900 | 196 | 200 | 199 | $3.23E + 24$ | $1.86E + 26$ |
| 144 | 5 | 15 | 18 | 35 | 904 | 133 | 198 | 197 | $2.79E + 24$ | $3.16E + 26$ |
| 145 | 5 | 15 | 18 | 35 | 905 | 167 | 200 | 183 | $3.50E + 24$ | $3.97E + 26$ |
| 146 | 5 | 14 | 18 | 35 | 911 | 195 | 199 | 199 | $4.07E + 24$ | $4.62E + 26$ |
| 147 | 5 | 14 | 18 | 35 | 910 | 200 | 196 | 195 | $4.18E + 24$ | $4.74E + 26$ |
| 148 | 5 | 15 | 18 | 35 | 904 | 156 | 195 | 191 | $3.27E + 24$ | $3.71E + 26$ |
| 149 | 5 | 15 | 18 | 35 | 904 | 135 | 200 | 198 | $2.83E + 24$ | $3.21E + 26$ |
| 150 | 5 | 13 | 18 | 35 | 916 | 259 | 188 | 188 | $5.38E + 24$ | $6.12E + 26$ |
| 151 | 7 | 15 | 18 | 35 | 1000 | 168 | 198 | 196 | $2.38E + 23$ | $8.66E + 24$ |
| 152 | 7 | 14 | 19 | 35 | 900 | 150 | 196 | 180 | $1.32E + 24$ | $4.65E + 25$ |
| 153 | 5 | 13 | 18 | 34 | 943 | 238 | 178 | 200 | $8.45E + 23$ | $9.70E + 25$ |
| 154 | 7 | 14 | 19 | 35 | 902 | 133 | 199 | 198 | $1.17E + 24$ | $4.12E + 25$ |
| 155 | 5 | 15 | 18 | 35 | 904 | 191 | 200 | 194 | $4.01E + 24$ | $4.54E + 26$ |
| 156 | 5 | 12 | 18 | 35 | 907 | 262 | 192 | 187 | $5.50E + 24$ | $6.23E + 26$ |
| 157 | 5 | 15 | 18 | 35 | 919 | 176 | 183 | 199 | $3.64E + 24$ | $4.15E + 26$ |
| 158 | 5 | 13 | 18 | 35 | 903 | 244 | 187 | 186 | $5.13E + 24$ | $5.81E + 26$ |
| 159 | 5 | 15 | 18 | 35 | 900 | 119 | 200 | 200 | $2.51E + 24$ | $2.83E + 26$ |
| 160 | 5 | 12 | 18 | 35 | 901 | 286 | 199 | 200 | $6.04E + 24$ | $6.82E + 26$ |
| 161 | 5 | 15 | 18 | 35 | 908 | 146 | 197 | 197 | $3.05E + 24$ | $3.46E + 26$ |
| 162 | 5 | 14 | 18 | 35 | 936 | 191 | 196 | 191 | $3.89E + 24$ | $4.46E + 26$ |
| 163 | 5 | 15 | 18 | 35 | 900 | 212 | 192 | 200 | $4.47E + 24$ | $5.05E + 26$ |
| 164 | 6 | 14 | 19 | 35 | 906 | 182 | 198 | 195 | $2.98E + 24$ | $1.73E + 26$ |
| 165 | 5 | 12 | 18 | 35 | 915 | 300 | 197 | 200 | $6.25E + 24$ | $7.10E + 26$ |
| 166 | 7 | 14 | 19 | 35 | 920 | 222 | 187 | 196 | $1.93E + 24$ | $6.82E + 25$ |
| 167 | 5 | 13 | 18 | 35 | 936 | 265 | 179 | 196 | $5.41E + 24$ | $6.20E + 26$ |
| 168 | 5 | 14 | 18 | 35 | 999 | 208 | 188 | 181 | $7.03E + 23$ | $8.23E + 25$ |
| 169 | 7 | 15 | 18 | 35 | 985 | 174 | 198 | 195 | $2.49E + 23$ | $9.04E + 24$ |
| 170 | 5 | 12 | 18 | 35 | 929 | 257 | 189 | 194 | $5.28E + 24$ | $6.03E + 26$ |
| 171 | 5 | 14 | 18 | 35 | 916 | 195 | 187 | 197 | $4.05E + 24$ | $4.61E + 26$ |
| 172 | 5 | 12 | 18 | 35 | 900 | 287 | 189 | 195 | $6.06E + 24$ | $6.85E + 26$ |
| 173 | 7 | 13 | 18 | 35 | 900 | 252 | 200 | 178 | $2.23E + 24$ | $7.83E + 25$ |
| 174 | 5 | 12 | 18 | 34 | 968 | 298 | 199 | 196 | $1.04E + 24$ | $1.20E + 26$ |
| 175 | 7 | 14 | 19 | 35 | 900 | 201 | 200 | 200 | $1.78E + 24$ | $6.24E + 25$ |
| 176 | 5 | 13 | 18 | 35 | 900 | 222 | 178 | 196 | $4.68E + 24$ | $5.29E + 26$ |
| 177 | 5 | 15 | 18 | 35 | 900 | 132 | 200 | 200 | $2.78E + 24$ | $3.14E + 26$ |
| 178 | 6 | 14 | 19 | 35 | 900 | 113 | 190 | 200 | $3.26E + 23$ | $1.88E + 25$ |
| 179 | 5 | 15 | 18 | 35 | 903 | 148 | 194 | 195 | $3.11E + 24$ | $3.52E + 26$ |
| 180 | 5 | 13 | 18 | 35 | 900 | 300 | 138 | 171 | $6.33E + 24$ | $7.16E + 26$ |
| 181 | 5 | 13 | 18 | 35 | 900 | 249 | 200 | 189 | $5.25E + 24$ | $5.94E + 26$ |
| 182 | 6 | 13 | 18 | 35 | 920 | 238 | 191 | 193 | $3.86E + 24$ | $2.24E + 26$ |
| 183 | 5 | 13 | 18 | 35 | 900 | 195 | 200 | 186 | $4.11E + 24$ | $4.65E + 26$ |
| 184 | 6 | 14 | 19 | 35 | 929 | 268 | 167 | 183 | $4.31E + 24$ | $2.51E + 26$ |

TABLE 7: Continued.

| Regime | Age at 1st thinning (years) | Age at 2nd thinning (years) | Age at 3rd thinning (years) | Clear felling age (years) | Initial planting density (stems ha$^{-1}$) | Stems thinned at 1st thinning (stems ha$^{-1}$) | Stems thinned at 2nd thinning (stems ha$^{-1}$) | Stems thinned at 3rd thinning (stems ha$^{-1}$) | Obj1- value prodn | Obj2-volume prodn |
|---|---|---|---|---|---|---|---|---|---|---|
| 185 | 5 | 15 | 18 | 35 | 905 | 147 | 194 | 194 | 3.08$E$ + 24 | 3.49$E$ + 26 |
| 186 | 6 | 15 | 18 | 35 | 903 | 148 | 200 | 194 | 2.43$E$ + 24 | 1.40$E$ + 26 |
| 187 | 5 | 15 | 18 | 35 | 903 | 126 | 200 | 199 | 2.65$E$ + 24 | 3.00$E$ + 26 |
| 188 | 6 | 13 | 18 | 35 | 905 | 274 | 185 | 191 | 4.50$E$ + 24 | 2.60$E$ + 26 |
| 189 | 6 | 15 | 18 | 34 | 1034 | 123 | 176 | 175 | 9.69$E$ + 21 | 5.85$E$ + 23 |
| 190 | 5 | 14 | 18 | 35 | 937 | 188 | 199 | 199 | 3.83$E$ + 24 | 4.39$E$ + 26 |
| 191 | 7 | 14 | 19 | 35 | 910 | 171 | 186 | 197 | 1.50$E$ + 24 | 5.28$E$ + 25 |
| 192 | 5 | 13 | 18 | 35 | 936 | 234 | 185 | 188 | 4.78$E$ + 24 | 5.47$E$ + 26 |
| 193 | 5 | 13 | 18 | 35 | 923 | 258 | 195 | 191 | 5.33$E$ + 24 | 6.08$E$ + 26 |
| 194 | 5 | 13 | 18 | 35 | 917 | 300 | 124 | 186 | 6.24$E$ + 24 | 7.09$E$ + 26 |
| 195 | 5 | 15 | 18 | 35 | 900 | 134 | 200 | 199 | 2.82$E$ + 24 | 3.19$E$ + 26 |
| 196 | 5 | 13 | 18 | 35 | 915 | 258 | 186 | 197 | 5.37$E$ + 24 | 6.10$E$ + 26 |
| 197 | 5 | 12 | 18 | 35 | 904 | 238 | 200 | 200 | 5.01$E$ + 24 | 5.66$E$ + 26 |
| 198 | 6 | 14 | 19 | 35 | 900 | 237 | 199 | 193 | 3.91$E$ + 24 | 2.26$E$ + 26 |
| 199 | 5 | 14 | 18 | 35 | 900 | 206 | 200 | 188 | 4.34$E$ + 24 | 4.91$E$ + 26 |
| 200 | 5 | 14 | 18 | 35 | 910 | 183 | 196 | 194 | 3.82$E$ + 24 | 4.34$E$ + 26 |
| 201 | 6 | 14 | 19 | 35 | 901 | 278 | 172 | 177 | 4.58$E$ + 24 | 2.65$E$ + 26 |
| 202 | 5 | 15 | 18 | 35 | 904 | 186 | 200 | 194 | 3.91$E$ + 24 | 4.42$E$ + 26 |
| 203 | 6 | 14 | 19 | 35 | 900 | 146 | 186 | 195 | 2.41$E$ + 24 | 1.39$E$ + 26 |
| 204 | 6 | 13 | 19 | 35 | 928 | 207 | 196 | 194 | 3.33$E$ + 24 | 1.94$E$ + 26 |
| 205 | 6 | 14 | 18 | 35 | 904 | 158 | 196 | 191 | 2.59$E$ + 24 | 1.50$E$ + 26 |
| 206 | 5 | 15 | 18 | 35 | 909 | 142 | 197 | 197 | 2.97$E$ + 24 | 3.36$E$ + 26 |
| 207 | 6 | 13 | 18 | 35 | 910 | 244 | 195 | 187 | 3.99$E$ + 24 | 2.31$E$ + 26 |
| 208 | 6 | 14 | 19 | 35 | 913 | 173 | 197 | 196 | 2.82$E$ + 24 | 1.63$E$ + 26 |
| 209 | 5 | 12 | 18 | 35 | 923 | 287 | 191 | 190 | 5.94$E$ + 24 | 6.76$E$ + 26 |
| 210 | 5 | 13 | 18 | 34 | 914 | 269 | 191 | 182 | 9.81$E$ + 23 | 1.11$E$ + 26 |
| 211 | 5 | 14 | 18 | 35 | 914 | 182 | 193 | 196 | 3.79$E$ + 24 | 4.30$E$ + 26 |
| 212 | 5 | 12 | 18 | 35 | 900 | 268 | 200 | 162 | 5.66$E$ + 24 | 6.40$E$ + 26 |
| 213 | 5 | 14 | 18 | 34 | 972 | 248 | 196 | 170 | 8.58$E$ + 23 | 9.96$E$ + 25 |
| 214 | 5 | 13 | 18 | 34 | 900 | 286 | 188 | 180 | 1.06$E$ + 24 | 1.19$E$ + 26 |
| 215 | 5 | 12 | 18 | 35 | 1055 | 255 | 178 | 196 | 8.26$E$ + 23 | 9.83$E$ + 25 |
| 216 | 5 | 14 | 18 | 34 | 1027 | 209 | 197 | 200 | 6.91$E$ + 23 | 8.16$E$ + 25 |
| 217 | 6 | 15 | 18 | 35 | 901 | 135 | 199 | 198 | 2.22$E$ + 24 | 1.28$E$ + 26 |
| 218 | 5 | 15 | 18 | 35 | 903 | 129 | 200 | 198 | 2.71$E$ + 24 | 3.07$E$ + 26 |
| 219 | 6 | 13 | 18 | 35 | 913 | 294 | 190 | 197 | 4.80$E$ + 24 | 2.78$E$ + 26 |
| 220 | 6 | 14 | 19 | 35 | 901 | 271 | 173 | 178 | 4.47$E$ + 24 | 2.58$E$ + 26 |
| 221 | 5 | 13 | 18 | 35 | 1033 | 185 | 181 | 195 | 1.07$E$ + 23 | 1.26$E$ + 25 |
| 222 | 5 | 14 | 18 | 35 | 922 | 215 | 188 | 196 | 4.44$E$ + 24 | 5.06$E$ + 26 |
| 223 | 6 | 13 | 18 | 35 | 900 | 282 | 178 | 200 | 4.66$E$ + 24 | 2.69$E$ + 26 |
| 224 | 5 | 12 | 18 | 35 | 936 | 244 | 186 | 188 | 4.98$E$ + 24 | 5.71$E$ + 26 |
| 225 | 7 | 14 | 19 | 35 | 903 | 172 | 199 | 199 | 1.51$E$ + 24 | 5.33$E$ + 25 |
| 226 | 7 | 14 | 19 | 35 | 918 | 154 | 199 | 199 | 1.34$E$ + 24 | 4.73$E$ + 25 |
| 227 | 5 | 13 | 18 | 34 | 913 | 300 | 190 | 184 | 1.10$E$ + 24 | 1.24$E$ + 26 |
| 228 | 5 | 12 | 18 | 35 | 993 | 272 | 192 | 197 | 5.29$E$ + 24 | 6.18$E$ + 26 |
| 229 | 7 | 14 | 19 | 35 | 919 | 233 | 183 | 200 | 2.02$E$ + 24 | 7.16$E$ + 25 |
| 230 | 5 | 13 | 18 | 35 | 924 | 206 | 187 | 190 | 4.25$E$ + 24 | 4.85$E$ + 26 |
| 231 | 7 | 13 | 18 | 35 | 900 | 278 | 192 | 153 | 2.46$E$ + 24 | 8.64$E$ + 25 |

Table 7: Continued.

| Regime | Age at 1st thinning (years) | Age at 2nd thinning (years) | Age at 3rd thinning (years) | Clear felling age (years) | Initial planting density (stems $ha^{-1}$) | Stems thinned at 1st thinning (stems $ha^{-1}$) | Stems thinned at 2nd thinning (stems $ha^{-1}$) | Stems thinned at 3rd thinning (stems $ha^{-1}$) | Obj1- value prodn | Obj2-volume prodn |
|---|---|---|---|---|---|---|---|---|---|---|
| 232 | 5 | 15 | 18 | 35 | 903 | 139 | 199 | 199 | $2.92E + 24$ | $3.30E + 26$ |
| 233 | 5 | 12 | 18 | 35 | 968 | 297 | 187 | 180 | $5.90E + 24$ | $6.83E + 26$ |
| 234 | 5 | 14 | 18 | 35 | 948 | 225 | 182 | 200 | $4.54E + 24$ | $5.23E + 26$ |
| 235 | 6 | 14 | 19 | 35 | 910 | 225 | 197 | 192 | $3.68E + 24$ | $2.13E + 26$ |
| 236 | 5 | 13 | 18 | 35 | 914 | 236 | 195 | 196 | $4.91E + 24$ | $5.59E + 26$ |
| 237 | 5 | 15 | 18 | 35 | 905 | 126 | 199 | 199 | $2.64E + 24$ | $2.99E + 26$ |
| 238 | 6 | 13 | 18 | 35 | 907 | 275 | 192 | 187 | $4.51E + 24$ | $2.61E + 26$ |
| 239 | 5 | 15 | 18 | 35 | 901 | 125 | 199 | 200 | $2.63E + 24$ | $2.97E + 26$ |
| 240 | 5 | 13 | 18 | 35 | 907 | 233 | 197 | 198 | $4.88E + 24$ | $5.54E + 26$ |
| 241 | 5 | 13 | 18 | 34 | 900 | 293 | 187 | 180 | $1.08E + 24$ | $1.22E + 26$ |
| 242 | 5 | 13 | 18 | 35 | 918 | 243 | 189 | 195 | $5.04E + 24$ | $5.74E + 26$ |
| 243 | 6 | 13 | 18 | 35 | 911 | 245 | 197 | 196 | $4.00E + 24$ | $2.32E + 26$ |
| 244 | 7 | 13 | 18 | 35 | 903 | 205 | 198 | 174 | $1.81E + 24$ | $6.36E + 25$ |
| 245 | 5 | 15 | 18 | 35 | 901 | 130 | 199 | 200 | $2.73E + 24$ | $3.09E + 26$ |
| 246 | 5 | 14 | 18 | 35 | 903 | 217 | 189 | 198 | $4.56E + 24$ | $5.17E + 26$ |
| 247 | 5 | 13 | 18 | 35 | 903 | 221 | 195 | 192 | $4.65E + 24$ | $5.26E + 26$ |
| 248 | 5 | 13 | 18 | 35 | 911 | 294 | 188 | 187 | $6.14E + 24$ | $6.98E + 26$ |
| 249 | 5 | 13 | 18 | 35 | 916 | 215 | 195 | 190 | $4.47E + 24$ | $5.08E + 26$ |
| 250 | 6 | 14 | 19 | 35 | 951 | 250 | 192 | 191 | $3.94E + 24$ | $2.32E + 26$ |
| 251 | 6 | 14 | 19 | 35 | 926 | 243 | 200 | 193 | $3.91E + 24$ | $2.28E + 26$ |
| 252 | 5 | 12 | 18 | 34 | 971 | 300 | 200 | 197 | $1.04E + 24$ | $1.21E + 26$ |
| 253 | 6 | 15 | 18 | 35 | 952 | 108 | 188 | 189 | $5.20E + 22$ | $3.06E + 24$ |
| 254 | 5 | 13 | 18 | 35 | 937 | 248 | 189 | 192 | $5.06E + 24$ | $5.80E + 26$ |
| 255 | 6 | 15 | 18 | 35 | 949 | 128 | 194 | 188 | $3.53E + 23$ | $2.07E + 25$ |
| 256 | 7 | 14 | 19 | 35 | 924 | 199 | 194 | 192 | $1.72E + 24$ | $6.10E + 25$ |
| 257 | 5 | 12 | 18 | 35 | 900 | 292 | 194 | 185 | $6.17E + 24$ | $6.97E + 26$ |
| 258 | 5 | 12 | 18 | 35 | 977 | 237 | 192 | 199 | $4.67E + 24$ | $5.42E + 26$ |
| 259 | 5 | 14 | 18 | 35 | 914 | 183 | 192 | 197 | $3.81E + 24$ | $4.33E + 26$ |
| 260 | 7 | 14 | 19 | 35 | 900 | 190 | 200 | 200 | $1.68E + 24$ | $5.90E + 25$ |
| 261 | 6 | 14 | 18 | 35 | 922 | 218 | 191 | 188 | $3.52E + 24$ | $2.05E + 26$ |
| 262 | 7 | 14 | 19 | 35 | 922 | 190 | 195 | 193 | $1.64E + 24$ | $5.83E + 25$ |
| 263 | 5 | 12 | 18 | 35 | 904 | 300 | 178 | 196 | $6.32E + 24$ | $7.15E + 26$ |
| 264 | 6 | 15 | 18 | 35 | 906 | 142 | 196 | 197 | $2.33E + 24$ | $1.35E + 26$ |
| 265 | 5 | 12 | 18 | 35 | 901 | 246 | 199 | 179 | $5.19E + 24$ | $5.87E + 26$ |
| 266 | 5 | 13 | 18 | 35 | 919 | 242 | 190 | 196 | $5.02E + 24$ | $5.71E + 26$ |
| 267 | 5 | 13 | 18 | 34 | 931 | 249 | 181 | 181 | $8.94E + 23$ | $1.02E + 26$ |
| 268 | 5 | 13 | 18 | 35 | 1031 | 224 | 189 | 200 | $7.38E + 23$ | $8.73E + 25$ |
| 269 | 6 | 15 | 18 | 35 | 949 | 158 | 200 | 194 | $4.36E + 23$ | $2.56E + 25$ |
| 270 | 5 | 13 | 18 | 35 | 938 | 245 | 200 | 181 | $4.99E + 24$ | $5.72E + 26$ |
| 271 | 5 | 15 | 18 | 35 | 910 | 170 | 197 | 197 | $3.55E + 24$ | $4.03E + 26$ |
| 272 | 5 | 13 | 18 | 35 | 905 | 255 | 196 | 192 | $5.36E + 24$ | $6.07E + 26$ |
| 273 | 7 | 14 | 19 | 35 | 919 | 162 | 198 | 198 | $1.41E + 24$ | $4.98E + 25$ |
| 274 | 6 | 13 | 18 | 35 | 908 | 246 | 197 | 191 | $4.03E + 24$ | $2.33E + 26$ |
| 275 | 6 | 13 | 18 | 35 | 900 | 252 | 197 | 168 | $4.16E + 24$ | $2.40E + 26$ |

Table 7: Continued.

| Regime | Age at 1st thinning (years) | Age at 2nd thinning (years) | Age at 3rd thinning (years) | Clear felling age (years) | Initial planting density (stems ha$^{-1}$) | Stems thinned at 1st thinning (stems ha$^{-1}$) | Stems thinned at 2nd thinning (stems ha$^{-1}$) | Stems thinned at 3rd thinning (stems ha$^{-1}$) | Obj1- value prodn | Obj2-volume prodn |
|---|---|---|---|---|---|---|---|---|---|---|
| 276 | 5 | 15 | 18 | 35 | 908 | 169 | 193 | 194 | $3.53E+24$ | $4.01E+26$ |
| 277 | 7 | 14 | 19 | 35 | 903 | 178 | 199 | 195 | $1.57E+24$ | $5.52E+25$ |
| 278 | 5 | 13 | 18 | 35 | 923 | 239 | 196 | 191 | $4.94E+24$ | $5.63E+26$ |
| 279 | 7 | 14 | 19 | 35 | 903 | 173 | 196 | 195 | $1.52E+24$ | $5.36E+25$ |
| 280 | 7 | 14 | 19 | 35 | 911 | 177 | 184 | 197 | $1.55E+24$ | $5.46E+25$ |
| 281 | 5 | 13 | 18 | 35 | 904 | 251 | 182 | 184 | $5.28E+24$ | $5.98E+26$ |
| 282 | 5 | 12 | 18 | 35 | 914 | 292 | 197 | 199 | $6.09E+24$ | $6.92E+26$ |
| 283 | 6 | 14 | 19 | 35 | 903 | 233 | 194 | 196 | $3.83E+24$ | $2.21E+26$ |
| 284 | 5 | 15 | 18 | 35 | 900 | 155 | 200 | 199 | $3.26E+24$ | $3.69E+26$ |
| 285 | 5 | 15 | 18 | 35 | 900 | 156 | 195 | 196 | $3.29E+24$ | $3.72E+26$ |
| 286 | 6 | 13 | 18 | 35 | 907 | 275 | 192 | 187 | $4.51E+24$ | $2.61E+26$ |
| 287 | 5 | 15 | 18 | 35 | 901 | 125 | 199 | 200 | $2.63E+24$ | $2.97E+26$ |
| 288 | 5 | 13 | 18 | 35 | 907 | 233 | 197 | 198 | $4.88E+24$ | $5.54E+26$ |
| 289 | 5 | 13 | 18 | 34 | 900 | 293 | 187 | 180 | $1.08E+24$ | $1.22E+26$ |
| 290 | 5 | 13 | 18 | 35 | 918 | 243 | 189 | 195 | $5.04E+24$ | $5.74E+26$ |
| 291 | 6 | 13 | 18 | 35 | 911 | 245 | 197 | 196 | $4.00E+24$ | $2.32E+26$ |
| 292 | 7 | 13 | 18 | 35 | 903 | 205 | 198 | 174 | $1.81E+24$ | $6.36E+25$ |
| 293 | 5 | 15 | 18 | 35 | 901 | 130 | 199 | 200 | $2.73E+24$ | $3.09E+26$ |
| 294 | 5 | 14 | 18 | 35 | 903 | 217 | 189 | 198 | $4.56E+24$ | $5.17E+26$ |
| 295 | 5 | 13 | 18 | 35 | 903 | 221 | 195 | 192 | $4.65E+24$ | $5.26E+26$ |
| 296 | 5 | 13 | 18 | 35 | 911 | 294 | 188 | 187 | $6.14E+24$ | $6.98E+26$ |
| 297 | 5 | 13 | 18 | 35 | 916 | 215 | 195 | 190 | $4.47E+24$ | $5.08E+26$ |
| 298 | 6 | 14 | 19 | 35 | 951 | 250 | 192 | 191 | $3.94E+24$ | $2.32E+26$ |
| 299 | 6 | 14 | 19 | 35 | 926 | 243 | 200 | 193 | $3.91E+24$ | $2.28E+26$ |
| 300 | 5 | 12 | 18 | 34 | 971 | 300 | 200 | 197 | $1.04E+24$ | $1.21E+26$ |
| 301 | 6 | 15 | 18 | 35 | 952 | 108 | 188 | 189 | $5.20E+22$ | $3.06E+24$ |
| 302 | 5 | 13 | 18 | 35 | 937 | 248 | 189 | 192 | $5.06E+24$ | $5.80E+26$ |
| 303 | 6 | 15 | 18 | 35 | 949 | 128 | 194 | 188 | $3.53E+23$ | $2.07E+25$ |
| 304 | 7 | 14 | 19 | 35 | 924 | 199 | 194 | 192 | $1.72E+24$ | $6.10E+25$ |
| 305 | 5 | 12 | 18 | 35 | 900 | 292 | 194 | 185 | $6.17E+24$ | $6.97E+26$ |
| 306 | 5 | 12 | 18 | 35 | 977 | 237 | 192 | 199 | $4.67E+24$ | $5.42E+26$ |
| 307 | 5 | 14 | 18 | 35 | 914 | 183 | 192 | 197 | $3.81E+24$ | $4.33E+26$ |
| 308 | 7 | 14 | 19 | 35 | 900 | 190 | 200 | 200 | $1.68E+24$ | $5.90E+25$ |
| 309 | 6 | 14 | 18 | 35 | 922 | 218 | 191 | 188 | $3.52E+24$ | $2.05E+26$ |
| 310 | 7 | 14 | 19 | 35 | 922 | 190 | 195 | 193 | $1.64E+24$ | $5.83E+25$ |
| 311 | 5 | 12 | 18 | 35 | 904 | 300 | 178 | 196 | $6.32E+24$ | $7.15E+26$ |
| 312 | 6 | 15 | 18 | 35 | 906 | 142 | 196 | 197 | $2.33E+24$ | $1.35E+26$ |
| 313 | 5 | 12 | 18 | 35 | 901 | 246 | 199 | 179 | $5.19E+24$ | $5.87E+26$ |
| 314 | 5 | 13 | 18 | 35 | 919 | 242 | 190 | 196 | $5.02E+24$ | $5.71E+26$ |
| 315 | 5 | 13 | 18 | 34 | 931 | 249 | 181 | 181 | $8.94E+23$ | $1.02E+26$ |
| 316 | 5 | 13 | 18 | 35 | 1031 | 224 | 189 | 200 | $7.38E+23$ | $8.73E+25$ |
| 317 | 6 | 15 | 18 | 35 | 949 | 158 | 200 | 194 | $4.36E+23$ | $2.56E+25$ |
| 318 | 5 | 13 | 18 | 35 | 938 | 245 | 200 | 181 | $4.99E+24$ | $5.72E+26$ |
| 319 | 5 | 15 | 18 | 35 | 910 | 170 | 197 | 197 | $3.55E+24$ | $4.03E+26$ |
| 320 | 5 | 13 | 18 | 35 | 905 | 255 | 196 | 192 | $5.36E+24$ | $6.07E+26$ |

Table 7: Continued.

| Regime | Age at 1st thinning (years) | Age at 2nd thinning (years) | Age at 3rd thinning (years) | Clear felling age (years) | Initial planting density (stems $ha^{-1}$) | Stems thinned at 1st thinning (stems $ha^{-1}$) | Stems thinned at 2nd thinning (stems $ha^{-1}$) | Stems thinned at 3rd thinning (stems $ha^{-1}$) | Obj1- value prodn | Obj2-volume prodn |
|---|---|---|---|---|---|---|---|---|---|---|
| 321 | 7 | 14 | 19 | 35 | 919 | 162 | 198 | 198 | 1.41*E* + 24 | 4.98*E* + 25 |
| 322 | 6 | 13 | 18 | 35 | 908 | 246 | 197 | 191 | 4.03*E* + 24 | 2.33*E* + 26 |
| 323 | 6 | 13 | 18 | 35 | 900 | 252 | 197 | 168 | 4.16*E* + 24 | 2.40*E* + 26 |
| 324 | 5 | 15 | 18 | 35 | 908 | 169 | 193 | 194 | 3.53*E* + 24 | 4.01*E* + 26 |
| 325 | 7 | 14 | 19 | 35 | 903 | 178 | 199 | 195 | 1.57*E* + 24 | 5.52*E* + 25 |
| 326 | 5 | 13 | 18 | 35 | 923 | 239 | 196 | 191 | 4.94*E* + 24 | 5.63*E* + 26 |
| 327 | 7 | 14 | 19 | 35 | 903 | 173 | 196 | 195 | 1.52*E* + 24 | 5.36*E* + 25 |
| 328 | 7 | 14 | 19 | 35 | 911 | 177 | 184 | 197 | 1.55*E* + 24 | 5.46*E* + 25 |
| 329 | 5 | 13 | 18 | 35 | 904 | 251 | 182 | 184 | 5.28*E* + 24 | 5.98*E* + 26 |
| 330 | 5 | 12 | 18 | 35 | 914 | 292 | 197 | 199 | 6.09*E* + 24 | 6.92*E* + 26 |
| 331 | 6 | 14 | 19 | 35 | 903 | 233 | 194 | 196 | 3.83*E* + 24 | 2.21*E* + 26 |
| 332 | 5 | 15 | 18 | 35 | 900 | 155 | 200 | 199 | 3.26*E* + 24 | 3.69*E* + 26 |
| 333 | 5 | 15 | 18 | 35 | 900 | 156 | 195 | 196 | 3.29*E* + 24 | 3.72*E* + 26 |
| 334 | 5 | 15 | 18 | 35 | 904 | 183 | 196 | 193 | 3.84*E* + 24 | 4.35*E* + 26 |
| 335 | 5 | 13 | 18 | 33 | 1035 | 256 | 200 | 186 | 1.47*E* + 23 | 1.74*E* + 25 |
| 336 | 6 | 14 | 18 | 35 | 907 | 223 | 194 | 198 | 3.65*E* + 24 | 2.11*E* + 26 |
| 337 | 5 | 15 | 18 | 35 | 905 | 173 | 200 | 182 | 3.63*E* + 24 | 4.11*E* + 26 |
| 338 | 5 | 12 | 18 | 35 | 900 | 269 | 193 | 188 | 5.68*E* + 24 | 6.42*E* + 26 |
| 339 | 7 | 14 | 19 | 34 | 962 | 137 | 190 | 196 | 2.00*E* + 23 | 7.20*E* + 24 |
| 340 | 5 | 13 | 18 | 35 | 920 | 254 | 173 | 197 | 5.26*E* + 24 | 5.99*E* + 26 |
| 341 | 7 | 14 | 19 | 35 | 900 | 163 | 200 | 200 | 1.44*E* + 24 | 5.06*E* + 25 |
| 342 | 5 | 15 | 18 | 35 | 906 | 157 | 196 | 195 | 3.29*E* + 24 | 3.73*E* + 26 |
| 343 | 6 | 13 | 18 | 35 | 914 | 267 | 186 | 181 | 4.35*E* + 24 | 2.52*E* + 26 |
| 344 | 6 | 14 | 19 | 35 | 963 | 194 | 200 | 188 | 3.02*E* + 24 | 1.78*E* + 26 |
| 345 | 6 | 14 | 19 | 35 | 903 | 130 | 198 | 199 | 2.14*E* + 24 | 1.23*E* + 26 |
| 346 | 7 | 14 | 19 | 35 | 900 | 157 | 200 | 200 | 1.39*E* + 24 | 4.87*E* + 25 |
| 347 | 5 | 13 | 18 | 35 | 966 | 257 | 176 | 176 | 5.11*E* + 24 | 5.92*E* + 26 |
| 348 | 6 | 14 | 19 | 35 | 904 | 178 | 189 | 193 | 2.92*E* + 24 | 1.69*E* + 26 |
| 349 | 5 | 14 | 18 | 35 | 959 | 216 | 199 | 195 | 4.32*E* + 24 | 4.99*E* + 26 |
| 350 | 7 | 14 | 19 | 35 | 956 | 247 | 188 | 197 | 2.07*E* + 24 | 7.45*E* + 25 |
| 351 | 5 | 15 | 18 | 35 | 906 | 138 | 200 | 198 | 2.89*E* + 24 | 3.28*E* + 26 |
| 352 | 6 | 14 | 19 | 35 | 912 | 236 | 191 | 193 | 3.85*E* + 24 | 2.23*E* + 26 |
| 353 | 5 | 14 | 18 | 35 | 908 | 224 | 192 | 192 | 4.69*E* + 24 | 5.32*E* + 26 |
| 354 | 5 | 15 | 18 | 35 | 900 | 158 | 189 | 178 | 5.82*E* + 23 | 6.59*E* + 25 |
| 355 | 7 | 14 | 19 | 35 | 900 | 116 | 200 | 196 | 1.79*E* + 23 | 6.30*E* + 24 |
| 356 | 5 | 13 | 18 | 35 | 911 | 273 | 188 | 187 | 5.70*E* + 24 | 6.48*E* + 26 |
| 357 | 5 | 15 | 18 | 35 | 900 | 152 | 199 | 198 | 3.20*E* + 24 | 3.62*E* + 26 |
| 358 | 6 | 13 | 18 | 35 | 905 | 288 | 183 | 191 | 4.73*E* + 24 | 2.74*E* + 26 |
| 359 | 5 | 15 | 18 | 35 | 904 | 165 | 194 | 192 | 3.46*E* + 24 | 3.92*E* + 26 |
| 360 | 6 | 14 | 19 | 35 | 905 | 269 | 176 | 182 | 4.42*E* + 24 | 2.55*E* + 26 |
| 361 | 5 | 15 | 18 | 35 | 907 | 152 | 196 | 196 | 3.18*E* + 24 | 3.61*E* + 26 |
| 362 | 7 | 14 | 19 | 35 | 900 | 142 | 181 | 188 | 2.19*E* + 23 | 7.71*E* + 24 |
| 363 | 5 | 12 | 18 | 35 | 900 | 286 | 194 | 185 | 6.04*E* + 24 | 6.83*E* + 26 |
| 364 | 7 | 14 | 19 | 34 | 961 | 206 | 196 | 198 | 3.01*E* + 23 | 1.08*E* + 25 |
| 365 | 5 | 12 | 18 | 35 | 913 | 251 | 190 | 196 | 5.24*E* + 24 | 5.95*E* + 26 |
| 366 | 5 | 12 | 18 | 35 | 1023 | 263 | 189 | 200 | 8.74*E* + 23 | 1.03*E* + 26 |

TABLE 7: Continued.

| Regime | Age at 1st thinning (years) | Age at 2nd thinning (years) | Age at 3rd thinning (years) | Clear felling age (years) | Initial planting density (stems $ha^{-1}$) | Stems thinned at 1st thinning (stems $ha^{-1}$) | Stems thinned at 2nd thinning (stems $ha^{-1}$) | Stems thinned at 3rd thinning (stems $ha^{-1}$) | Obj1- value prodn | Obj2-volume prodn |
|---|---|---|---|---|---|---|---|---|---|---|
| 367 | 5 | 14 | 18 | 35 | 910 | 193 | 193 | 192 | 4.03$E$ + 24 | 4.57$E$ + 26 |
| 368 | 5 | 13 | 18 | 35 | 1009 | 227 | 192 | 191 | 7.62$E$ + 23 | 8.94$E$ + 25 |
| 369 | 5 | 15 | 18 | 35 | 903 | 125 | 200 | 199 | 2.62$E$ + 24 | 2.97$E$ + 26 |
| 370 | 6 | 14 | 19 | 35 | 900 | 299 | 191 | 194 | 4.94$E$ + 24 | 2.85$E$ + 26 |
| 371 | 5 | 13 | 18 | 35 | 934 | 254 | 172 | 196 | 5.19$E$ + 24 | 5.95$E$ + 26 |
| 372 | 5 | 13 | 18 | 35 | 917 | 284 | 186 | 186 | 5.90$E$ + 24 | 6.71$E$ + 26 |
| 373 | 6 | 14 | 19 | 34 | 900 | 157 | 185 | 200 | 4.53$E$ + 23 | 2.61$E$ + 25 |
| 374 | 6 | 14 | 19 | 35 | 939 | 228 | 188 | 188 | 3.63$E$ + 24 | 2.12$E$ + 26 |
| 375 | 6 | 14 | 18 | 35 | 900 | 201 | 199 | 195 | 3.31$E$ + 24 | 1.91$E$ + 26 |
| 376 | 6 | 13 | 18 | 35 | 907 | 237 | 193 | 185 | 3.89$E$ + 24 | 2.25$E$ + 26 |
| 377 | 5 | 13 | 18 | 35 | 905 | 291 | 183 | 186 | 6.12$E$ + 24 | 6.93$E$ + 26 |
| 378 | 6 | 14 | 19 | 35 | 929 | 209 | 187 | 191 | 3.36$E$ + 24 | 1.96$E$ + 26 |
| 379 | 5 | 15 | 18 | 35 | 910 | 178 | 191 | 194 | 3.72$E$ + 24 | 4.22$E$ + 26 |
| 380 | 5 | 13 | 18 | 35 | 910 | 290 | 178 | 194 | 6.07$E$ + 24 | 6.88$E$ + 26 |
| 381 | 6 | 14 | 19 | 35 | 900 | 145 | 194 | 186 | 2.39$E$ + 24 | 1.38$E$ + 26 |
| 382 | 5 | 14 | 18 | 35 | 916 | 199 | 198 | 184 | 4.13$E$ + 24 | 4.70$E$ + 26 |
| 383 | 5 | 14 | 18 | 35 | 906 | 171 | 197 | 196 | 3.58$E$ + 24 | 4.06$E$ + 26 |
| 384 | 5 | 13 | 18 | 34 | 950 | 239 | 195 | 194 | 8.43$E$ + 23 | 9.71$E$ + 25 |
| 385 | 5 | 12 | 18 | 35 | 901 | 283 | 199 | 200 | 5.97$E$ + 24 | 6.75$E$ + 26 |
| 386 | 5 | 13 | 18 | 35 | 924 | 299 | 191 | 170 | 6.17$E$ + 24 | 7.04$E$ + 26 |
| 387 | 6 | 13 | 18 | 35 | 945 | 260 | 196 | 185 | 4.12$E$ + 24 | 2.42$E$ + 26 |
| 388 | 6 | 15 | 18 | 35 | 900 | 153 | 200 | 187 | 2.52$E$ + 24 | 1.45$E$ + 26 |
| 389 | 6 | 14 | 19 | 35 | 903 | 139 | 198 | 199 | 2.28$E$ + 24 | 1.32$E$ + 26 |
| 390 | 5 | 13 | 18 | 34 | 932 | 253 | 177 | 178 | 9.07$E$ + 23 | 1.04$E$ + 26 |
| 391 | 5 | 15 | 18 | 35 | 903 | 124 | 200 | 199 | 2.60$E$ + 24 | 2.95$E$ + 26 |
| 392 | 6 | 14 | 19 | 35 | 901 | 140 | 199 | 200 | 2.30$E$ + 24 | 1.33$E$ + 26 |
| 393 | 5 | 14 | 18 | 35 | 907 | 168 | 197 | 194 | 3.52$E$ + 24 | 3.99$E$ + 26 |
| 394 | 6 | 14 | 19 | 35 | 962 | 286 | 197 | 199 | 4.47$E$ + 24 | 2.64$E$ + 26 |
| 395 | 7 | 14 | 19 | 35 | 1016 | 154 | 195 | 200 | 2.15$E$ + 23 | 7.88$E$ + 24 |
| 396 | 5 | 14 | 18 | 35 | 922 | 223 | 200 | 199 | 4.61$E$ + 24 | 5.25$E$ + 26 |
| 397 | 6 | 14 | 19 | 35 | 992 | 275 | 200 | 187 | 4.19$E$ + 24 | 2.50$E$ + 26 |
| 398 | 6 | 15 | 18 | 35 | 903 | 142 | 197 | 196 | 2.33$E$ + 24 | 1.35$E$ + 26 |
| 399 | 5 | 13 | 18 | 35 | 916 | 245 | 194 | 197 | 5.09$E$ + 24 | 5.79$E$ + 26 |
| 400 | 5 | 14 | 18 | 35 | 903 | 172 | 189 | 190 | 3.61$E$ + 24 | 4.09$E$ + 26 |
| 401 | 6 | 14 | 18 | 35 | 942 | 246 | 199 | 180 | 3.91$E$ + 24 | 2.29$E$ + 26 |
| 402 | 5 | 14 | 18 | 35 | 905 | 170 | 196 | 194 | 3.57$E$ + 24 | 4.04$E$ + 26 |
| 403 | 5 | 15 | 18 | 35 | 900 | 170 | 200 | 193 | 3.58$E$ + 24 | 4.05$E$ + 26 |
| 404 | 5 | 14 | 18 | 35 | 910 | 244 | 191 | 186 | 5.10$E$ + 24 | 5.79$E$ + 26 |
| 405 | 6 | 14 | 19 | 35 | 923 | 271 | 183 | 190 | 4.38$E$ + 24 | 2.55$E$ + 26 |
| 406 | 5 | 15 | 18 | 35 | 900 | 127 | 200 | 200 | 2.67$E$ + 24 | 3.02$E$ + 26 |
| 407 | 5 | 15 | 18 | 35 | 909 | 181 | 194 | 195 | 3.78$E$ + 24 | 4.29$E$ + 26 |
| 408 | 6 | 14 | 19 | 35 | 969 | 209 | 199 | 200 | 3.24$E$ + 24 | 1.92$E$ + 26 |
| 409 | 5 | 12 | 18 | 35 | 930 | 280 | 169 | 191 | 5.75$E$ + 24 | 6.57$E$ + 26 |
| 410 | 5 | 14 | 18 | 35 | 900 | 176 | 198 | 200 | 3.71$E$ + 24 | 4.19$E$ + 26 |

Table 7: Continued.

| Regime | Age at 1st thinning (years) | Age at 2nd thinning (years) | Age at 3rd thinning (years) | Clear felling age (years) | Initial planting density (stems ha$^{-1}$) | Stems thinned at 1st thinning (stems ha$^{-1}$) | Stems thinned at 2nd thinning (stems ha$^{-1}$) | Stems thinned at 3rd thinning (stems ha$^{-1}$) | Obj1- value prodn | Obj2-volume prodn |
|---|---|---|---|---|---|---|---|---|---|---|
| 411 | 5 | 13 | 18 | 34 | 907 | 259 | 189 | 176 | $9.50E+23$ | $1.08E+26$ |
| 412 | 7 | 14 | 19 | 35 | 901 | 193 | 199 | 180 | $1.70E+24$ | $5.99E+25$ |
| 413 | 5 | 14 | 18 | 35 | 913 | 197 | 195 | 188 | $4.10E+24$ | $4.66E+26$ |
| 414 | 7 | 15 | 18 | 35 | 973 | 185 | 197 | 195 | $2.68E+23$ | $9.67E+24$ |
| 415 | 5 | 12 | 18 | 35 | 900 | 277 | 199 | 200 | $5.85E+24$ | $6.61E+26$ |
| 416 | 5 | 15 | 18 | 35 | 904 | 143 | 199 | 197 | $3.00E+24$ | $3.40E+26$ |
| 417 | 6 | 13 | 18 | 35 | 911 | 268 | 193 | 191 | $4.38E+24$ | $2.54E+26$ |
| 418 | 5 | 13 | 18 | 34 | 902 | 261 | 193 | 191 | $9.62E+23$ | $1.09E+26$ |
| 419 | 5 | 15 | 18 | 35 | 913 | 174 | 198 | 194 | $3.62E+24$ | $4.12E+26$ |
| 420 | 5 | 13 | 18 | 35 | 914 | 253 | 190 | 190 | $5.27E+24$ | $5.99E+26$ |
| 421 | 5 | 14 | 18 | 34 | 908 | 224 | 192 | 184 | $8.20E+23$ | $9.31E+25$ |
| 422 | 5 | 12 | 18 | 35 | 901 | 273 | 190 | 195 | $5.76E+24$ | $6.51E+26$ |
| 423 | 5 | 13 | 18 | 35 | 926 | 176 | 193 | 194 | $3.62E+24$ | $4.13E+26$ |
| 424 | 6 | 14 | 19 | 35 | 915 | 191 | 195 | 194 | $3.11E+24$ | $1.80E+26$ |
| 425 | 5 | 15 | 18 | 35 | 900 | 124 | 200 | 200 | $2.61E+24$ | $2.95E+26$ |
| 426 | 7 | 13 | 18 | 35 | 901 | 220 | 194 | 174 | $1.94E+24$ | $6.83E+25$ |
| 427 | 6 | 14 | 19 | 35 | 900 | 128 | 200 | 200 | $2.11E+24$ | $1.22E+26$ |
| 428 | 5 | 13 | 18 | 35 | 915 | 232 | 196 | 195 | $4.83E+24$ | $5.49E+26$ |
| 429 | 5 | 15 | 18 | 35 | 911 | 154 | 199 | 193 | $3.21E+24$ | $3.65E+26$ |
| 430 | 5 | 13 | 18 | 35 | 901 | 199 | 200 | 186 | $4.19E+24$ | $4.74E+26$ |
| 431 | 5 | 14 | 18 | 35 | 906 | 194 | 196 | 194 | $4.07E+24$ | $4.61E+26$ |
| 432 | 5 | 12 | 18 | 35 | 918 | 276 | 186 | 198 | $5.73E+24$ | $6.52E+26$ |
| 433 | 7 | 14 | 19 | 35 | 923 | 135 | 200 | 200 | $1.17E+24$ | $4.14E+25$ |
| 434 | 5 | 14 | 18 | 35 | 910 | 232 | 189 | 189 | $4.85E+24$ | $5.50E+26$ |
| 435 | 7 | 14 | 19 | 35 | 938 | 211 | 191 | 197 | $1.80E+24$ | $6.42E+25$ |
| 436 | 6 | 14 | 18 | 34 | 935 | 185 | 187 | 188 | $5.17E+23$ | $3.02E+25$ |
| 437 | 5 | 13 | 18 | 35 | 994 | 242 | 188 | 200 | $4.70E+24$ | $5.49E+26$ |
| 438 | 6 | 15 | 18 | 35 | 946 | 108 | 197 | 190 | $2.99E+23$ | $1.75E+25$ |
| 439 | 5 | 15 | 18 | 35 | 911 | 153 | 199 | 197 | $3.19E+24$ | $3.62E+26$ |
| 440 | 6 | 13 | 18 | 35 | 947 | 224 | 199 | 199 | $3.54E+24$ | $2.08E+26$ |
| 441 | 5 | 13 | 18 | 34 | 951 | 195 | 192 | 196 | $6.87E+23$ | $7.91E+25$ |
| 442 | 7 | 13 | 18 | 35 | 901 | 258 | 198 | 163 | $2.28E+24$ | $8.01E+25$ |
| 443 | 5 | 15 | 18 | 35 | 913 | 155 | 194 | 195 | $3.23E+24$ | $3.67E+26$ |
| 444 | 7 | 14 | 19 | 35 | 903 | 161 | 196 | 181 | $1.42E+24$ | $4.99E+25$ |
| 445 | 6 | 14 | 18 | 35 | 919 | 222 | 194 | 188 | $3.60E+24$ | $2.09E+26$ |
| 446 | 5 | 15 | 18 | 35 | 933 | 215 | 199 | 181 | $4.40E+24$ | $5.03E+26$ |
| 447 | 6 | 13 | 18 | 35 | 906 | 206 | 195 | 190 | $3.38E+24$ | $1.95E+26$ |
| 448 | 5 | 15 | 18 | 35 | 909 | 132 | 199 | 200 | $2.76E+24$ | $3.13E+26$ |
| 449 | 5 | 15 | 18 | 35 | 910 | 180 | 193 | 196 | $3.76E+24$ | $4.27E+26$ |
| 450 | 6 | 13 | 18 | 35 | 914 | 280 | 194 | 199 | $4.56E+24$ | $2.65E+26$ |
| 451 | 5 | 12 | 18 | 35 | 953 | 284 | 200 | 194 | $5.72E+24$ | $6.58E+26$ |
| 452 | 5 | 13 | 18 | 35 | 935 | 241 | 191 | 191 | $4.92E+24$ | $5.64E+26$ |
| 453 | 6 | 15 | 18 | 35 | 904 | 176 | 194 | 194 | $2.89E+24$ | $1.67E+26$ |
| 454 | 6 | 14 | 19 | 35 | 931 | 213 | 187 | 191 | $3.41E+24$ | $1.99E+26$ |
| 455 | 5 | 15 | 18 | 35 | 900 | 123 | 200 | 200 | $2.59E+24$ | $2.93E+26$ |

Table 7: Continued.

| Regime | Age at 1st thinning (years) | Age at 2nd thinning (years) | Age at 3rd thinning (years) | Clear felling age (years) | Initial planting density (stems ha$^{-1}$) | Stems thinned at 1st thinning (stems ha$^{-1}$) | Stems thinned at 2nd thinning (stems ha$^{-1}$) | Stems thinned at 3rd thinning (stems ha$^{-1}$) | Obj1- value prodn | Obj2-volume prodn |
|---|---|---|---|---|---|---|---|---|---|---|
| 456 | 5 | 12 | 18 | 35 | 946 | 265 | 200 | 194 | 5.37$E+24$ | 6.17$E+26$ |
| 457 | 5 | 13 | 18 | 35 | 918 | 243 | 189 | 195 | 5.04$E+24$ | 5.74$E+26$ |
| 458 | 5 | 14 | 18 | 35 | 935 | 216 | 187 | 188 | 4.41$E+24$ | 5.05$E+26$ |
| 459 | 5 | 13 | 18 | 35 | 920 | 253 | 174 | 197 | 5.24$E+24$ | 5.97$E+26$ |
| 460 | 5 | 13 | 18 | 35 | 929 | 237 | 200 | 195 | 4.87$E+24$ | 5.56$E+26$ |
| 461 | 5 | 15 | 18 | 35 | 902 | 136 | 182 | 200 | 5.00$E+23$ | 5.66$E+25$ |
| 462 | 5 | 15 | 18 | 35 | 904 | 135 | 198 | 198 | 2.83$E+24$ | 3.21$E+26$ |
| 463 | 5 | 13 | 18 | 35 | 911 | 210 | 197 | 194 | 4.38$E+24$ | 4.98$E+26$ |
| 464 | 5 | 13 | 18 | 35 | 920 | 229 | 196 | 192 | 4.74$E+24$ | 5.40$E+26$ |
| 465 | 5 | 12 | 18 | 35 | 919 | 292 | 185 | 200 | 6.06$E+24$ | 6.90$E+26$ |
| 466 | 5 | 13 | 18 | 35 | 913 | 260 | 192 | 192 | 5.42$E+24$ | 6.16$E+26$ |
| 467 | 6 | 14 | 19 | 35 | 921 | 164 | 197 | 196 | 2.65$E+24$ | 1.54$E+26$ |
| 468 | 6 | 15 | 18 | 35 | 901 | 133 | 199 | 198 | 2.19$E+24$ | 1.26$E+26$ |
| 469 | 5 | 14 | 18 | 35 | 915 | 194 | 191 | 196 | 4.03$E+24$ | 4.59$E+26$ |
| 470 | 5 | 13 | 18 | 35 | 921 | 240 | 199 | 193 | 4.97$E+24$ | 5.66$E+26$ |
| 471 | 5 | 14 | 18 | 35 | 906 | 218 | 194 | 197 | 4.57$E+24$ | 5.18$E+26$ |
| 472 | 7 | 13 | 18 | 35 | 901 | 249 | 197 | 168 | 2.20$E+24$ | 7.73$E+25$ |
| 473 | 6 | 14 | 19 | 35 | 916 | 184 | 189 | 197 | 2.99$E+24$ | 1.73$E+26$ |
| 474 | 6 | 15 | 18 | 35 | 906 | 141 | 196 | 197 | 2.31$E+24$ | 1.34$E+26$ |
| 475 | 6 | 14 | 19 | 34 | 901 | 142 | 192 | 200 | 4.09$E+23$ | 2.36$E+25$ |
| 476 | 6 | 15 | 18 | 35 | 952 | 126 | 194 | 187 | 3.47$E+23$ | 2.04$E+25$ |
| 477 | 6 | 14 | 19 | 35 | 903 | 155 | 195 | 190 | 2.55$E+24$ | 1.47$E+26$ |
| 478 | 5 | 15 | 18 | 35 | 913 | 146 | 196 | 197 | 3.04$E+24$ | 3.45$E+26$ |
| 479 | 6 | 14 | 19 | 35 | 906 | 164 | 185 | 193 | 2.69$E+24$ | 1.55$E+26$ |
| 480 | 5 | 13 | 18 | 35 | 969 | 274 | 200 | 200 | 5.44$E+24$ | 6.30$E+26$ |
| 481 | 5 | 14 | 18 | 35 | 909 | 212 | 193 | 187 | 4.43$E+24$ | 5.03$E+26$ |
| 482 | 5 | 12 | 18 | 35 | 936 | 285 | 192 | 187 | 5.83$E+24$ | 6.67$E+26$ |
| 483 | 5 | 14 | 18 | 35 | 903 | 211 | 191 | 192 | 4.44$E+24$ | 5.02$E+26$ |
| 484 | 5 | 14 | 18 | 35 | 904 | 174 | 193 | 192 | 3.65$E+24$ | 4.14$E+26$ |
| 485 | 5 | 13 | 18 | 35 | 1001 | 263 | 190 | 200 | 5.08$E+24$ | 5.95$E+26$ |
| 486 | 5 | 12 | 18 | 35 | 903 | 292 | 187 | 196 | 6.15$E+24$ | 6.96$E+26$ |
| 487 | 5 | 15 | 18 | 35 | 912 | 181 | 194 | 195 | 3.77$E+24$ | 4.28$E+26$ |
| 488 | 5 | 12 | 18 | 35 | 900 | 284 | 189 | 184 | 6.00$E+24$ | 6.78$E+26$ |
| 489 | 6 | 14 | 19 | 35 | 903 | 175 | 198 | 198 | 2.88$E+24$ | 1.66$E+26$ |
| 490 | 5 | 15 | 18 | 35 | 900 | 171 | 200 | 193 | 3.60$E+24$ | 4.07$E+26$ |
| 491 | 6 | 14 | 19 | 35 | 906 | 186 | 186 | 194 | 3.05$E+24$ | 1.76$E+26$ |
| 492 | 5 | 15 | 18 | 35 | 908 | 137 | 199 | 198 | 2.86$E+24$ | 3.25$E+26$ |
| 493 | 5 | 14 | 18 | 35 | 910 | 197 | 190 | 192 | 4.11$E+24$ | 4.67$E+26$ |
| 494 | 5 | 13 | 18 | 35 | 912 | 272 | 185 | 187 | 5.68$E+24$ | 6.45$E+26$ |
| 495 | 5 | 12 | 18 | 35 | 906 | 300 | 181 | 197 | 6.30$E+24$ | 7.14$E+26$ |
| 496 | 6 | 14 | 19 | 35 | 900 | 212 | 197 | 195 | 3.50$E+24$ | 2.02$E+26$ |
| 497 | 6 | 14 | 18 | 35 | 916 | 228 | 193 | 182 | 3.71$E+24$ | 2.15$E+26$ |
| 498 | 7 | 14 | 19 | 35 | 929 | 150 | 198 | 197 | 1.29$E+24$ | 4.58$E+25$ |
| 499 | 5 | 13 | 18 | 35 | 913 | 274 | 149 | 190 | 5.71$E+24$ | 6.49$E+26$ |
| 500 | 5 | 13 | 18 | 35 | 902 | 226 | 200 | 193 | 4.76$E+24$ | 5.38$E+26$ |

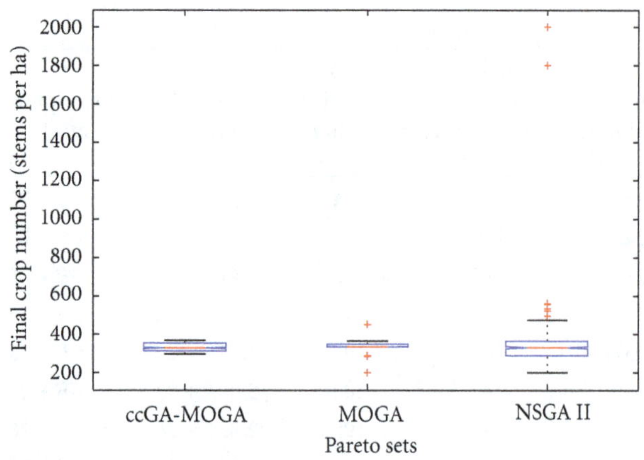


FIGURE 8: Final crop number box plots from the ccGA-MOGA, MOGA, and NSGA II Pareto sets.

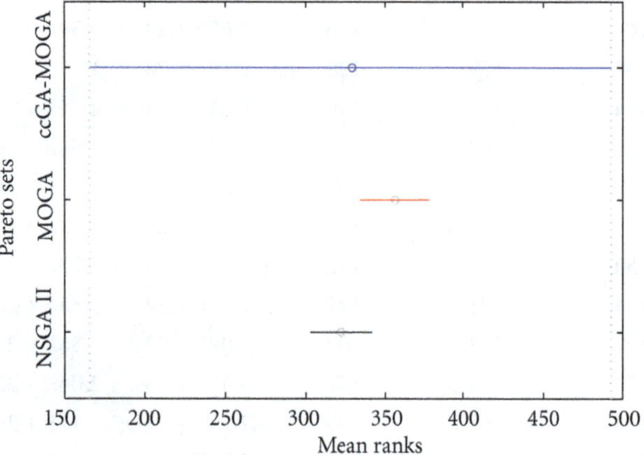


FIGURE 9: Wilcoxon rank sum test for the final crop numbers from the ccGA-MOGA, MOGA, and NSGA II Pareto sets.

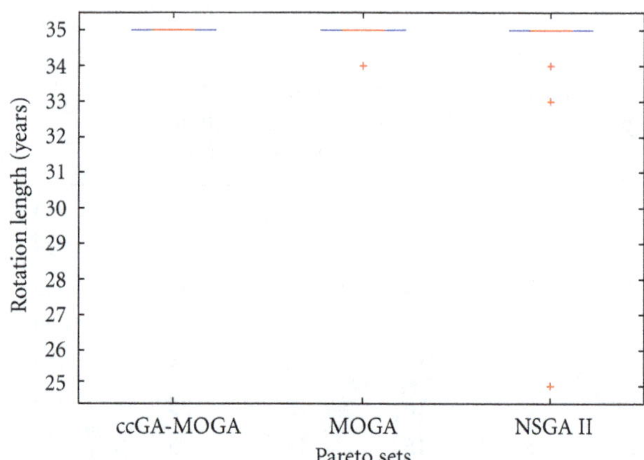


FIGURE 10: Rotation length box plots from the ccGA-MOGA, MOGA, and NSGA II Pareto sets.

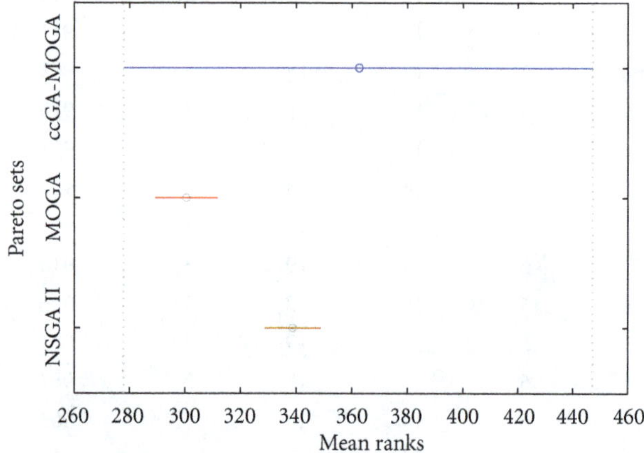


FIGURE 11: Wilcoxon rank sum test for the rotation lengths from the ccGA-MOGA, MOGA and NSGA II Pareto sets.

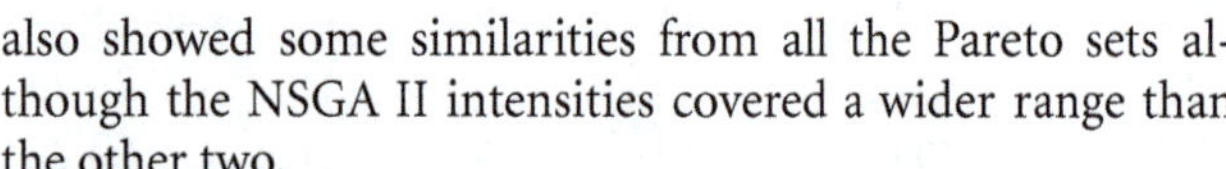

also showed some similarities from all the Pareto sets although the NSGA II intensities covered a wider range than the other two.

The timings for the second thinnings were similar for the MOGA and NSGA II, whereas the intensities from all the Pareto sets were also the same, with NSGA II, again, showing a wider range of intensities.

The timings for the third thinnings showed dissimilarities between MOGA and NSGA II but showing thinning intensity similarities from all the three Pareto sets, with both MOGA and NSGA II showing a wider range of thinning intensities.

## 5. Discussion

The previous results suggest that the nondominated solutions from the three Pareto sets belong to the same family of solutions. It is clear that the Pareto front is disjointed with many topologies and the differences observed in the decision variables (i.e., initial planting stocking; final crop number; rotation length; frequency, timing, and intensity of thinning) are mere reflections of the different topologies of the Pareto front or sub-domains (niches) of the nondominated solutions. It also seems that the computational efficiency of MOGA or ccGA-MOGA is due to the fact that the Fonseca and Fleming ranking scheme attempts to only predict part of the Pareto front, whereas NSGA II seems to approximate all the Pareto points, hence the differences in the turnaround times as shown in Table 3. This is despite employing parallel computing capability for the MOGA and NSGA II algorithms. It is also important to note that, with just an Intel Core 2 Duo processor, it may not be possible to experience appreciable improvements in the computational efficiencies.

The question that lingers with the Fonseca and Fleming ranking scheme is how it determines the subdomain of the Pareto front to approximate and whether it is possible for the user to have control on which subdomain of the Pareto front to estimate. The same question can be raised about the NSGA II algorithm, whether it can be made to concentrate on estimating Pareto points on a specific subdomain rather than all the Pareto points, making the algorithm more computationally efficient. Although the answer to the above question is beyond the scope of this paper, it does seem obvious that MOGA and NSGA II can be applied for different output and computational requirements. Where computational efficiency is of paramount importance, MOGA would be the appropriate choice but at the expense of approximating only a part of the Pareto front. The user will also not have the ability to choose the subregion on the Pareto front to estimate. Situations with large problems characterised by many objectives (greater than three) and a large set of decision variables may benefit from the application of MOGA. However, when it is vitally important to approximate the entire Pareto front, NSGA II would be ideal, but with a computational overhead. Note also that most research in evolutionary algorithms has centered on dealing with 2 or 3 objective functions and beyond that is still an area of active research [39].

A larger set of nondominated solutions from the search space implies greater flexibility in decision making in sifting through a plethora of alternatives based on prevailing preferences and values of the decision maker(s). From a silvicultural management perspective, this is good news for the forester, as such flexibility at a stand level facilitates implementing a mosaic of thinning regimes across an estate to meet sawlog and pulpwood demands from markets that are plagued with uncertainty.

The ability of NSGA II to maintain diversity in the search space to approximate an entire Pareto front lies in the way it does its fitness assignment (through fitness sharing) as alluded to earlier. The crowding model in NSGA II has been demonstrated to work, which encouraged a good spread of solutions over all the subdomains or topologies of the Pareto front. Although MOGA also has the ability to form niches and distribute the population over the Pareto-optimal region, its fitness sharing is done in the objective function domain. This maintains diversity in the objective function values and may not necessarily do so for the decision variables, which are important to the decision maker(s) [16]. An attempt to understand this further led to running MOGA and NSGA II models over 1000 generations. MOGA converged to a single point on the Pareto front, whereas NSGA II kept all its 500 Pareto points. As for the ccGA-MOGA, which is just a modified MOGA with 5 subpopulations linked through migration instead of one large population [2], diversity was maintained through the subpopulations [5] and convergence to a single Pareto point after 1000 generations was avoided. To construct a better MOGA that will not converge to a single point on the Pareto front, several niching and elitist models have to be considered, and Obayashi et al. [40] reviewed some of these models.

At this stage there is no similar study in this same problem domain of optimal silvicultural regimes that can be compared against the results obtained here. However, there is work in other disciplines that compare well with some of the findings in this paper.

(a) Ducheyne et al. [41] found out NSGA II to produce better results than MOGA although the latter had a more evenly spread-out Pareto front, for a biobjective forest estate management problem where the estate consisted of 295 stands (or management units). The problem was run for only 50 generations, leaving questions as to whether the results would have been any different if the problem was run over more generations. The Pareto front was also smooth and continuous, a sharp contrast to the one in this paper;

(b) Zitzler et al. [15] provided a systematic comparison of various approaches to biobjective function optimisation using carefully chosen test functions. MOGA and the first-generation NSGA were among some of these approaches. Some of the test functions produced smooth well-distributed nondominated sets and the most difficult ones, disjointed Pareto fronts. NSGA was ranked better than MOGA for performance.

(c) Sağ and Çunkaş [42] also compared various approaches that included NSGA, NSGA II, MOGA, VEGA, SPEA, SPEA2 (second-generation SPEA), NPGA, and PESA, using test functions for multiobjective optimisation. Their general findings were that while the NSGA-II, SPEA2, and PESA could keep track of all the feasible solutions found during the optimisation, the number of Pareto-optimal solutions of other algorithms remained poor. Focusing on a more difficult problem that had two objective functions and two constraints where the Pareto set was discontinuous and divided into 5 regions, NSGA-II, SPEA2 and PESA kept track of all the feasible solutions. MOGA was poor at finding points at the extreme ends of the curve giving a good sampling of solutions at the midsection of the curve. However, the number of Pareto-optimal solutions remained poor for MOGA.

This paper has attempted to provide an insight into some of the well-established algorithms for estimating MOEA Pareto fronts at a stand level. The intention was not to minimise or disregard the important work already done in other disciplines. Instead, the aim was to bring forest analysts close to the MOEA community so that both can interact and mutually benefit. In the near future, if forest analysts take on board some of the findings presented in this paper, then the goals of this paper will have been accomplished.

## 6. Conclusion

The results clearly indicate that, for the stand-level optimal thinning regime problem, the NSGA II is capable of approximating an entire Pareto front at the expense of computational efficiency and that MOGA and ccGA-MOGA, though computationally efficient, will crowd solutions and also estimate only a part of the Pareto front. The key to NSGA II wider coverage in estimating the entire Pareto front lies in its ability of information sharing across individual potential solutions and maintaining diversity in the search space in collectively satisfying the two objectives, sawlog production and pulpwood production. For a better MOGA that will approximate an entire Pareto front, a niching and elitist model will have to be included that will provide a means for controlling the number, location, extent, and distribution of solutions in the search space. In its current form, MOGA suffers from genetic drift and will converge to a single point on the Pareto front if run for many generations. However, with the island model version, ccGA-MOGA, where the population is divided into subpopulations, genetic drift may be minimised although not sufficiently enough to avert estimation for only a subregion of the Pareto front. The final take-home message here is that, for a MOEA, it is important to choose an approach fit for purpose as in finding the appropriate balance between computational efficiency and the extent of the Pareto front to be estimated.

## Acknowledgments

This project was funded by the New Zealand Foundation for Research Science and Technology under Contract no. C04X0806, "Protecting and Enhancing the Environment through Forestry," and the contribution from Future Forest Research Ltd is gratefully acknowledged. The author is also thankful to Dr. Peter Clinton (Scion) and Dr. Ruth Falshaw (Scion) for their final comments on this paper.

## References

[1] O. Chikumbo, R. H. Bradbury, and S. Davey, "Large-scale ecosystem management as a complex systems problem: multi-objective optimisation with spatial constraints," in *Applied Complexity—From Neural Nets to Managed Landscapes*, S. Halloy and T. Williams, Eds., pp. 124–155, Institute for Crop & Food Research Ltd, Christchurch, New Zealand, 2000.

[2] O. Chikumbo and I. Nicholas, "Efficient thinning regimes for *Eucalyptus fastigata*: multi-objective stand-level optimisation using the island model genetic algorithm," *Ecological Modelling*, vol. 222, no. 10, pp. 1683–1695, 2011.

[3] M. J. Osborne and A. Rubenstein, *A Course in Game Theory (p7)*, MIT Press, Cambridge, MA, USA, 1994.

[4] P. Siarry, A. Pétrowski, and M. Bessaou, "A multipopulation genetic algorithm aimed at multimodal optimization," *Advances in Engineering Software*, vol. 33, no. 4, pp. 207–213, 2002.

[5] O. Chikumbo, "Exploration and exploitation in function optimization using stochastic generate-and-test algorithms," in *Proceedings of the International Conference on Genetic and Evolutionary Methods*, H. R. Arabnia and A. M. G. Solo, Eds., pp. 22–27, CSREA Press, Las Vegas, Nev, USA, July 2009.

[6] C. M. Fonseca and P. J. Fleming, "Genetic algorithms for multiple objective optimization: formulation, discussion and generalization," in *Proceedings of the 5th International Conference on Genetics Algorithms*, S. Forrest, Ed., Morgan Kaufmann Publishers, San Mateo, Calif, USA, 1993.

[7] L. T. Bui, J. Branke, and H. A. Abbass, "Multi-objective optimisation for dynamic environments," in *Proceedings of the Congress on Evolutionary Computation*, pp. 2349–2356, IEEE Press, Edinburgh, UK, 2005.

[8] J. D. Schaffer, "Multiple objective optimisation with vector evaluated genetic algorithms," in *Genetic Algorithms and their Applications: Proceedings of the 1st International Conference on Genetic Algorithms*, L. Erlbaum, Ed., pp. 93–100, 1985.

[9] H. Polheim, "GEATbx: introduction, evolutionary algorithms: overview, methods and operators," 2006, http://www.geatbx.com/.

[10] J. H. Holland, *Adaptation in Natural and Artificial Systems*, The University of Michigan Press, Ann Arbor, Mich, USA, 1975.

[11] I. Rechenberg, *Evolutionsstrategie—Optimierung Technischer Systeme Nach Prinzipien der Biologischen Evolution*, Frommann-Holzboog, Stuttgart, Germany, 1973.

[12] H.-P. Schwefel, *Numerical Optimisation of Computer Models*, John Wiley & Sons, Chichester, UK, 1981.

[13] L. J. Fogel, A. J. Owens, and M. J. Walsh, "Intelligent decision making through a simulation of evolution," *Behavioral Science*, vol. 11, no. 4, pp. 253–272, 1966.

[14] C. A. Coello Coello, "20 Years of evolutionary multi-objective optimization: what has been done and what remains to be done," in *Computational Intelligence: Principles and Practice*, G. Y. Yen and D. B. Fogel, Eds., pp. 73–78, IEEE Computational Intelligence Society, 2006.

[15] E. Zitzler, K. Deb, and L. Thiele, "Comparison of multiobjective evolutionary algorithms: empirical results," *Evolutionary Computation*, vol. 8, no. 2, pp. 173–195, 2000.

[16] N. Srinivas and K. Deb, "Multi-objective optimisation using non-dominated sorting in genetic algorithms," *Evolutionary Computation*, vol. 2, no. 3, pp. 221–248, 1994.

[17] F. Y. Edgeworth, *Mathematical Psychics*, P. Keagan, London, UK, 1881.

[18] V. Pareto, *Cours D'Economie Politique*, vol. 1-2, F. Rouge, Lausanne, Switzerland, 1896.

[19] P. Hajela and C. Y. Lin, "Genetic search strategies in multicriterion optimal design," *Structural Optimization*, vol. 4, no. 2, pp. 99–107, 1992.

[20] J. Horn, N. Nafpliotis, and D. E. Goldberg, "A niched Pareto genetic algorithm for multi-objective optimisation," in *Proceedings of the 1st IEEE Conference on Evolutionary Computation, IEEE World Congress on Computational Computation*, vol. 1, pp. 82–87, IEEE Services Centre, Piscataway, NJ, USA, 1994.

[21] C. A. Coello Coello, "Using the min-max method to solve multi-objective optimization problems with genetic algorithms," in *Proceedings of the 6th Ibero-American Conference on AI: Progress in Artificial Intelligence (IBERAMIA '98)*, pp. 303–313, Springer, London, UK, 1998.

[22] E. Zitzler and L. Thiele, "An evolutionary algorithm for multi-objective optimization: the strength Pareto approach," Tech. Rep. 43, Computer Engineering and Networks Laboratory, Swiss Federal Institute of Technology, Zurich, 1998.

[23] D. W. Corne, J. D. Knowles, and M. J. Oates, "The Pareto envelope-based selection Algorithm for multi-objective optimisation," in *Proceedings of the Parallel Problem Solving from Nature 6th Conference*, M. Schoenauer, K. Deb, G. Rudolph, X. Yao, E. Lutton, and J. J. Merelo, Eds., pp. 839–849, Paris, France, 2000.

[24] D. J. Knowles and D. Corne, "The Pareto archived evolution strategy: a new baseline algorithm for Pareto multi-objective optimisation," in *Proceedings of the Congress on Evolutionary Computation (CEC '99)*, P. J. Angeline, Z. Michalewicz, M. Schoenauer, X. Yao, and A. Zalzala, Eds., vol. 1, pp. 98–105, 1999.

[25] J. A. Nelder, "New kinds of systematic designs for spacing experiments," *Biometrics*, vol. 18, pp. 283–307, 1962.

[26] J. H. Mayo and T. J. Straka, "The holding value premium in standing timber valuation," *Appraisal Journal*, vol. 37, no. 1, pp. 98–106, 2005.

[27] L. Ljung, *System Identification: Theory for the User*, Prentice Hall, Upper Saddle River, NJ, USA, 1987.

[28] O. Chikumbo, I. M. Y. Mareels, and B. J. Turner, "Predicting stand basal area in thined stands using a dynamical model,"

*Forest Ecology and Management*, vol. 116, no. 1–3, pp. 175–187, 1999.

[29] O. Chikumbo, "An optimal regime model using competitive co-evolutionary genetic algorithms," in *Proceedings of the International Joint Conference on Computational Intelligence*, A. Dourado, A. Rosa, and K. Madani, Eds., pp. 210–217, Institute for Systems and Technologies of Information, Control and Communication (ISTICC), Madeira, Portugal, October 2009.

[30] D. E. Goldberg, *Genetic Algorithms in Search, Optimisation and Machine Learning*, Addison-Wesley, Reading, Mass, USA, 1989.

[31] D. E. Goldberg and J. Richardson, "Genetic algorithms with sharing for multimodal function optimization," in *Proceedings of the 2nd International Conference on Genetic Algorithms and their Applications*, J. Grefenstette, Ed., pp. 41–49, Lawrence Erlbaum Associates, Cambridge, Mass, USA, 1987.

[32] C. A. Coello Coello, *An empirical study of evolutionary techniques for multi-objective optimization in engineering design*, Ph.D. thesis, Tulane University, New Orleans, La, USA, 1996.

[33] C. A. Coello Coello, "A short tutorial on evolutionary multiobjective optimisation," in *Proceedings of the 1st International Conference on Evolutionary Multi-Criterion Optimisation*, E. Zitzler, K. Deb, L. Thiele, C. A. Coello Coello, and D. Corne, Eds., Lecture Notes in Computer Science No. 1993, pp. 21–40, Springer, 2001.

[34] K. Deb, A. Pratap, S. Agarwal, and T. Meyarivan, "A fast and elitist multiobjective genetic algorithm: NSGA-II," *IEEE Transactions on Evolutionary Computation*, vol. 6, no. 2, pp. 182–197, 2002.

[35] MathWorks, 2011, http://www.mathworks.com/help/techdoc/index.html.

[36] O. Chikumbo and I. M. Y. Mareels, "Predicting terminal time and final crop number for a forest plantation stand: Pontryagin's maximum principle," in *Ecosystems and Sustainable Development*, E. Tiezzi, C. A. Brebbia, and J. L. Uso, Eds., vol. 2 of *ISBN 1-85312-834-X*, pp. 1227–1237, WIT Press, 2003.

[37] W. Kruskal and W. A. Wallis, "Use of ranks in one-criterion analysis of variance," *Journal of the American Statistical Association*, vol. 47, pp. 583–621, 1952.

[38] J. D. Gibbons, *Nonparametric Statistical Inference*, Marcel Dekker, New York, NY, USA, 2nd edition, 1985.

[39] J. Wallenius, J. S. Dyer, P. C. Fishburn, R. E. Steuer, S. Zionts, and K. Deb, "Multiple criteria decision making, multiattribute utility theory: recent accomplishments and what lies ahead," *Management Science*, vol. 54, no. 7, pp. 1336–1349, 2008.

[40] S. Obayashi, S. Takahashi, and Y. Takeguchi, "iching and elitist models for MOGAs," in *Proceedings of the 5th International Conference Parallel Problem Solving From Nature (PPSN '98)*, A. E. Eiben, M. Schoenauer, and H. P. Schwefel, Eds., pp. 260–269, 1998.

[41] E. Ducheyne, R. R. De Wulf, and B. De Baets, "Bi-objective genetic algorithms for forest management planning: a comparative study," in *Proceedings of the Genetic and Evolutionary Computation Conference*, L. Spector, E. D. Goodman, A. Wu et al., Eds., pp. 63–66, Morgan Kaufmann Publishers, July 2001.

[42] T. Sağ and M. Çunkaş, "A tool for multiobjective evolutionary algorithms," *Advances in Engineering Software*, vol. 40, no. 9, pp. 902–912, 2009.

# 10

# Introduction of *Eucalyptus* spp. into the United States with Special Emphasis on the Southern United States

**R. C. Kellison,[1] Russ Lea,[1,2] and Paul Marsh[3]**

[1] *North Carolina State University, Raleigh, NC, USA*
[2] *CEO NEON, Inc., Boulder, CO, USA*
[3] *1433 Lutz Avenue, Raleigh, NC 27607, USA*

Correspondence should be addressed to R. C. Kellison; bkelliso@bellsouth.net

Academic Editor: John Stanturf

Introduction of *Eucalyptus* spp. into the United States from Australia on a significant scale resulted from the gold rush into California in 1849. Numerous species were evaluated for fuel, wood products, and amenity purposes. The first recorded entry of eucalyptus into the southern United Stated was in 1878. Subsequent performance of selected species for ornamental purposes caused forest industry to visualize plantations for fiber production. That interest led the Florida Forestry Foundation to initiate species-introduction trials in 1959. The results were sufficiently promising that a contingent of forest products companies formed a cooperative to work with the USDA Forest Service, Lehigh Acres, FL, USA, on genetic improvement of selected species for fiber production. The Florida initiative caused other industrial forestry companies in the upper South to establish plantations regardless of the species or seed source. The result was invariably the same: failure. Bruce Zobel, Professor of Forestry, North Carolina State University, initiated a concerted effort to assess the potential worth of eucalyptus for plantation use. The joint industrial effort evaluated 569 sources representing 103 species over a 14-year period. The three levels of testing, screening, in-depth, and semioperational trials led to identification of some species and sources that offered promise for adaptation, but severe winter temperatures in late 1983 and early 1984 and 1985 terminated the project. Despite the failed attempt valuable silvicultural practices were ascertained that will be beneficial to other researchers and practitioners when attempts are again made to introduce the species complex into the US South.

## 1. Introduction

More than 500 *Eucalyptus* spp. (Myrtaceae) are indigenous to Australia and the bordering islands of Polynesia [1]. They occur in environments from 10°N to 44°S latitude (Mindanao Island, Philippines through Tasmania, Australia), from sea level to 2000 meters elevation (snow line) and from 10 (Northern Territory, Australia) to 375 centimeters of rainfall (Papua New Guinea). These vast differences in climate have allowed a great diversity to develop within the *Eucalyptus* genus. The inherent diversity has resulted in successful introduction of many of the species, for landscape, fuelwood and timber purposes, to areas within the tropical, subtropical, and warm temperate zones of the world [2].

As with other plants and animals, introduction of eucalypts to areas of the world where they are not indigenous sometimes allows for performance that is greatly superior to that exhibited in their native habitat. Reasons for the differences in performance include favorable climatic and edaphic conditions and the general lack of pests in the new environment. Notable examples of successful species introductions include *E. grandis*, *E. urophylla*, and their hybrid (Brazil, Colombia, Venezuela, Republic of Congo, Zimbabwe, South Africa), *E. globulus* (Chile, Portugal, Spain, southern California (USA)), *E. camaldulensis* (Israel, Nigeria, Morocco, India, Northern California (USA)), and *E. viminalis* (Argentina, Brazil, Georgia (formerly part of USSR)) [2].

Long before the species generated so much enthusiasm for plantation forestry in parts of the world, other than North America, attempts were made to introduce selected species into California. The occasion was the gold rush of 1849. The influx of a half million people resulted in a shortage of foodstuff and supplies essential for survival and development (http://ceres.ca.gov/ceres/calweb/geology/goldrush.html).

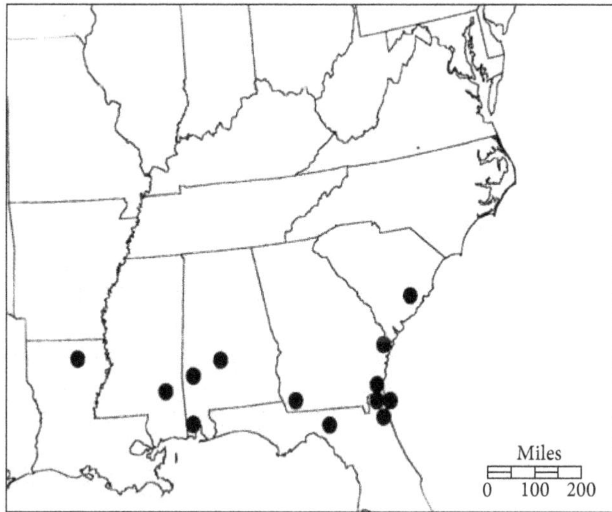


FIGURE 1: Locations for Hardwood Research Cooperative's eucalyptus evaluation trials.

This being the time before completion of the transcontinental railroad with its terminus in California [3] trade was established with offshore countries, especially Australia, New Zealand, and Chile. Wheat (*Triticum* spp.) for the making of flour was high on the list of imports. Schooners were soon plying the waters between domestic and foreign ports for the exchange of goods and services [4]. That trade ostensibly allowed for the exchange of seeds from forest tree species, which eventually accounted for extensive plantations of Monterey pine (*Pinus radiata*) in the three countries and for the introduction of eucalyptus species into California as well as into other parts of the world.

Interest in eucalyptus wood in California resulted from observations that the schooners from Australia were far superior to those manufactured from native wood in the United States. Inquiry about the wood source identified blue gum (*E. globulus*) as the favored species. That resulted in test planting of blue gum and other eucalyptus species from seedlings provided by the Golden Gate Nursery in San Francisco as early as 1853 [5]. The performance of a selected lot of those species so impressed landowners and wood users that additional nurseries joined the fray. By 1870, the State Board of Agriculture was espousing the need for artificial forests, with special emphasis on eucalyptus to cover areas of barren terrain and to produce wood for the manufacture of wagons, carriages, and agriculture implements. A few years later the Central Pacific Railroad was championing the need for eucalyptus timber for railroad ties. Furniture manufacturers visualized the added source of wood for their needs. Others envisioned the resource for desirable urban shade trees, markets for ointments and scented products from the processing of the odiferous eucalyptus leaves, and honey from bee pollination of the nectar-rich flowers [6, 7].

The State Board of Forestry became involved in the eucalyptus introduction program by 1887. That organization established experiment stations at six locations (Santa Monica, Chico, Merced, Hesperia, San Jacinto, and Lake Hemet) within the state. The primary purpose of the stations was to urge landowners in the respective areas to establish plantations of selected eucalyptus species. A few years later the State Board of Forestry was disbanded and the experiment stations and related activities became the responsibility of the College of Agriculture, University of California at Berkeley [5]. Planting of eucalyptus for its many potential uses resulted in a major expansion program during the 1905–1912 era. The residuals of that program are the large trees that are today scattered across the landscape of the Golden State.

TABLE 1

| Code | Cold damage |
|---|---|
| 0 | Destroyed |
| 1 | 0–25% crown functional |
| 2 | 25–50% crown functional |
| 3 | 50–75% crown functional |
| 4 | Greater than 75% crown functional |
| 5 | Functionally undamaged |
| 9 | Missing or dead before freeze |

As with any new venture with promises, the ills of Pandora's Box came slithering out. And so it was with the wonders of eucalyptus. The railroad companies learned that the ties made from eucalyptus wood were unacceptable because the spikes needed for holding the rails in place lost their grip. Similarly the furniture manufacturers and ship builders found that instability of the fast-growing timber was a bane to their needs. Even others visualized environmental constraints on the landscape from the escape of the "noxious weed" (http://www.columbia.edu/itc/cerc/danoff-burg/invasion_bio/inv_spp_summ/Eucalyptus_globulus.html), depletion of soil nutrients and ground water, and housing of insects and diseases that would be detrimental to native forests. Those limitations, both real and imagined, largely resulted in a hiatus of eucalyptus plantations for decades [5].

With the advent of new forest products and energy needs, interest was renewed in the introduction of eucalyptus species

Table 2: Sample output from the simulator of the North Carolina State University Hardwood Cooperative eucalypts seed source evaluation program at 24 months from planting.

| Seed source[a] | Survival (%) | | Height (ft.) | | DBH (in.) | | Cold-hardiness | | Bole form[b] | | Index score[c] |
|---|---|---|---|---|---|---|---|---|---|---|---|
| | Mean % | Std. err. | Mean | Std. err. | Mean | Std. err. | Mean | Std. err. | Mean | Std. err. | |
| DAL008 | 30 | 0.01 | 12.3 | 0.09 | 1.6 | 0.02 | 66 | 1.46 | 3.1 | 0.04 | 44.2 |
| MAC017 | 30 | 0.01 | 9.1 | 0.08 | 0.9 | 0.01 | 58 | 1.22 | 4.0 | 0.04 | 40.6 |
| DAL007 | 30 | 0.01 | 7.2 | 0.08 | 0.6 | 0.01 | 47 | 1.05 | 2.1 | 0.05 | 34.7 |
| RUB003 | 30 | 0.01 | 6.3 | 0.04 | 0.4 | 0.01 | 49 | 1.77 | 2.8 | 0.06 | 34.6 |
| CAP014 | 30 | 0.01 | 4.5 | 0.04 | 0.2 | 0.01 | 69 | 1.41 | 3.2 | 0.04 | 33.2 |
| VIM078 | 30 | 0.01 | 5.6 | 0.38 | 0.5 | 0.02 | 49 | 3.47 | 2.7 | 0.19 | 31.8 |
| BRI001 | 30 | 0.01 | 5.2 | 0.05 | 0.4 | 0.01 | 51 | 1.69 | 1.8 | 0.05 | 30.5 |
| RUB004 | 30 | 0.01 | 4.8 | 0.84 | 0.7 | 0.02 | 21 | 4.82 | 1.0 | 0.19 | 26.5 |

[a]See Table 7 for species code.
[b]Based on a score of 0 to 5 with the higher score being more desirable.
[c]Higher score being more desirable.

Table 3: Comparison of actual versus predicted development—age 6 months.

| Seed source | Survival (%) | | Mean height (m) | |
|---|---|---|---|---|
| | Actual | Predicted | Actual | Predicted |
| VIM078 | 84 | 85 | 0.97 | 0.60 |
| DAL008 | 93 | 91 | 1.09 | 0.77 |
| NOV012 | 86 | 91 | 0.65 | 0.49 |
| MAC017 | 87 | 77 | 0.73 | 0.69 |
| BRI001 | 89 | 79 | 0.69 | 0.49 |
| DAL007 | 90 | 89 | 0.57 | 0.57 |
| RUB004 | 89 | 80 | 0.73 | 0.77 |
| CAP014 | 71 | 87 | 0.57 | 0.40 |
| RUB003 | 77 | 84 | 0.65 | 0.57 |

Table 4: Comparison of actual versus predicted development—age 12 months.

| Seed source | Survival (%) | | Mean height (m) | |
|---|---|---|---|---|
| | Actual | Predicted | Actual | Predicted |
| DAL008 | 72 | 72 | 1.98 | 1.86 |
| NOV012 | 67 | 72 | 0.97 | 1.25 |
| VIM078 | 62 | 60 | 0.73 | 1.21 |
| MAC017 | 32 | 48 | 1.66 | 1.54 |
| BRI001 | 53 | 49 | 0.93 | 0.97 |
| DAL007 | 56 | 67 | 1.38 | 1.26 |
| RUB003 | 54 | 56 | 0.89 | 1.17 |
| RUB004 | 38 | 45 | 1.13 | 1.50 |
| CAP014 | 58 | 64 | 0.73 | 0.85 |

Table 5: Comparison of actual versus predicted development—age 24 months.

| Seed source | Survival (%) | | Mean height (m) | |
|---|---|---|---|---|
| | Actual | Predicted | Actual | Predicted |
| VIM078 | 45 | 42 | 2.23 | 2.31 |
| DAL008 | 45 | 48 | 4.17 | 4.66 |
| NOV012 | 58 | 47 | 2.96 | 3.40 |
| MAC017 | 46 | 38 | 2.79 | 3.44 |
| BRI001 | 38 | 33 | 2.02 | 2.02 |
| DAL007 | 36 | 36 | 2.59 | 2.75 |
| RUB004 | 6 | 7 | 2.55 | 2.71 |
| CAP014 | 51 | 42 | 1.62 | 1.74 |
| RUB003 | 27 | 23 | 2.19 | 2.43 |

Table 6

| Source | CSIRO Seed lot no. |
|---|---|
| VIM 077 | 9217 |
| VIM 069 | 8978 |
| VIM 063 | 10726 |
| VIM 061 | 10836 |
| VIM 113 | 11978 |
| CAP 013 | 5317 |
| DAL 008 | 8847 |
| MAC 010 | S10942 |
| MAC 016 | 12084 |

into California after a lapse of nearly half a century. One such event was in the 1950s and 1960s when fiberboard became a commodity in home construction and furniture manufacture. The second event resulted from the oil embargo instituted by the Organization of Petroleum Exporting Countries (OPEC) in the late 1970s [5]. Both of these events proved to be a "flash in the pan." The first one was exposed when it was learned that the manufactured board could be made from wood residue from other manufacturing processes, and the second one took its leave when oil became readily available at relatively low prices. Due to the volatile conditions in the Middle East and the relatively high price of crude oil another effort to grow eucalyptus was undertaken in California in the 1990s, but the effort came with financial and environmental scrutiny. The complications proved greater than the benefits [5].

Table 7: Eucalyptus seed lots tested by species and number of installation in the Lower Coastal Plain of the US South by members of the Hardwood Research Cooperative.

| Species code | Species name | No. of installations by species | No. of seed lots tested |
|---|---|---|---|
| AGL | AGGLOMERATA | 5 | 12 |
| AGR | AGGREGATA | 1 | 1 |
| ALP | ALPINA | 2 | 2 |
| AMP | AMPFOLIA | 4 | 12 |
| AND | ANDREWSII | 2 | 2 |
| BAD | BADJENSIS | 5 | 11 |
| BAN | BANCROFTII | 0 | 0 |
| BIC | BICOSTATA | 7 | 29 |
| BLA | BLAKELYI | 4 | 14 |
| BOT | BOTRYOIDES | 2 | 9 |
| BRI | BRIDGESIANA | 2 | 13 |
| CAA | CAMPANULATA | 4 | 6 |
| CAE | CAMERONII | 1 | 1 |
| CAL | CALIGINOSA | 2 | 2 |
| CAM | CAMALDULENSIS | 7 | 17 |
| CAP | CAMPHORA | 17 | 68 |
| CHA | CHAPMANIANA | 3 | 5 |
| CIN | CINEREA | 5 | 24 |
| CIT | CITRIODORA | 1 | 1 |
| COC | COCCIFERA | 2 | 2 |
| COR | CORDATA | 0 | 0 |
| CRE | CRENULATA | 3 | 16 |
| CYP | CYPELLOCARPA | 5 | 16 |
| DAL | DALRYMPLEANA | 33 | 63 |
| DEA | DEANEI | 2 | 12 |
| DEB | DEBEUZEVILLEI | 2 | 2 |
| DEL | DELEGATENSIS | 10 | 15 |
| DIR | DIVERSICOLOR | 0 | 0 |
| DIV | DIVES | 2 | 2 |
| DRX | DELEGATENSIS X | 1 | 1 |
| DUN | DUNNII | 3 | 15 |
| ELA | ELATA | 2 | 2 |
| EUG | EUGENIODES | 2 | 4 |
| EXP | EXPELLACARPA | 1 | 1 |
| FAS | FASTIGATA | 4 | 7 |
| FRA | FRAXINOIDES | 4 | 6 |
| GLA | GLAUCESCENS | 3 | 11 |
| GLO | GLOBULUS | 1 | 3 |
| GRA | GRANDIS | 1 | 1 |
| GUN | GUNNII | 2 | 7 |
| JOH | JONSTONII | 4 | 10 |
| KAR | KARTZOFFIANA | 1 | 5 |
| KYB | KYBEANENSIS | 1 | 1 |
| LAE | LAEVOPINEA | 4 | 4 |
| LEU | LEUCOXYLON | 2 | 3 |

Table 7: Continued.

| Species code | Species name | No. of installations by species | No. of seed lots tested |
|---|---|---|---|
| MAC | MACARTHURII | 22 | 90 |
| MAI | MAIDENII | 0 | 0 |
| MAL | MALACOXYLON | 1 | 2 |
| MAN | MANNIFERA | 1 | 6 |
| MAR | MACRORHYNCHA | 1 | 1 |
| MAT | MARGINATA | 0 | 0 |
| MEL | MELLIODORA | 3 | 11 |
| MIC | MICROCORYS | 2 | 6 |
| MIT | MITCHELLIANA | 2 | 3 |
| MOO | MOOREI | 1 | 1 |
| MOR | MORRISBYI | 2 | 2 |
| MUL | MULLERIANA | 1 | 1 |
| NEG | NEGLECTA | 2 | 4 |
| NIP | NIPHOPHILIA | 1 | 1 |
| NIT | NITENS | 40 | 154 |
| NOV | NOVA-ANGLICA | 14 | 57 |
| NTD | NITIDA | 0 | 0 |
| NUM | NUMEROSA | 0 | 0 |
| NVX | NOVA-ANGLICA X | 2 | 3 |
| OBL | OBLIQUA | 4 | 4 |
| ORE | OREADEA | 5 | 7 |
| OVA | OVATA | 8 | 23 |
| PAL | PALIFORMIS | 1 | 1 |
| PAR | PARVIFOLIA | 4 | 12 |
| PAU | PAUCIFLORA | 3 | 3 |
| PER | PERRINIANA | 4 | 9 |
| PIL | PILULARIS | 1 | 1 |
| PIP | PIPERITA | 2 | 2 |
| POL | POLVERULENTA | 0 | 0 |
| POY | POLYANTHEMOS | 2 | 7 |
| PSE | PSEUDOGLOBULUS | 1 | 1 |
| PUL | PULCHELLA | 1 | 1 |
| PUV | PULVERULENTA | 3 | 18 |
| QUA | QUADRANGULATA | 1 | 4 |
| RAD | RADIATA | 5 | 7 |
| REG | REGNANS | 6 | 6 |
| RES | RESINIFERA | 0 | 0 |
| RIS | RISDONII | 1 | 1 |
| ROB | ROBERTSONII | 2 | 2 |
| ROD | RODWAYII | 1 | 3 |
| ROS | ROSSII | 2 | 2 |
| ROU | ROBUSTA | 1 | 1 |
| RUB | RUBIDA | 3 | 15 |
| SAL | SALIGNA | 1 | 5 |
| SIB | SIEBERIANA | 0 | 0 |

Table 7: Continued.

| Species code | Species name | No. of installations by species | No. of seed lots tested |
|---|---|---|---|
| SID | SIDEROXYLON | 2 | 4 |
| SIE | SIEBERI | 2 | 2 |
| SMI | SMITHII | 3 | 10 |
| STE | STELLULATA | 1 | 3 |
| STJ | ST JOHNII | 1 | 1 |
| STN | STENOSTOMA | 2 | 3 |
| TEN | TENUIRAMIS | 1 | 1 |
| TER | TERETICORMIS | 4 | 16 |
| UNK | UNKNOWN | 1 | 1 |
| URN | URNIGERA | 3 | 14 |
| URO | UROPHYLLA | 1 | 7 |
| VIM | VIMINALIS | 134 | 575 |
| YOU | YOUMANII | 1 | 2 |

## 2. Historical Perspective of Eucalyptus Introduction to the South

The outstanding performance of eucalyptus in some of the exotic environments in offshore countries has often impressed foresters and their superiors from the warm temperate zones of the southern United States. The logic was that the growth rate of the eucalyptus would be far superior to that of indigenous plantations hardwoods of cottonwood (*Populus deltoides*), sweetgum (*Liquidambar styraciflua*), sycamore (*Platanus occidentalis*), and other species. The added impetus for the introduction of eucalyptus for plantation use is the excellent wood properties (high fiber count) of the myriad species for pulp and paper manufacture [8].

Species of eucalyptus were introduced to the South as early as 1878 [9], but no significant commercial plantations were established through the first half of the 20th century. Even though forest-based organizations such as St. Regis Paper Company and Champion International Corporation trial planted eucalyptus in Florida and Texas, respectively, in the 1950s and 1960s, most plantings before 1970 were of small scale for windbreaks, ornamentals, and shade trees in central and Southern Florida and Texas [10].

In 1959, the Florida Forests Foundation initiated research on eucalyptus as a potential source of hardwood pulpwood on rangeland and other low-quality sites in the area generally south of Tampa, Florida. Their research was absorbed by the USDA Forest Service and the Florida Division of Forestry in 1968. In the early 1970s, a eucalypts research cooperative was formed by Buckeye Cellulose Corporation, Brunswick Pulp Land Company, Container Corporation of America, Hudson Pulp and Paper Company, ITT Rayonier and St. Regis Paper Company, in conjunction with Lykes Brothers Land Company, to provide financial and research support to the USDA Forest Service scientists at Lehigh Acres, Florida [9]. This effort led to the selection of *E. grandis*, *E. robusta*, *E. camaldulensis,* and *E. tereticornis* and to the development of cultural practices by which to raise seedlings and establish commercial plantations over parts of South Florida. A significant part of this effort was four phases of genetic improvement of *E. grandis* by Dr. Carlyle Franklin and Mr. George Meskimen of the USDA Forest Service, and to lesser levels of genetic improvement of other species [9].

## 3. North Carolina State University Hardwood Research Cooperative Becomes Involved

In the 1950s and 1960s several southern forest-based companies other than Champion and St. Regis tried to introduce eucalyptus for plantation use. The scenario followed a similar line: a newly appointed vice president would become infatuated with the performance of eucalyptus in an exotic environment such as Brazil. The ensuing directive was to plant eucalyptus seedlings on company land, often in the hundreds of acres. The lack of adaptation of the exotic species was invariably the same: failure. The event became so repeatable that it was labeled the "seven-year silly cycle." The seven-year cycle spanned the approximate time before one vice president succeeded another.

By 1971, Bruce Zobel and others at the North Carolina State University decided to evaluate the introduction of eucalyptus into the southern US on a scientific manner. Working with company members of the Hardwood Research Cooperative, the plan was to systematically evaluate eucalyptus species and sources to determine their adaptability [11]. By 1978, the industrial members of the Florida group united with the Hardwood Cooperative in pursuit of the goal. The eucalyptus dream was pursued until 1985 when the 14-year effort came to an end, following severe freezes on December 24, 1983, January 20, 1984, and January 9, 1985.

## 4. Purpose of This Report

The initial purpose of this report was to provide a record of the species and sources that were evaluated, method of evaluation, location of the tests, cause of failure, and recommendations for subsequent research on eucalypts for use in southern forestry. The rationale was that, with time, people involved in the project would have forgotten the particulars, and for their successors to be unaware that such an exhaustive effort ever took place. To repeat an adage: "Those that ignore the past are condemned to repeat it."

## 5. The Method

During the 14-year period, 569 eucalyptus seed sources representing 103 species (Table 7) were planted in 141 different tests throughout the southern USA The general geographic locations of the study sites are depicted in Figure 1.

The seed material was obtained from the indigenous range of the species in Australia and neighboring islands as well as from exotic plantations. The emphasis on obtaining seeds from exotic plantations outside of Australia was because of the supposition that there would have been a genetic selection for adaptability to foreign sites. In addition, the International Paper Company deployed Research Forester Ron Hunt to Australia to collaborate with the Tree Seed Centre of the Commonwealth Scientific and Industrial Research Organization (CSIRO) in making seed collections of eucalyptus species that appeared to have good potential for the southern United States [11]. The Hardwood Research Cooperative served as the clearing house for the distribution of seeds and plant material to cooperators, for plot design, and for collection and analysis of data from test plantings.

To evaluate the material, 3 types of tests were installed: (1) screening trials that consisted of 6-tree row plots replicated 6 times, (2) in-depth trials that consisted of 4 replications of 25-tree plots of the most promising species/sources identified in the screening trials, and (3) semioperational trials of 0.4 to 2 hectares that consisted of the most promising species/sources identified in the in-depth trials.

All new seed lots were exposed to the screening trial, at two or more locations (sometimes as many as eight) depending upon the amount of seed available. The initial trials were installed from North Carolina, Tennessee, and Arkansas southward. It soon became evident, however, that the harsher winters in the northern tier of the southern states left no room for selection at the species, source, or tree level because all plants were badly damaged or killed by freezing temperatures. The decision was then made to restrict trial plantings to the USDA Plant Hardiness Zone VIII and higher (http://www.usna.usda.gov/Hardzone/ushzmap.html), which encompasses an area from Charleston, SC, southward along the coast to Savannah, GA, thence westward along the upper fringes of the Lower Coastal Plain through Georgia, Alabama, Mississippi, Louisiana, and into East Texas.

*5.1. Screening Trials.* Evaluation of the screening trials was performed after the first growing season, to determine survival and potential growth before the onset of winter. A second evaluation was made at about the time of bud break in the spring of the second growing season which allowed for assessment of cold damage incurred during the winter. The third assessment was made at the beginning of the third growing season. The only data recorded at these measurement periods were tree height and survival. A subjective judgment was made on cold tolerance. Any source exhibiting acceptable performance levels for these two traits and being relatively free of cold damage was designated for "in-depth" planting. Most screening trials were abandoned after the second growing season since the seed sources with below-average survival and cold tolerance would not significantly change rank after two growing seasons. Survival rates, inclusive of cold damage, from the screening trials were low: over 80% of the sources had lower than 40 percent survival at the end of two years. The remaining 113 sources were advanced to the in-depth trials.

*5.2. In-Depth Trials.* The 25-tree square plot design used for the in-depth trials allowed assessment of the 16 interior trees without interference from the adjacent plot of trees of a separate source. These replicated trials were installed at two to eight locations, depending upon the availability of seeds and seedlings.

In similar fashion as that described for the screening trials, the first measurements from the in-depth trials were of height and survival at the end of the first growing season. Cold damage was assessed at the beginning of the following growing season and every two years thereafter until the end of the study, using the scale in Table 1.

In addition to survival, total height, and cold tolerance, measurements were made of diameter at breast height, and bole form was subjectively ranked.

Some of the species and sources were extremely susceptible to the swings in winter temperatures and were killed during the first growing season. Others, however, were little affected by the fluctuating temperatures for the first several years. The species with initial cold tolerance nearly always exhibited a slower growth and poorer form than did those most susceptible to freezing temperatures. One such species was the *E. camphora*. Other species and sources, however, such as *E. viminalis*, *E. macarthurii,* and *E. nova-anglica* were highly variable in their performance. Some showed a good growth and form along with a good cold tolerance. Others with a good growth and form were very susceptible to freezing temperatures, and still others showed combinations of these traits. It was these lots that commanded attention because the opportunity ostensibly existed to select and breed for the desired traits.

Analysis of the data proved difficult because none of the 569 seed sources were planted on more than 20 percent of the sites and no site had more than 16 percent of the sources. This limitation arose because of the need to distribute the work load among cooperators; thus, the consignment of 20 seed lots, for example, going to Cooperator A for trial establishment was often very different from those going to Cooperator B. The rather complicated statistical analysis that was developed to make comparisons consisted of three parts: (1) using statistical regression, predictive equations were formulated for each of the measured variables, (2) a linear composite of these variables was created that permitted comparison of the seed sources over locations and time, and (3) a multivariate computer simulation, involving the predictive estimators, was used to compute the mean and variance of the linear composite for each source at each location. The simulator allowed an index score for a species/source to be determined across locations (Table 2).

The simulator was validated by comparing the actual survival and mean height of a specific species/source in a specific test to its predicted value. Tables 3, 4 and 5 show these values for nine of the species/sources at 6, 12, and 24 months. In the absence of cold damage or other unexpected events, the simulator could predict with confidence the species/sources that had the greatest potential for plantation use in the southern United States.

Since the severe freezes of 1983, 1984, and 1985 killed or damaged the plant material beyond salvage, the results of the statistical analyses are moot and therefore will be omitted.

*5.3. Semioperational Trials.* An analysis of the in-depth trials helped to identify nine sources of eucalyptus that were superior in performance to all the other sources evaluated. (ses Table 6).

These sources were planted in 10 demonstration trials across the South in 1981 but were severely damaged or killed by the severe freezes of December 24, 1983, January 20, 1984, and January 9, 1985. This list is different from the one issued in 1978 in which Hunt and Zobel [11] identified *E. viminalis*, *E. nova-anglica*, *E. macarthurii*, and *E. camphora* as the most promising species. The difference between the two lists is that the present one is the result of statistical analysis whereas the former one was based on observation. Again, the difference is moot because all species, regardless of the method of selection, were equally vulnerable to the severe freezes of 1983 to 1985.

## 6. Lessons Learned

Among the lessons learned from the 14 years of testing of eucalypts for use in the southern United States are the following.

*6.1. Eucalyptus Have Indeterminate Growth.* No amount of field testing will identify a species or source that will have a universal adaptability to the fluctuating winter temperatures in the southern United States. The reason is that *Eucalyptus* spp. have an indeterminate type of growth that results from their naked buds [2]. As opposed to indigenous species of cottonwood, sweetgum, sycamore, loblolly pine (*Pinus taeda*), and so forth that have covered buds, eucalyptus begins cell division during the winter whenever the ambient nighttime air temperature exceeds 6°C for six consecutive days. A sudden drop in temperature to near the freezing point during such an event is enough to cause freeze damage to the active tissue. The indigenous species, with their covered buds, do not begin cell activity under such circumstances until a combination of temperature and photoperiod trigger the cambial activity.

*6.2. Variation in Cold Tolerance.* A considerable variation in cold tolerance exists within every species and source of eucalyptus. Among those with greatest cold tolerance, the opportunity exists to select the trees most resistant to freeze damage and to breed for increased tolerance Meskimen and Franklin [12]. Progress was made during our 14 years of trial and error with *E. viminalis,* whereby the best trees for cold tolerance, growth, and form were selected for inclusion in a proposed clonal seed orchard at Fort Green Springs, Florida, on lands of the Container Corporation of America. The concept was to interbreed the selected trees and subject their progeny to freezing temperatures from which the survivors would again be selected for their tolerance to cold. This procedure was to be repeated until a genetic line with the desired traits would be available for plantation use.

The weakness of the plan was that *E. viminalis* and the other most promising species for the Florida environment flowered in February, much as they do in their native habitat. Even at the southerly location of central Florida, seed set was scanty at best and usually nonexistent because the flowers were susceptible to winter frosts. A further limitation to viable seed set was the hibernation of pollinating insects, especially honeybees, during that season of the year.

To overcome these limitations, consideration was given to vegetatively propagating the selected trees for shipment to a frost-free environment in South America. Cooperation with Carton de Colombia and Carton de Venezuela, sister companies of the Container Corporation of America was considered. The concept was to produce seeds from intercrossing the selections in the foreign environment and to screen the seedlings for cold tolerance in the Northern Hemisphere. Graduate Assistant Michael Cunningham of North Carolina State University, with funding from the USDA Forest Service and Container Corporation of America, was successful in cloning the selections by tissue culture [13], but the severe freezes of the mid-1980s destroyed the last vestiges of the project.

*6.3. Soil Suitability.* Planting trials were established on the variety of soil conditions in the Lower Coastal Plain, from sand ridges to imperfectly drained soils. The best performance was obtained on sandy loam soils with a rooting depth of 45 to 90 centimeters. Imperfectly drained soils, inclusive of bedding, were poorly suited to acceptable performance of eucalypts after about two growing seasons. Similarly, the excessively drained soils proved inopportune because of nutrient and moisture deficiency [8].

*6.4. Containerized versus Bareroot Seedlings.* The initial trials were installed exclusively with containerized seedlings. Well-balanced seedlings with a root-shoot ratio of about 0.6 gave superior results. Containerized seedlings can be planted with good results any time during the growing season if there is sufficient soil moisture. Later studies, conducted by the International Paper Company and Brunswick Pulp Land Company, showed that bareroot seedlings performed equally as well or better than the containerized ones if the seedlings were planted immediately after being lifted from the nursery bed after the last spring frost [2].

*6.5. Weed Competition. Eucalyptus* species do not tolerate weed competition, either for sunlight or below ground resources during their first two years after planting. The lack of efficacious and safe silvichemicals during the period of the early trials made competition control a difficult challenge. Cold tolerance of adapted eucalypts is positively correlated with vigor. Trees that are stressed by weed competition or lack of nutrients are more susceptible to freeze damage than those without the limitations [14, 15].

*6.6. Response to Added Nutrients.* Eucalypt seedlings responded to added nutrients, especially nitrogen, in every trial in which fertilizers were added [16]. Soils with phosphorus deficiency showed that plant growth declined as soon as the second year after planting. Potassium and magnesium deficiencies were also observed at some of the planting sites. Best responses were obtained from soils of pH 5.5 to 6.0 [14].

*6.7. Tree Spacing.* Spacing for the test plantings was 3.5 × 3.2 m. For test purposes this spacing was sufficient, but observations from commercial plantations in Brazil, South Africa, and elsewhere suggest that tree spacing of 4 × 4 m or something similar is required for rotations longer than 8 to 10 years. Because nearly all eucalyptus species are intolerant of overlapping crowns, thinning is required for optimum growth of closely spaced trees for rotations longer than 10 years [8, 17].

*6.8. Greenhouse Studies for Cold Tolerance Poorly Correlated to Field Studies.* Results from cold tolerance studies, conducted in greenhouses and phytotrons, were poorly correlated to field studies because of the highly variable performance of eucalyptus in the experimental plantings across the South [18]. Some species/sources were subject to the first winter freezes, whereas others performed well for a number of years before the combination of variables developed to damage or kill the trees. Freezing temperatures in combination with frigid chill factors had a particularly devastating effect. In 1978, for example, an uncommonly heavy snow of 60 cm that blanketed the Eastern Seaboard resulted in the death of the plant material extending above the snow line. Foliage below the line survived without damage because of the insulation effect of snow.

*6.9. Performance Modeling Failed.* Performance modeling of the best species/source has proven to be an impossible task because a combination of factors would prove fatal to the trees in one year and a separate set of factors would prove equally as destructive in another year. High hopes were generated for the success of the project when certain trees, especially of *E. viminalis*, would escape the ravages of winter in the midst of other species/sources that succumbed to the freezing temperatures. Some of those trees grew to be 25 meters tall and 25 cm diameter at breast height in 6 to 8 years. In a subsequent winter, however, they too would be killed, even when the temperatures were less severe than those of the previous winter.

*6.10. Fluctuations in Winter Temperatures Prove Fatal.* Outside of the severe freezes that occurred in 1983, 1984, and 1985, the greatest damage always resulted when there were great fluctuations in temperature within one or two days. Temperatures in excess of 15°C in midwinter are occasionally followed by temperature drops of 12°C or more within a 24-hour period as a result of cold fronts that originate in northern latitudes and sweep through the Great Plains to the southern United States. The active cambial growth of the eucalypts, with their naked buds, is extremely vulnerable to such fluctuations. Under such circumstances, freeze damage can happen to some species/sources of eucalypts at temperatures slightly above 0°C.

*6.11. Wood Properties Prove Acceptable.* Four- to six-year-old trees of *E. viminalis* produced wood with specific gravity, fiber dimensions, and pulp and paper properties similar to that of sweetgum of 15 years of age [19, 20]. Fluff pulp is especially desired from eucalypts because of the high fiber to vessel ratio.

*6.12. Inhospitable Environment for Plants.* The southern United States is an inhospitable environment for the growing of exotic forest trees. The area has been compared with that of Siberia, without its severity, for its great fluctuations in winter temperature. This phenomenon prohibits the successful establishment of forest trees from most other parts of the world where the mean winter temperature can be even lower than that in the southern United States. The temperatures in those areas, however, decline steadily from summer highs to winter lows in the absence of fluctuating warm and cold temperatures. That is why seed sources of thriving plantations of *Eucalyptus* spp. from other areas of the world, such as *E. dunnii* from southern Brazil, *E. deanii* from France, and *E. viminalis* from Georgia (formerly the Soviet Union), failed when introduced to the United States.

*6.13. Molecular Genetics to Aid Species Introduction.* In combination with warming trends, the successful introduction of eucalypts to the southern United States will be aided by the use of molecular genetics. Work has been accomplished on the gene mapping of *E. grandis*, *E. globulus* and their hybrid by the Biotechnology Program at the North Carolina State University [12]. Quantitative trait loci (QTL) have been identified for tree growth, and speculation exists that a similar phenomenon exists for oliogenic inheritance for cold tolerance.

## 7. Summary

The outstanding performance of eucalyptus species growing in foreign countries caused industrial foresters to establish semioperational plantations in the southern United States in the 1950s and 1960s. Failure of the plantations, due primarily to cold damage, resulted in establishment of research trials across the landscape from the Atlanta Ocean to areas west of the Mississippi River. Seed sources for the trials were obtained from natural stands in Australia and from exotic plantations from much of the Tropics and Warm Temperate Zones of the world where the species were growing. Over a 14-year span, 569 seed sources, representing 103 species were evaluated for industrial use. The species/sources showing most promise in screening trials were further tested in in-depth trials, and those of greatest promise in in-depth trials were subjected to semioperational trials.

Statistical analyses of the collected tree data consisted of three steps: (1) using regression and analysis, predictive equations were formulated for each of the measured variables, (2) a linear composite of these variables was created that permitted comparison of the seed sources over locations and time, and (3) a multivariate computer simulation, involving the predictive estimators, was used to compute the mean and variance of the linear composite for each source at each location. The simulator allowed an index score for species/sources to be determined across locations.

Regardless of the growth potential of selected species/ sources the effort came to naught because of susceptibility to severe winter temperatures in late 1983 and early 1984 and 1955. Since the initiative lost its appeal more than 25 years ago, temperatures have apparently moderated enough across the South for establishment of new introductory trials. With the experience gained from the past endeavor, we urge present-day investigators to concentrate their efforts on soils of sandy clay loam and clay loam and to avoid imperfectly drained soils and those of excessive soil drainage (high sand content). It is also highly recommended that the plantations be kept free from weed competition for at least the first two growing seasons and that attention be given to soil fertility. Seedling quality for bareroot planting should be in the range of 0.6 for root-shoot ratio, and such seedlings should be planted in early spring, but after the last frost. Containerized seedlings can be planted at any time of the year whenever there is adequate soil moisture. Fall planting of containerized seedlings must occur in time for root establishment before first frosts.

## Disclosure

The basic part of this paper was written in 1985. It was never published but was committed to the files so that the results could be shared whenever there was a renewed effort to introduce eucalyptus to the U.S. South. The authors of the 27 years ago paper have revised it to coincide with today's needs as well as to shed light on the introduction of eucalyptus to California.

## References


[1] Introduction to *Eucalyptus*. Blog posted by Phytopath, 2010.

[2] J. Davidson, "Ecological Aspects of *Eucalyptus* Plantations," FAO Corporate Document Repository. Proceedings of Regional Expert Consultation o Eucalyptus. Vol.1, 1979.

[3] D. H. Bain, *Empire Express; Building the First Transcontinental Railroad*, Viking Penguin, 1999.

[4] S. Villalobos, S. Osvaldo, S. Fernando, and E. Patricio, History de Chile. Editorial Universitaria, Chile. pp. 406–413, 1974.

[5] R. L. Santos, *The Eucalyptus of California*, Alley-Cass, Denair, Calif, USA, 1997.

[6] G. M. Groenendaal, "*Eucalyptus* helped solve a timber problem," in *Proceedings of the Workshop on Eucalyptus in California*, pp. 1853–1880, Sacramento, Calif, USA, 1983.

[7] R. Crawford, "The Way We Were: Eucalyptus Trees Have Deep Roots in California's History," The San Diego Union-Tribune, 2008.

[8] C. Foelkel, "The *Eucalyptus* fibers and the pulp quality requirements for paper manufacturing," 2007, http://www.eucalyptus.com.br/.

[9] T. F. Geary, G. F. Meskimen, and E. C. Franklin, "Growing Eucalypts in Florida for industrial wood production," Tech. Rep. SE-23, Southeastern Forest Experiment Station, USDA Forest Service, Asheville, NC, USA, 1983.

[10] R. J. Moultin, "Tree planting in the United States, 1997," *Tree Planters' Notes*, vol. 49, no. 1, pp. 5–15, 1999.

[11] R. Hunt and B. Zobel, "Frost-hardy *Eucalyptus* grow well in the Southeast," *Southern Journal of Applied Forestry*, vol. 2, no. 1, pp. 6–10, 1978.

[12] A. A. Myburg, A. R. Griffin, R. R. Sederoff, and R. W. Whetten, "Comparative genetic linkage maps of *Eucalyptus grandis, Eucalyptus globulus* and their F1 hybrid based on a double pseudo-backcross mapping approach," *Theoretical and Applied Genetics*, vol. 107, no. 6, pp. 1028–1042, 2003.

[13] M. Cunningham and R. Mott, "Micropropagation of *Eucalyptus* viminalis," in *Proceedings of the Forestry Conference*, p. 8, Division of Continuing Education, Louisiana State University, Baton Rouge, La, USA, 1985.

[14] G. Meskimen, "Fertilizer tablets stimulate eucalyptus in Florida trial," 1971USDA Forest Service Research Note SE-162, Lehigh Acres, Fla, USA, 12, 1971.

[15] A. P. G. Schonau, R. Verloren von Themaat, and D. L. Boden, "The importance of complete site preparation and fertilizing in the establishment of *Eucalyptus grandis*," *South African Journal of Forestry*, vol. 116, pp. 1–10, 1981.

[16] N. F. Barros, "Some aspects of *Eucalyptus* fertilization: A review of current literature as a contribution to future Eucalyptus research and planting in Florida," Florida Agric. Exp. Stations, Institute of Food and Agricultural Sciences, University of Florida, Gainesville, 1977.

[17] G. Meskimen and E. C. Franklin, "Spacing *Eucalyptus grandis* in southern Florida: a question of merchantable versus total volume," *Southern Journal of Applied Forestry*, vol. 1, pp. 3–5, 1978.

[18] E. B. Schultz, *Artificial cold hardiness of Eucalyptus [M.S. thesis]*, North Carolina State University, Raleigh, NC, USA, 1979.

[19] E. C. Franklin, "Yield and properties of pulp from eucalypt wood grown in Florida," *Tappi*, vol. 60, no. 6, pp. 65–67, 1977.

[20] G. O. Otegebe and R. C. Kellison, "Genetics of wood and barfk characteristics of *Eucalyptus viminalis*," *Silva Genetica*, vol. 29, no. 1, pp. 27–30, 1980.

# 11

# Inconsistent Growth Response to Fertilization and Thinning of Lodgepole Pine in the Rocky Mountain Foothills Is Linked to Site Index

**Bradley D. Pinno,[1,2] Victor J. Lieffers,[1] and Simon M. Landhäusser[1]**

[1] *Alberta School of Forest Science and Management, Faculty of Agricultural Life and Environmental Sciences, University of Alberta, Edmonton, AB, Canada T6G 2H1*
[2] *Natural Resources Canada, Canadian Forest Service, Northern Forestry Centre, 5320 122 Street, Edmonton, AB, Canada T6H 3S5*

Correspondence should be addressed to Bradley D. Pinno, bpinno@nrcan.gc.ca

Academic Editor: John Sessions

Fertilization of conifers often results in highly variable growth responses across sites which are difficult to predict. The goal of this study was to predict the growth response of lodgepole pine (*Pinus contorta var. latifolia*) crop trees to thinning and fertilization using basic site and foliar characteristics. Fifteen harvest-origin stands along the foothills of the Rocky Mountains of Alberta were subjected to six treatments including two levels of thinning (thinning to 2500 stems per hectare and a control) and three types of fertilization (nitrogen-only fertilization, complete fertilization including nitrogen with added P, K, S, Mg, and B, and no fertilization). After three growing seasons, the growth response and foliar status of the crop trees were examined and this response was related to site and foliar characteristics. There was a small and highly variable additive response to fertilization and thinning; diameter growth of crop trees increased relative to the controls an average of 0.3 cm with thinning, 0.3 cm with either N-only or complete fertilization and 0.6 cm when thinned and fertilized. The increase in diameter growth with thinning and nitrogen-only fertilization was positively related to site index but not to any other site factors or pretreatment foliar variables such as nutrient concentrations, ratios, or thresholds.

## 1. Introduction

Lodgepole pine (*Pinus contorta var. latifolia* Loudon) is the dominant tree species in the foothills of Alberta and is capable of growing on a wide range of site types. Precommercial thinning of juvenile high density lodgepole pine stands can be used to avoid stand repression [1] and increase the growth of individual trees [2, 3]. Fertilization is used to increase both individual tree growth and total stand volume [4]. Fertilization of lodgepole pine in North America usually focuses on nitrogen (N) but limitations of other nutrients, including sulfur (S), phosphorus (P), boron (B), and zinc (Zn), have been identified in some sites in British Columbia [5, 6].

Fertilization of lodgepole pine and other conifers has been extensively studied around the world and a common finding has been that, on average, fertilizing conifer stands result in a significant increase in growth but there is usually high variability across sites. For example, five-year stem growth of *Pinus sylvestris* increased on average 45% after fertilization, but the growth response ranged from 11–104% across 28 sites in Scandinavia [7] with no obvious connection between growth response and site characteristics.

For lodgepole pine, pretreatment foliar nutrient concentrations, their ratios with foliar N [8–10] and adequate foliar nutrient concentrations [11] have shown promise as diagnostic tools to predict site response to fertilization. For example, pretreatment foliar sulfate concentration and N/S ratios were successful in predicting lodgepole pine growth response to N and N + S fertilization in British Columbia [4]. Other site variables including site index and soil type have been used to predict the response to midrotation fertilization

Table 1: Site properties for each of the 15 stands.

| Site number | Site index (m @ 50 years) | Age (at breast height) | Density (stems $ha^{-1}$) | Basal area ($m^2$ $ha^{-1}$) | Elevation (m) | Ecological subregion |
|---|---|---|---|---|---|---|
| 1 | 20.3 | 18 | 8000 | 23.0 | 1238 | Upper foothills |
| 2 | 20.4 | 16 | 11040 | 24.6 | 1281 | Upper foothills |
| 3 | 19.5 | 9 | 5740 | 5.8 | 1197 | Upper foothills |
| 4 | 21.7 | 6 | 5060 | 3.9 | 1104 | Lower foothills |
| 5 | 19.7 | 22 | 8300 | 17.0 | 1064 | Lower foothills |
| 6 | 21.6 | 21 | 8667 | 29.9 | 1041 | Lower foothills |
| 7 | 19.3 | 12 | 11160 | 15.3 | 1341 | Upper foothills |
| 8 | 21.4 | 7 | 6533 | 8.3 | 1255 | Upper foothills |
| 9 | 18.2 | 15 | 2420 | 11.1 | 1346 | Lower foothills |
| 10 | 16.6 | 19 | 3027 | 11.2 | 1473 | Upper foothills |
| 11 | 18.7 | 4 | 9200 | 1.2 | 1480 | Upper foothills |
| 12 | 20.8 | 18 | 6960 | 21.9 | 1084 | Upper foothills |
| 13 | 18.8 | 18 | 9280 | 15.3 | 1169 | Upper foothills |
| 14 | 18.4 | 18 | 14640 | 22.1 | 1208 | Upper foothills |
| 15 | 22.5 | 9 | 10080 | 14.9 | 1096 | Lower foothills |

of loblolly pine (*Pinus taeda*) in the southern United States [12], and these types of relationships between fertilization response and basic site and foliar characteristics may be applicable in Alberta.

Our study examines tree growth in relation to both thinning and fertilization applied to harvest origin lodgepole pine stands across a range of sites within the same ecological region. We examined site and foliar characteristics that might be used to predict response to treatment. Previous lodgepole pine thinning and fertilization studies have generally only examined a single site [1, 3] or compared stands from different ecological regions [13]. Our approach uses sites from a 350 km north-south transect across most of Alberta's foothills region and allows us to examine the differential growth benefits of thinning and fertilization over a large number of sites differing in productivity.

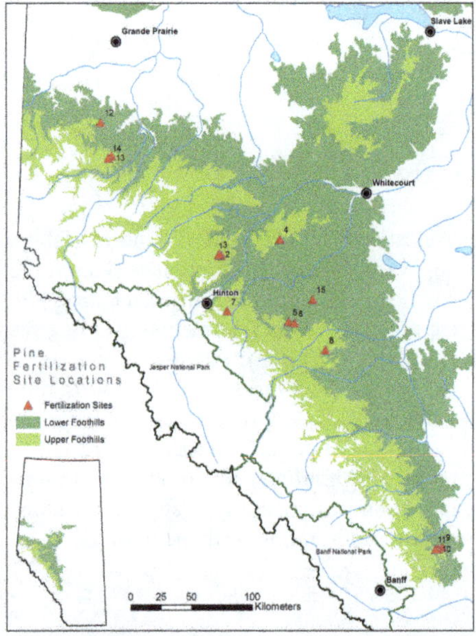


Figure 1: Map of the study region in the foothills of Alberta.

## 2. Methods

We studied 15 relatively pure lodgepole pine stands in the foothills of the Rocky Mountains in Alberta, Canada (Figure 1). Stands were of harvest origin, ranging in breast height age from 6 to 22 years, and were dominated by lodgepole pine (pine made up over 95% of the stand density in all but three of the stands). Site index, elevation, age, density, and other characteristics are given in Table 1. Within each stand, six 200 $m^2$ square plots were established and randomly assigned to a thinning (2 levels) and fertilization treatment (3 levels). The thinning treatments were no thinning and a low thinning to a density of 2500 stems per hectare with all deciduous trees removed. The fertilization treatments were control (no fertilization), N-only fertilization, and complete fertilizer blend with N and added P, K, S, magnesium (Mg), and B (Table 2).

Within each plot, all conifer trees were measured before treatment for height and diameter at breast height (DBH). Thinning and fertilization treatments were carried out by hand in May of 2006. All trees were remeasured the following winter and again after three growing seasons. After the first and third growing seasons, foliar samples were also collected during the winter from the upper third of the crown of three dominant or codominant lodgepole pine trees. A sample of 100 needle fascicles of the youngest age class was isolated from each tree and the dry weight was determined after drying at 68°C. Samples were ground and pooled for foliar nutrient analysis, including foliar N, P, K, S, calcium (Ca), Mg, sulfate ($SO_4$), B, copper (Cu), Zn, manganese (Mn), and iron (Fe) concentrations. Nitrogen concentration was determined colourimetrically using an autoanalyzer after digestion with $H_2SO_4$ while K, Ca, Mg,

TABLE 2: Fertilizer formulations for N-only and complete fertilizers.

| | Ingredient | Nutrients ($kg\ ha^{-1}$) | | | | | |
|---|---|---|---|---|---|---|---|
| | | N | P | K | S | Mg | B |
| N-only fertilizer | Urea | 300 | | | | | |
| | Total | 300 | 0 | 0 | 0 | 0 | 0 |
| Cbmplete fertilizer | Urea | 251 | | | | | |
| | Monoammonium phosphate | 49 | 100 | | 7 | | |
| | Muriate of potash | | | 46 | | | |
| | Sulphate potassium magnesia | | | 54 | 68 | 33 | |
| | Borate granular | | | | | | 3 |
| | Total | 300 | 100 | 100 | 75 | 33 | 3 |

Cu, Zn, and Mn were determined by atomic absorption after the same digestion. The azomethine-H method was used to determine B concentration after dry ashing. Available $SO_4$ was determined colourimetrically on a HI-bismuth reducible distillate after 0.1 N HCl extraction. Active Fe concentration was determined by atomic absorption after 1 N HCl extraction.

Growth rate per tree was determined for each plot based on the largest 12 stems (600 stems per hectare). This approach was taken because a snow storm damaged 9 of the 15 of the stands, with damage concentrated on the medium- and smaller-sized trees [14]. Only in the thinning + fertilization treatments were some of the larger trees affected. By concentrating on the largest trees, the impact of the snow damage on growth responses can be greatly reduced and still allows for meaningful comparisons among treatments. Analyzing the growth response of the largest trees in stands has been done previously [1, 15] and is relevant as these trees can be considered the crop trees that will likely survive to final harvest. Growth increment was calculated as tree size (DBH and volume) after three growing seasons minus the initial tree size prior to treatment. Stem volume was calculated using a taper volume equation developed for the area [16]. Foliar N mass per 100 fascicles was determined by multiplying N concentration by the mass of 100 fascicles. Foliar N uptake as a result of the treatments was estimated by subtracting the foliar N mass per 100 fascicles of the control plot from the foliar N mass of the treatment plots.

Statistical analysis involved comparing growth and foliar characteristics among treatments using two-way ANOVAs ($3 \times 2$) blocked by site. Tukey's HSD test was used to further examine differences among treatment levels. To evaluate the differential response to fertilization and thinning between sites, differential growth increment (treatment growth-control growth) was regressed against site conditions, including site index, density, elevation, and age, and foliar properties, including foliar mass, nutrient concentrations, ratios, uptake, and adequate nutrient values, of the control plots. Statistical analysis was conducted using JMP 8.0 (SAS Institute Inc., Cary, NC, USA).

## 3. Results

Thinning and fertilization had an additive impact on individual tree growth of the 600 crop trees $ha^{-1}$ after the third growing season, increasing diameter growth from 1.4 cm in the control plots to an average of 1.7 in the thinned or fertilized plots to greater than 2.0 cm in the thinned + fertilized plots (Figure 2(a)). However, the differential diameter growth increment was highly variable ranging from −0.12 to 1.27 cm diameter growth across treatments. Relative to the control plots, this corresponds to a diameter growth increase of 22% with thinning, 24% with N-only fertilization, 25% with complete fertilization and 47% with thinning and fertilization combined. For volume growth, only the thinning + fertilization treatments resulted in significantly greater growth than the controls (Figure 2(b)). Overall, there was a positive effect of thinning and fertilization on both diameter and volume increment, but there was no difference in average growth response between N-only and complete fertilization either with or without thinning.

Initial foliar N concentrations ranged from 1.02–1.23% N. The first year after treatment, foliar N concentration increased with complete fertilization (both with and without thinning) and thinning + N-only fertilization reaching an average of 1.48% N, while the N-only treatment without thinning was 1.35% N and the control fertilization plots (either unthinned or thinned) were less than 1.14% N (Figure 3(a)). After three growing seasons, foliar N concentrations were not significantly different from the control plots in all but the unthinned + complete fertilization plots (Figure 3(a)).

Foliar mass responded strongly to the combination of thinning and fertilization the first year after treatment (Figure 3(b)). Thinning or N-only fertilization alone did not increase foliar mass while the highest foliar mass was in the thinning + complete fertilization treatment. After three growing seasons, foliar mass was not significantly different from the control plots in any of treatments. Crop tree foliar nitrogen uptake increased with fertilization with the greatest average uptake in the thinning + complete fertilization treatment (Figure 4). There was great variability among sites, however, with 4 of the 15 sites showing no foliar N uptake with N-only fertilization alone, 8 of the 15 sites showing no foliar N uptake with thinning only and many other treatment units showing very little foliar N uptake.

After one growing season, foliar P concentration increased in response to complete fertilization (Table 3). Foliar S concentration increased with complete fertilization

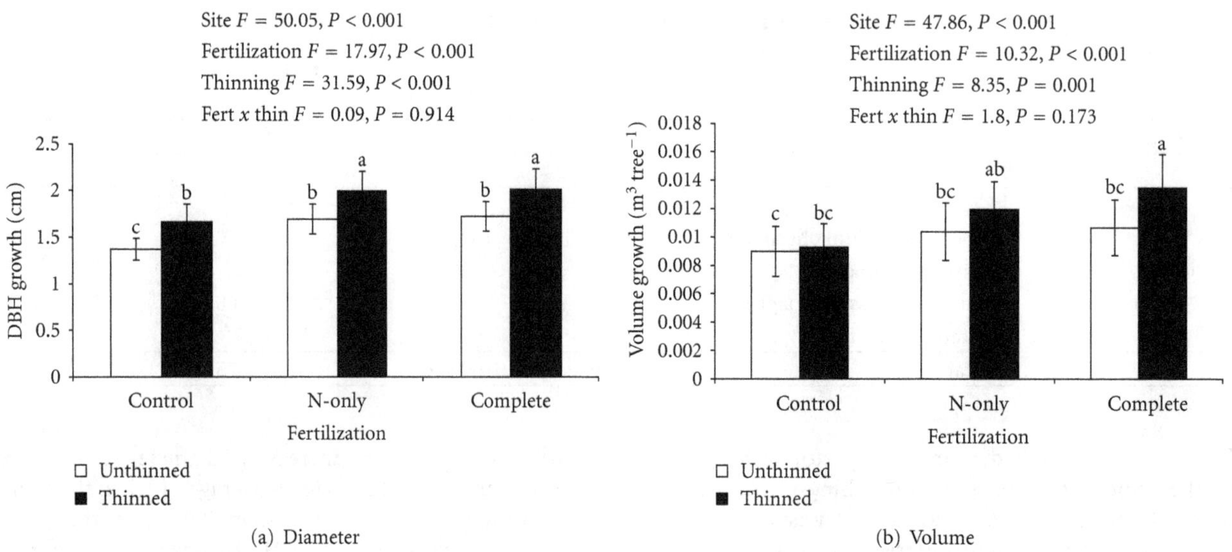


(a) Diameter (b) Volume

FIGURE 2: Three-year crop tree diameter (a) and volume (b) growth in relation to thinning and fertilization treatments. Letters represent differences in total growth among the six treatments.

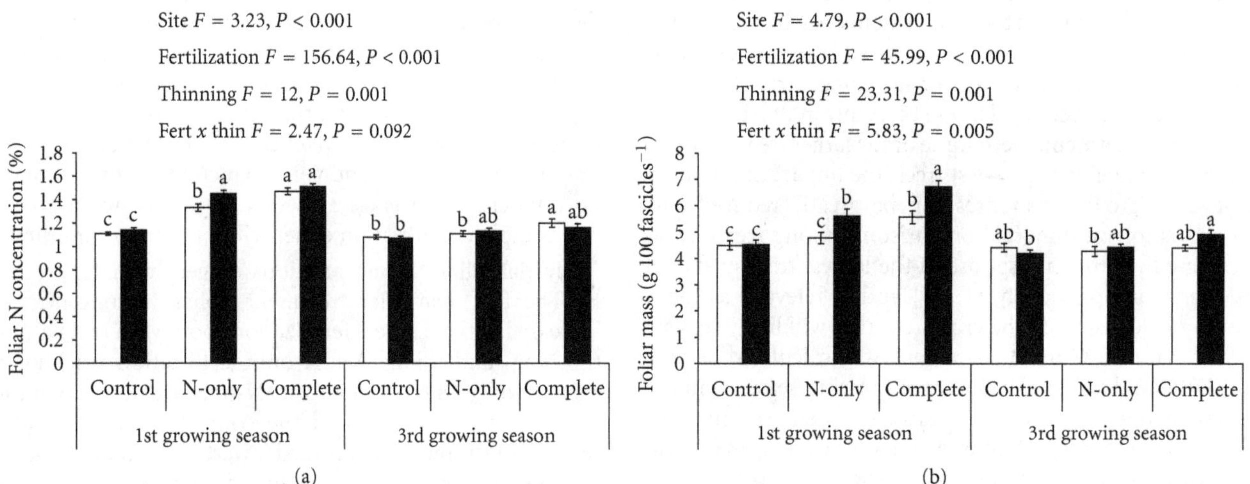


(a) (b)

FIGURE 3: Response of (a) foliar nitrogen concentration and (b) mass of 100 fascicles to thinning and fertilization treatments after 1 and 3 growing seasons. White bars represent unthinned stands and black bars represent thinned stands. Different letters represent significant differences among treatments for each growing season. ANOVA statistics are given for the 1st growing season data.

but $SO_4$ concentration decreased with N-only fertilization. Foliar base cation concentrations (K, Ca, and Mg) did not respond significantly to either fertilization or thinning but foliar K concentration tended to increase with complete fertilization while Mg tended to decrease with all fertilizer treatments. Foliar B concentration increased with complete fertilization but tended to decrease with N-only fertilization. Foliar Fe decreased with complete fertilization while Zn, Cu and Mn did not respond to the treatments. After three growing seasons, only B and $SO_4$ concentrations were still significantly different from the controls.

We could not detect any correlations between growth differential (treatment growth—control growth) and prethinning density, elevation, age, foliar mass, foliar nutrient concentrations, nutrient ratios, or nutrient thresholds of the controls. Further, the growth differential was not related to estimated foliar N uptake among sites for any of the treatments (Figure 4). We did, however, find that diameter growth differential increased with site index in the thinning only, N-only fertilization, and thinning + N-only fertilization treatments (Figures 5(a) and 5(b)); there was no correlation of diameter growth differential with site index in the complete fertilization treatments (Figure 5(c)).

## 4. Discussion

Thinning and fertilization produced an additive growth response in lodgepole pine with the best growth occurring when plots were both thinned and fertilized—diameter growth after three growing seasons increased on average

Table 3: Foliar nutrient concentrations after the first growing season in relation to thinning and fertilization treatments. Letters represent significant differences between treatments.

| Thinning | Fertilization | P (%) | K (%) | S (%) | Ca (%) | Mg (%) | $\text{So}_4^-$ (ppm) | B (ppm) | Cu (ppm) | Zn (ppm) | Fe (ppm) | Mn (ppm) |
|---|---|---|---|---|---|---|---|---|---|---|---|---|
| Unthinned | Control | $0.132^{c}$ | 0.409 | $0.089^{bc}$ | 0.187 | 0.082 | $127.7^{a}$ | $11.16^{b}$ | 3.49 | 45.6 | $53.3^{ab}$ | 370.3 |
| | N-only | $0.133^{bc}$ | 0.407 | $0.088^{c}$ | 0.182 | 0.075 | $48.7^{b}$ | $8.67^{b}$ | 3.29 | 43.8 | $47.7^{bc}$ | 360.7 |
| | Complete | $0.149^{a}$ | 0.435 | $0.108^{a}$ | 0.189 | 0.075 | $94.9^{a}$ | $34.29^{a}$ | 3.23 | 42.7 | $44.5^{c}$ | 360.7 |
| Thinned | Control | $0.135^{bc}$ | 0.413 | $0.091^{bc}$ | 0.211 | 0.084 | $121.2^{a}$ | $11.78^{b}$ | 3.54 | 45.4 | $54.9^{a}$ | 400.8 |
| | N-only | $0.143^{ab}$ | 0.405 | $0.096^{b}$ | 0.187 | 0.077 | $48.6^{b}$ | $8.93^{b}$ | 3.77 | 44.7 | $48.9^{abc}$ | 360.5 |
| | Complete | $0.150^{a}$ | 0.447 | $0.111^{a}$ | 0.189 | 0.075 | $97.8^{a}$ | $32.10^{a}$ | 3.48 | 43.5 | $46.6^{c}$ | 378.0 |

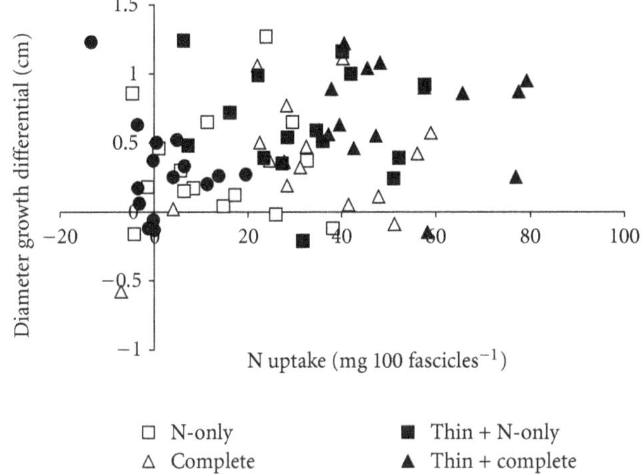


Figure 4: Diameter growth differential of thinning and fertilization treatments relative to the controls in relation to nitrogen uptake by the foliage in year 1 (N content of treated trees—N content of control trees). Diameter growth differential is the difference between DBH growth after three years of the treated stand and DBH growth of the associated control stand.

0.3 cm with thinning, 0.3 cm with either N-only or complete fertilization, and 0.6 cm with thinning and fertilization combined. This additive crop tree growth response to thinning and fertilization has been recorded before in lodgepole pine stands [1, 3, 13]; however, the real story of our study relates to our inability to diagnose which stands would respond to treatment, particularly for the complete fertilization treatment.

All stands had low foliar N concentrations prior to treatment (average 1.1%) compared to the adequate value of 1.35% N [11]. We therefore expected a greater growth increase as a result of the fertilization. Further, many of our sites showed little or no positive growth response to our treatments which is a disappointing result given that all of the sites were below the critical level of foliar N prior to treatment and the high rate of fertilizer applied (300 kg N ha$^{-1}$ along with other nutrients). This inconsistent response to fertilization is similar to what has been found in other conifer fertilization studies (e.g., in lodgepole pine [4, 17], *Picea glauca* [18], *Picea abies* [19], *Pseudotsuga menziesii* [20], and mixed conifer stands of the Pacific Northwest [21]).

Foliar nutrient concentrations and ratios have been successfully used in predicting the growth response of lodgepole pine to fertilization [4, 6] but did not work in our study. We believe that the main reason that these techniques were not useful in our study is that in many of the sites the nutrients supplied by the fertilizer were not successfully taken up by the trees. These diagnostic tests can only work well if the tree actually takes up the nutrients. In our study, at 4 of the 15 sites, trees showed no uptake of N in the N-only treatment and another 4 sites had only a small amount of uptake.

The uptake of N may be related to the type of fertilization applied with the complete formulation resulting in greater N uptake, even though the total amount of N applied was the same between the N-only and complete fertilization treatments. This increased N uptake with complete fertilization could simply be related to increased tree growth stimulating greater uptake of N. The addition of other potentially limiting nutrients can also increase N uptake as has been seen previously in *Eucalyptus grandis*, where fertilization with P increased N absorption through a mechanism not related to increased N demand [22]. The difference in fertilizer formulation may have also affected N uptake; in the N-only fertilizer urea was the single source for N while in the complete fertilizer 49 kg N ha$^{-1}$ was derived from monoammonium phosphate. This direct addition of $NH_4$ could have made N more readily available for uptake since the N from monoammonium phosphate may have been less likely to be volatilized than the N derived from urea [23] in these forests with thick organic layers. It is likely that ammonium nitrate is a better formulation than urea for N fertilization in boreal forests [24].

Even when N was increased in the foliage, there were sometimes poor tree growth responses and we suggest that these were related to internal nutrient imbalances. With the addition of N-only fertilizer, macronutrient imbalances can be induced in lodgepole pine [4] and other conifers [25]. Micronutrient deficiencies (Cu and Zn) have also developed after repeated N fertilization of lodgepole pine, thereby limiting potential growth [5]. In our study, the idea that nutrient imbalances limit growth response is supported by the fact that only the N-only and thinning-only treatments had a positive relationship between growth differential and

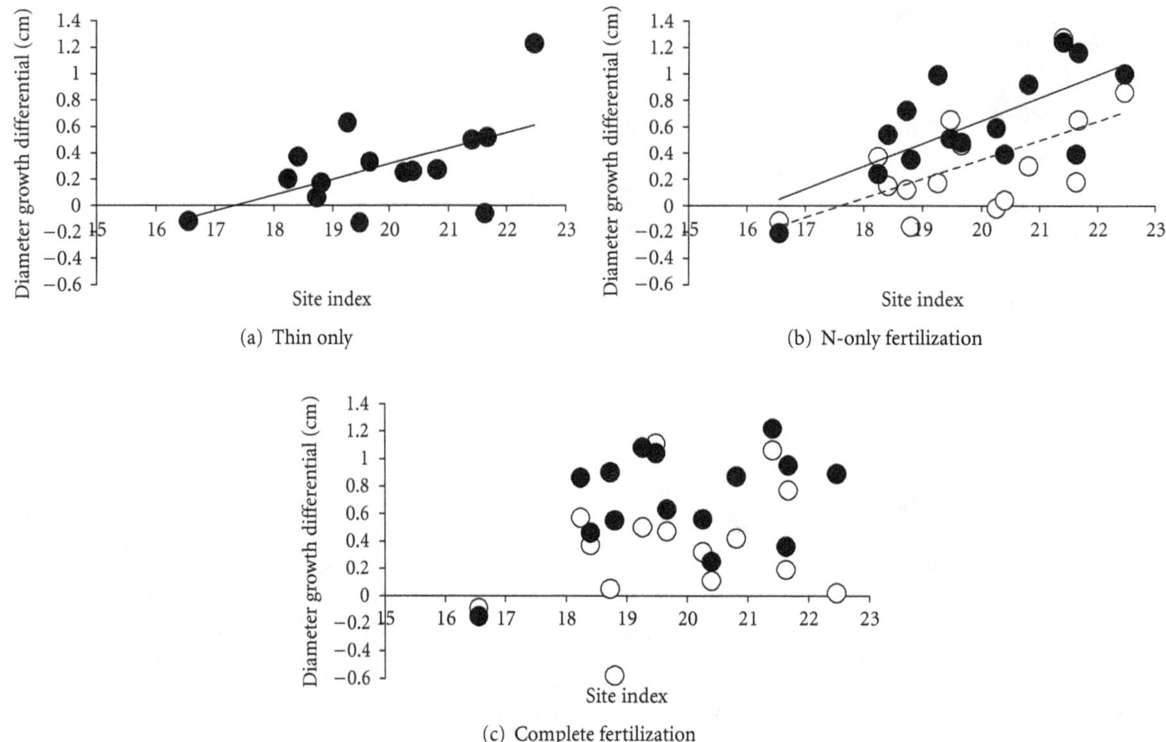


(a) Thin only

(b) N-only fertilization

(c) Complete fertilization

Figure 5: Diameter growth differential of thinning and fertilization treatments relative to site index of the control stands. The response variable, diameter growth differential, is the difference between DBH growth after three years of the treated stand and DBH growth of the associated control stand. White dots represent unthinned plots and black dots represent thinned plots. (a) Thin only, $P = 0.032$, $r^2 = 0.254$, (b) N-only, $P = 0.020$, $r^2 = 0.303$, N-only + thinning, $P = 0.003$, $r^2 = 0.469$, and (c) complete fertilization, $P = 0.334$, complete + thinning, $P = 0.110$.

site index. We argue that sites with low site index are likely limited by several nutrients so thinning or N-only fertilization will not result in increased growth. On better-quality sites, other nutrients such as P, S, and micronutrients, are likely in higher supply so N-only fertilization and thinning will increase growth without causing internal nutrient imbalances or inducing other nutrient deficiencies. In contrast, the differential growth response to complete fertilizer was not related to site index, likely because any potential nutrient imbalances were eliminated.

The link between site index and growth response to silvicultural treatments has been varied. For example, growth response to competition control in hybrid poplar plantations was positively correlated to site productivity [26], while in jack pine plantations the growth response to site preparation treatments was negatively correlated to site productivity [27]. The positive relationship between lodgepole pine growth response to thinning and N-only fertilization and site index has not previously been documented but appears to be related to site nutrient availability and nutrient imbalances.

In summary, the combination of fertilization and thinning is likely to result in the greatest growth response of lodgepole pine crop trees. On better quality sites it may be possible to use N-only fertilizer but the poorer sites may also need other nutrients in order to stimulate a growth response. We recommend caution in extrapolating the growth results in our study to yield at the stand level because our study focused on the response of only the largest trees in the stand, that is, the crop trees. Future work should concentrate on identifying specific fertilizer formulations, particularly the forms of N, and their delivery methods to enhance nutrient uptake of trees in the field.

## Acknowledgments

The authors thank Dick Dempster for directing the establishment and measurement of the study plots. Funding was provided by the Foothills Growth and Yield Association, the Forest Resource Improvement Association of Alberta, and the National Sciences and Engineering Research Council of Canada.

## References

[1] C. Farnden and L. Herring, "Severely repressed lodgepole pine responds to thinning and fertilization: 19-year results," *Forestry Chronicle*, vol. 78, no. 3, pp. 404–414, 2002.

[2] X. Liu, U. Silins, V. J. Lieffers, and R. Man, "Stem hydraulic properties and growth in lodgepole pine stands following thinning and sway treatment," *Canadian Journal of Forest Research*, vol. 33, no. 7, pp. 1295–1303, 2003.

[3] R. C. Yang, "Foliage and stand growth responses of semimature lodgepole pine to thinning and fertilization," *Canadian Journal of Forest Research*, vol. 28, no. 12, pp. 1794–1804, 1998.

[4] R. P. Brockley, "Using foliar variables to predict the response of lodgepole pine to nitrogen and sulphur fertilization," *Canadian Journal of Forest Research*, vol. 30, no. 9, pp. 1389–1399, 2000.

[5] I. G. Amponsah, P. G. Comeau, R. P. Brockley, and V. J. Lieffers, "Effects of repeated fertilization on needle longevity, foliar nutrition, effective leaf area index, and growth characteristics of lodgepole pine in interior British Columbia, Canada," *Canadian Journal of Forest Research*, vol. 35, no. 2, pp. 440–451, 2005.

[6] R. E. Carter and R. P. Brockley, "Boron deficiencies in British Columbia: diagnosis and treatment evaluation," *Forest Ecology and Management*, vol. 37, no. 1–3, pp. 83–94, 1990.

[7] U. Sikström, H. Ö. Nohrstedt, F. Pettersson, and S. Jacobson, "Stem-growth response of *Pinus sylvestris* and *Picea abies* to nitrogen fertilization as related to needle nitrogen concentration," *Trees*, vol. 12, no. 4, pp. 208–214, 1998.

[8] R. P. Brockley, "Response of thinned, immature lodgeopole pine to nitrogen and boron fertilization," *Canadian Journal of Forest Research*, vol. 20, no. 5, pp. 579–585, 1990.

[9] R. P. Brockley and F. J. Sheran, "Foliar nutrient status and fascicle weight of lodgepole pine after nitrogen and sulphur fertilization in the interior of British Columbia," *Canadian Journal of Forest Research*, vol. 24, no. 4, pp. 792–803, 1994.

[10] B. E. Kishchuk and R. P. Brockley, "Sulfur availability on lodgepole pine sites in British Columbia," *Soil Science Society of America Journal*, vol. 66, no. 4, pp. 1325–1333, 2002.

[11] R. P. Brockley, "Foliar sampling guidelines and nutrient interpretative criteria for lodgepole pine," Extension Note 52, B.C. Ministry of Forests, Victoria, Canada, 2001.

[12] T. R. Fox, H. L. Allen, T. J. Albaugh, R. Rubilar, and C. A. Carlson, "Tree nutrition and forest fertilization of pine plantations in the southern United States," *Southern Journal of Applied Forestry*, vol. 31, no. 1, pp. 5–11, 2007.

[13] P. M. F. Lindgren, T. P. Sullivan, D. S. Sullivan, R. P. Brockley, and R. Winter, "Growth response of young lodgepole pine to thinning and repeated fertilization treatments: 10-year results," *Forestry*, vol. 80, no. 5, pp. 587–611, 2007.

[14] F. P. Teste and V. J. Lieffers, "Snow damage in lodgepole pine stands brought into thinning and fertilization regimes," *Forest Ecology and Management*, vol. 261, no. 11, pp. 2096–2104, 2011.

[15] D. P. Blevins, C. E. Prescott, H. L. Allen, and T. A. Newsome, "The effects of nutrition and density on growth, foliage biomass, and growth efficiency of high-density fire-origin lodgepole pine in central British Columbia," *Canadian Journal of Forest Research*, vol. 35, no. 12, pp. 2851–2859, 2005.

[16] S. M. Huang, "Ecologically based individual tree volume estimation for major *Alberta tree species*," Report #1, Alberta Sustainable Resource Development, Edmonton, Canada, 1994.

[17] B. E. Kishchuk, G. F. Weetman, R. P. Brockley, and C. E. Prescott, "Fourteen-year growth response of young lodgepole pine to repeated fertilization," *Canadian Journal of Forest Research*, vol. 32, no. 1, pp. 153–160, 2002.

[18] R. F. Sutton, "White spruce establishment: initial fertilization, weed control, and irrigation evaluated after three decades," *New Forests*, vol. 9, no. 2, pp. 123–133, 1995.

[19] K. Dralle and J. B. Larsen, "Growth response to different types of NPK-fertilizer in Norway spruce plantations in Western Denmark," *Plant and Soil*, vol. 168-169, no. 1, pp. 501–504, 1995.

[20] D. Binkley and P. Reid, "Long-term responses of stem growth and leaf area to thinning and fertilization in a Douglas-fir plantation," *Canadian Journal of Forest Research*, vol. 14, no. 5, pp. 656–660, 1984.

[21] R. Rose and J. S. Ketchum, "Interaction of vegetation control and fertilization on conifer species across the Pacific Northwest," *Canadian Journal of Forest Research*, vol. 32, no. 1, pp. 136–152, 2002.

[22] C. Graciano, J. F. Goya, J. L. Frangi, and J. J. Guiamet, "Fertilization with phosphorus increases soil nitrogen absorption in young plants of *Eucalyptus grandis*," *Forest Ecology and Management*, vol. 236, no. 2-3, pp. 202–210, 2006.

[23] D. E. Kissel, M. L. Cabrera, and R. B. Ferguson, "Reactions of ammonia and urea hydrolysis products with soil," *Soil Science Society of America Journal*, vol. 52, no. 6, pp. 1793–1796, 1988.

[24] R. F. Fisher and D. Binkley, *Ecology and Management of Forest Soils*, John Wiley & Sons, New York, NY, USA, 3rd edition, 2000.

[25] M. T. Garrison, J. A. Moore, T. M. Shaw, and P. G. Mika, "Foliar nutrient and tree growth response of mixed-conifer stands to three fertilization treatments in Northeast Oregon and North central Washington," *Forest Ecology and Management*, vol. 132, no. 2-3, pp. 183–198, 2000.

[26] B. D. Pinno and N. Bélanger, "Competition control in juvenile hybrid poplar plantations across a range of site productivities in central Saskatchewan, Canada," *New Forests*, vol. 37, no. 2, pp. 213–225, 2009.

[27] F. Marquis and D. Paré, "The role of permanent site factors in the assessment of soil treatment effects: a case study with a site preparation trial in jack pine plantations on glacial outwashes," *Canadian Journal of Soil Science*, vol. 89, no. 1, pp. 81–91, 2009.

# 12

# Assessing the Invasion Risk of *Eucalyptus* in the United States Using the Australian Weed Risk Assessment


**Doria R. Gordon,[1] S. Luke Flory,[2] Aimee L. Cooper,[2] and Sarah K. Morris[2]**

[1] *The Nature Conservancy and Department of Biology, University of Florida, P.O. Box 118526, Gainesville, FL 32611, USA*
[2] *Agronomy Department, University of Florida, P.O. Box 110500, Gainesville, FL 32611, USA*

Correspondence should be addressed to Doria R. Gordon, dgordon@tnc.org



Academic Editor: Matias Kirst



Many agricultural species have undergone selection for traits that are consistent with those that increase the probability that a species will become invasive. However, the risk of invasion may be accurately predicted for the majority of plant species tested using the Australian Weed Risk Assessment (WRA). This system has been tested in multiple climates and geographies and, on average, correctly identifies 90% of the major plant invaders as having high invasion risk, and 70% of the noninvaders as having low risk. We used this tool to evaluate the invasion risk of 38 *Eucalyptus* taxa currently being tested and cultivated in the USA for pulp, biofuel, and other purposes. We predict 15 taxa to have low risk of invasion, 14 taxa to have high risk, and 9 taxa to require further information. In addition to a history of naturalization and invasiveness elsewhere, the traits that significantly contribute to a high invasion risk conclusion include having prolific seed production and a short generation time. Selection against these traits should reduce the probability that eucalypts cultivated in the USA will become invasive threats to natural areas and agricultural systems.


## 1. Introduction

Global travel and trade have resulted in unprecedented introductions of nonnative species [1, 2]. Furthermore, many agronomic and silvicultural species are being selected and bred for rapid growth, high fecundity, and tolerance to a wide range of climatic and environmental conditions [3, 4]. These traits are the same as those of many invasive species, thus there is increasing concern that cultivated species may become invasive [5]. Because biological invasions can have considerable ecological and economic costs [5, 6], accurately predicting which introduced species are likely to become invasive can have significant benefits.

As the majority of the ecological and economic impacts are caused by a relatively small proportion of nonnative species that become harmful invaders [4, 7], tools that differentiate this group from the non-invasive majority are critical. One such tool, the Australian Weed Risk Assessment (hereafter WRA) [8], was developed in Australia and has been used for regulatory purposes for over a decade. This system has now been tested in temperate, tropical, island, and continental geographies and appears to have comparable accuracy across regions [9]. On average across these tests, the WRA correctly identified 90% of the harmful plant invaders as of high invasion risk and 70% of the noninvaders as having low risk [9]. Roughly 10% of each of the noninvaders and the harmful invaders were misclassified, with the remainder requiring further evaluation. This WRA discriminates between invaders and noninvaders independently of the proportion of species in either category, which is important because the true proportions are unknown but are clearly dominated by noninvaders [9, 10]. While concerns about this base-rate, misclassifications and bias have been raised about the WRA and other predictive tools [11], the ecological and economic value of prevention supports the implementation of a proactive approach [12]. The cost savings for Australia associated with implementation of the WRA were conservatively estimated to save up to US$1.67 billion over 50 years [13].

*Eucalyptus* species (Myrtaceae) are widely cultivated in subtropical and tropical regions for reforestation, production of timber, pulp, and other forest products, and increasingly, as potential bioenergy feedstocks [14, 15]. Several species, hybrids, and genotypes (hereafter "taxa") of eucalypts show rapid growth across a wide range of environments [16]. Selection and genetic modification are increasing that range, focused on a potential need for 20 million tons/year from *Eucalyptus* for pulp and biofuel production in the Southern USA alone by 2022 [17]. As a result, 5,000 to 10,000 ha/year may be converted to commercial *Eucalyptus* plantations in this region [17].

More than 200 *Eucalyptus* taxa have been screened for cultivation outside their native range over the last 180 years [18]. Only a few of these have become harmful invaders, including *E. globulus* [15, 19], *E. megacornuta* [20], *E. camaldulensis* [21, 22], *E. grandis* [21, 22], *E. conferruminata* [15, 22], *E. robusta* [15], and *E. diversicolor* [22]. *Corymbia citriodora* (*C. maculata*), *E. cinerea*, *E. cladocalyx*, *E. tereticornis*, and *E. saligna* have also been identified as invasive [15, 22].

The relatively small proportion of eucalypts introduced that has become invasive (~5%) may reflect the frequency and extent of cultivation, number of propagules introduced, or geographic range of introduction (propagule pressure *sensu* [27]), rather than increased likelihood of invasion risk of these taxa [15]. Three of the species identified as invaders above are among the four species (*E. camaldulensis*, *E. globulus*, *E. grandis*, and *E. urophylla*) and their hybrids that represent 80% of global *Eucalyptus* plantations [14]. The majority of taxa that have become invasive belong to the subgenus *Symphyomyrtus* [15], so there may be specific biological traits that influence invasion risk. As cultivation of eucalypts is anticipated to increase in the USA, predicting and avoiding those taxa that are likely to become expensive and damaging invasive species would be beneficial.

Given the increasing focus on eucalypts for pulp and bioenergy crop production in both the USA and elsewhere [14, 16], we have selected a suite of *Eucalyptus* taxa for evaluation using the WRA. As these taxa have already been introduced into the USA and received some testing for forestry production, they do not represent a random sample within this genus. Despite their intended use for forest products and biomass, we hypothesized that high and low risk taxa may be identified using the WRA. Assuming that some of these taxa are likely to be cultivated despite the potential for substantial external costs associated with unintended escape and invasion, we identified the traits that were most closely associated with the high risk invasion group. We suggest that selection against those traits could reduce invasion risk as cultivation of eucalypts in the USA increases.

## 2. Methods

We selected 38 *Eucalyptus* taxa (Table 1) that had previously been evaluated using the WRA in Hawaii, the Pacific, or Florida for which the results are available online (http://www.hear.org/pier/, http://plants.ifas.ufl.edu/assessment/predictive_response_forms.html). Inclusion in previous assessments likely indicates current or historic interest in those taxa for cultivation. These assessments were conducted at the regional scale across limited environmental conditions, but they provided an initial source of the literature and data for each taxon. Furthermore, we found more recent data to address several of the questions than were available when the original assessments were completed.

We followed the published guidance available for use of the WRA [28], which was modified for application in the USA [29]. Like the original WRA, this system has 49 questions that address historical, biogeographical, and biological traits of the species. Responses to the questions result in points ranging from −3 to 5, with the majority ranging from −1 for negative responses to 1 for positive responses [8]. At least 10 questions from the three categories of questions with specified distribution must be addressed for completion of the WRA. The points are summed for a total score with the corresponding conclusions: scores below one indicate that the species has a low risk of being invasive; scores of one through six indicate that further evaluation is necessary before risk level may be concluded; scores above six indicate the species has a high risk of becoming invasive [8]. We used the secondary screen for species requiring further evaluation [23] to resolve the risk level where possible.

Sources of information included primary literature from forestry, biological and invasive species references, floras and websites for different regions of the world, and the U.S. Department of Agriculture (USDA) Germplasm Resources Information Network [30] and USDA PLANTS [31] databases. Natural distribution of eucalypts was determined from Australian floras [32, 33]. All data, including evidence and references used to develop scores, are archived and available (http://plants.ifas.ufl.edu/assessment/predictive_response_forms.html).

We assumed that any published information for a species or hybrid applied to the taxon of that name. Unless we had specific information about seed dispersal, we assumed that all taxa had negligible wind dispersal, but could be water dispersed [15]. Additionally, we assumed that seeds are not dispersed through animal ingestion because *Eucalyptus* seed does not survive the alimentary canal [34]. We also answered negatively about presence of a persistent propagule bank [15] for all taxa. Responses to all other questions were based on taxon-specific data; questions were left blank if no data were available.

The WRA results allowed us to identify *Eucalyptus* taxa with the lowest and the highest risk for invasion. We used regression analyses (Proc REG, SAS Enterprise, 2010) to investigate whether the total WRA score was dependent on the number of questions answered for each taxon independently and all taxa combined. We also identified the questions that differentiated taxa with low and evaluate further conclusions (scores $\leq 6$) from high risk taxa (scores $> 6$) using Welch's $t$-tests for samples with unequal variances (Proc $t$-test, SAS Enterprise, 2010).

## 3. Results

We were able to complete the WRA for all taxa, answering an average of 28 questions (range: 23–35). The scores varied from –3 to 18 across taxa (Figure 1). Out of the 38 taxa,

Table 1: Taxa assessed using the Weed Risk Assessment (WRA) modified for the USA [22], with WRA results from other assessments for comparison.

| No.[a] | Species | Subgenus | USA score | Other published score(s) | Risk level |
|---|---|---|---|---|---|
| 1 | *Eucalyptus dorrigoensis* | *Symphyomyrtus* | −3 | — | Low |
| 2 | *Eucalyptus dunnii* | *Symphyomyrtus* | −2 | 0[b] | Low |
| 3 | *Eucalyptus salubris* | *Symphyomyrtus* | −1 | −3[b], −2[c] | Low |
| 4 | *Eucalyptus amplifolia* | *Symphyomyrtus* | 0 | 2[d] | Low |
| 5 | *Eucalyptus benthamii* | *Symphyomyrtus* | 0 | — | Low |
| 6 | *Eucalyptus stoatei* | *Symphyomyrtus* | 0 | −2[b] | Low |
| 7 | *Eucalyptus cloeziana* | *Idiogenes* | 1 | −1[b] | Low[e] |
| 8 | *Eucalyptus nitens* | *Symphyomyrtus* | 1 | — | Low[e] |
| 9 | *Eucalyptus smithii* | *Symphyomyrtus* | 1 | — | Low[e] |
| 10 | *Eucalyptus caesia* | *Symphyomyrtus* | 1 | 0[b] | Evaluate |
| 11 | *Eucalyptus gardneri* | *Symphyomyrtus* | 2 | 0[b] | Low[e] |
| 12 | *Eucalyptus gunnii* | *Symphyomyrtus* | 2 | — | Low[e] |
| 13 | *Eucalyptus erythrocorys* | *Eudesmia* | 2 | 6[b] | Evaluate |
| 14 | *Eucalyptus platypus* | *Symphyomyrtus* | 2 | 0[b] | Evaluate |
| 15 | *Eucalyptus pellita* | *Symphyomyrtus* | 3 | 3[b] | Evaluate |
| 16 | *Eucalyptus kruseana* | *Symphyomyrtus* | 4 | 0[b] | Low[e] |
| 17 | *Eucalyptus macrocarpa* | *Symphyomyrtus* | 4 | 3[b] | Low[e] |
| 18 | *Eucalyptus urograndis (=E. grandis X E. urophylla)* | *Symphyomyrtus* | 4 | — | Evaluate |
| 19 | *Eucalyptus microcorys* | *Alveolata* | 5 | 1[b], 0[c] | Low[e] |
| 20 | *Eucalyptus torquata* | *Symphyomyrtus* | 5 | −1[b] | Low[e] |
| 21 | *Eucalyptus intermedia (=Corymbia intermedia)* | *Corymbia* | 5 | 1[b] | Evaluate |
| 22 | *Eucalyptus yarraensis* | *Symphyomyrtus* | 5 | 1[b] | Evaluate |
| 23 | *Eucalyptus crebra* | *Symphyomyrtus* | 6 | −1[b], −1[c] | Evaluate |
| 24 | *Eucalyptus macarthurii* | *Symphyomyrtus* | 6 | — | Evaluate |
| 25 | *Eucalyptus cinerea* | *Symphyomyrtus* | 7 | 4[b] | High |
| 26 | *Eucalyptus paniculata* | *Symphyomyrtus* | 7 | 11[b], 6[c] | High |
| 27 | *Eucalyptus sideroxylon* | *Symphyomyrtus* | 7 | 2[b] | High |
| 28 | *Eucalyptus urophylla* | *Symphyomyrtus* | 7 | 4[b] | High |
| 29 | *Eucalyptus deglupta* | *Symphyomyrtus* | 8 | 2[b] | High |
| 30 | *Eucalyptus saligna* | *Symphyomyrtus* | 9 | 7[b] | High |
| 31 | *Eucalyptus grandis* | *Symphyomyrtus* | 10 | 11[b], 8[d] | High |
| 32 | *Eucalyptus tereticornis* | *Symphyomyrtus* | 10 | 5[b] | High |
| 33 | *Eucalyptus viminalis* | *Symphyomyrtus* | 10 | — | High |
| 34 | *Eucalyptus robusta* | *Symphyomyrtus* | 11 | 3[b], −1[c] | High |
| 35 | *Eucalyptus citriodora (=Corymbia citriodora)* | *Corymbia* | 12 | 9[b], 6[c] | High |
| 36 | *Eucalyptus torelliana (=Corymbia torelliana)* | *Corymbia* | 13 | 4[b] | High |
| 37 | *Eucalyptus camaldulensis* | *Symphyomyrtus* | 18 | 12[d] | High |
| 38 | *Eucalyptus globulus* | *Symphyomyrtus* | 18 | 10[b] | High |

[a]The taxon number corresponds to the number on the bars in Figure 1.
[b][23, 24]: Hawaii and Pacific.
[c][25]: Tanzania.
[d][26]: U.S.
[e]Outcome determined after use of the secondary screen [23].

15 (39%) were determined to have a low probability of invasion, 14 (37%) were predicted to have high probability of invasion, and 9 (24%) required further information (Table 1), even after use of the secondary screen. Several of our predictions (33%) were different from WRA results found previously for the USA and other regions (Table 1), reflecting both the availability of new published data and differences in how the WRA was implemented. All our scores were higher than those found by earlier assessments. In several cases, we found new evidence that a taxon has naturalized beyond its native range (e.g., for *E. robusta, E. deglupta, and E. tereticornis*), explaining a difference of 2 to 4 points. Other authors also responded negatively (incurring negative points) to some questions when they found no

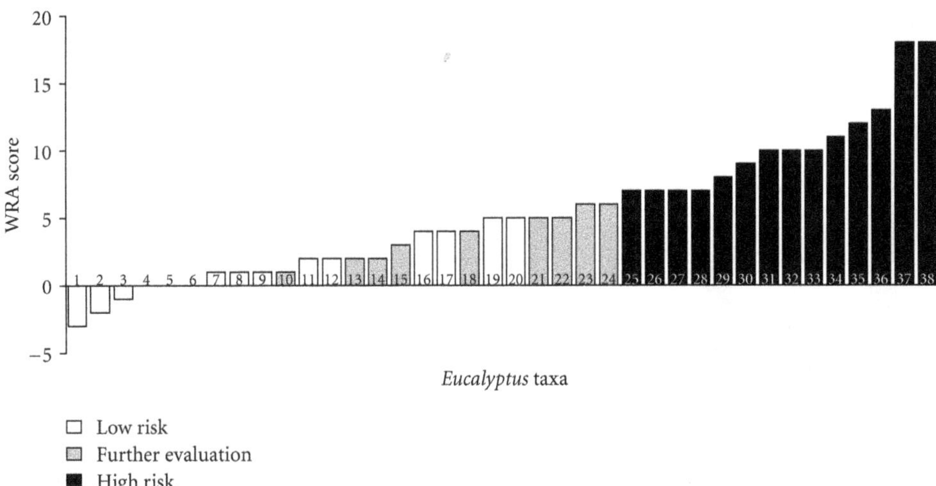


FIGURE 1: Distribution of Weed Risk Assessment (WRA) scores of 38 *Eucalyptus* taxa evaluated for the USA. Scores < 1 suggest that the taxon is a low risk for invasion; scores of 1–6 indicate that the taxon requires further evaluation unless a secondary screen [23] allowed resolution to a risk outcome; scores > 6 suggest that the taxon is a high risk for invasion [8]. See Table 1 for identification of the taxon associated with each score. Numbers within bars correspond to numbers in the first column of Table 1.

affirmative data. In this case, we were more likely to leave the question unanswered (0 points) following the WRA guidance [28]. We have not presented Florida WRA results in this comparison (http://plants.ifas.ufl.edu/assessment/) as our group conducted those regional analyses.

Across all taxa, total scores were dependent on the number of questions answered ($n = 38$, $P = 0.03$, $r^2 = 0.13$; Figure 2). However, this relationship was not found for taxa predicted to be of high risk ($n = 14$, $P = 0.10$, $r^2 = 0.20$) or for those needing further evaluation or low risk for invasion ($n = 24$, $P = 0.21$, $r^2 = 0.07$). Taxa for which we could answer high numbers of questions but were designated as low risk included *E. dunnii* (31 questions answered), *E. macarthurii* (31), *E. nitens* (32), *E. benthamii* (34), and *E. gunnii* (34).

Not surprisingly, traits associated with whether the taxon has been introduced and invasive elsewhere (naturalized beyond its native range, invasive in disturbed, agricultural, or natural areas) were disproportionately ($P < 0.0001$, $P = 0.0009$, $P = 0.04$, and $P = 0.02$, resp.) associated with high risk taxa (Figure 3). Other traits contributing to the invasive conclusion included short generation time ($df = 16$, $t = 2.13$, $P = 0.045$) and prolific seed production ($df = 5$, $t = 3.16$, $P = 0.025$) (Figure 3). Absence of a specific pollinator requirement ($P = 0.08$) and seed dispersal by animals (invertebrates; seed not ingested) ($P = 0.08$) also contributed to prediction of high risk for invasion.

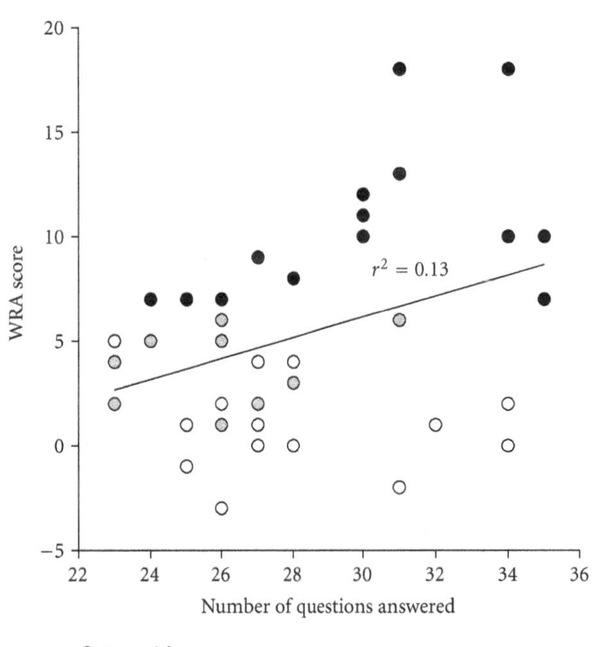


FIGURE 2: Relationship between the number of questions answered and the WRA score for *Eucalyptus* taxa assessed.

## 4. Discussion

The majority of eucalypts assessed was determined to present a low risk for invasion or required further evaluation (collectively, 63%). Conversely, over a third of the taxa (37%) were predicted to pose a high invasion risk. While this percentage would be unexpectedly high if the taxa had been randomly selected from all possible eucalypts [7], higher proportions are not unusual for forestry species cultivated over large areas. For example, 24% of the species introduced to Australia for forestry have naturalized, and 17% have become harmful invaders [35]. Forestry species represent 13% and 24% of the invasive species flora in North America and Europe, respectively [4]. Twenty percent (22/110) of *Pinus* species are invasive outside their native ranges [4]. Overall, tree species with multiple uses are disproportionately likely to be invasive [4].

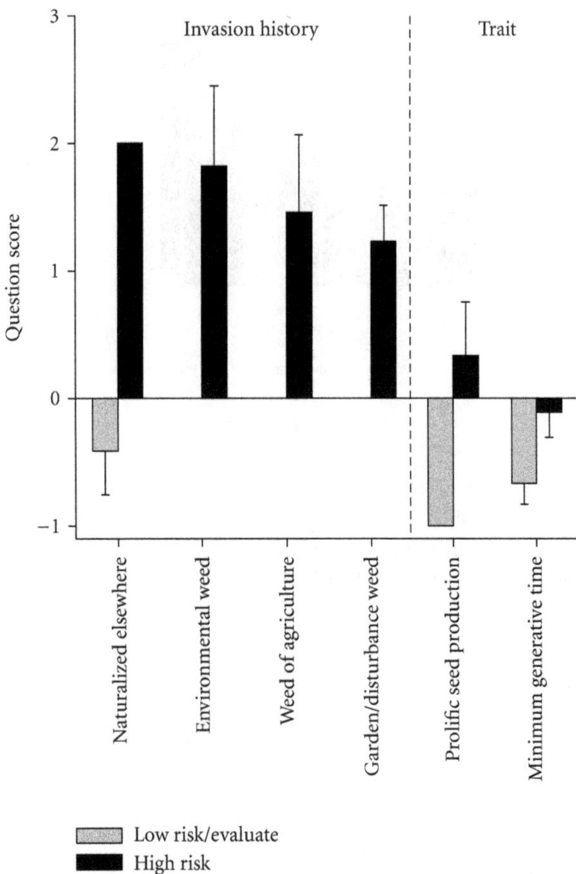


FIGURE 3: WRA questions on invasion history elsewhere or specific traits with responses (mean ± SE) that significantly ($P < 0.05$) differentiated taxa that scored ≤6 (low risk or requiring further evaluation) from >6 (high risk).

Of the *Eucalyptus* taxa we evaluated, more were predicted to have a high risk of invasion than are currently recognized as harmful invaders. This result may be due to the lag time often observed with invasion [36]. The only study that has quantified the lag time in invasive tree species found that it takes an average of 170 years from the time of their introduction to identification as invasive [37]. That lag can result from several factors, from intrinsic rates of population growth, to selection for tolerance to the new environment, to changes in climate or other environmental or biotic characteristics [37]. Propagule pressure has been demonstrated to influence the lag time and the probability of invasion as more genotypes are introduced into more environments, increasing the opportunity for taxa to encounter a situation favorable for population growth [27, 38].

Invasion by eucalypts has been contrasted with that by pines as likely resulting more from propagule pressure than from biological traits [15]. While the four globally most extensively cultivated taxa: *E. globulus, E. camaldulensis, E. grandis*, and *E. tereticornis* [15, 17] had high risk outcomes, so did a number of other taxa. *Eucalyptus urograndis*, the other most frequently cultivated taxon [17], requires further evaluation before risk can be assessed. Significant reliance of the WRA on whether the taxon is invasive outside its native range (Figure 3) supports the contribution of propagule pressure to invasion risk. However, traits of specific taxa and characteristics of introduction sites may also be critical.

While our evaluation was conducted at the national scale because eucalypts are cultivated in multiple states and territories, the majority of new cultivation is likely to be across the southeastern states [17]. Thus, as discussed previously, a more regional assessment may provide greater resolution of these outcomes for different species (Florida WRA results by region at http://plants.ifas.ufl.edu/assessment/). Differences in phenology, age at reproductive maturity, seed viability, and cold tolerance will certainly impact the potential invasiveness of species and genotypes. As the acreage planted in *Eucalyptus* increases, the potential for spread from plantations will be better understood. Moreover, the active selection for genotypes that are cold tolerant and have desirable growth and wood characteristics (e.g., [14]) means that although some species have been introduced for many years, novel genotypes with unknown invasiveness are being propagated. As a result, the list of *Eucalyptus* taxa currently considered invasive in the USA may not be indicative of the long-term invasion risks from this genus.

The differences in WRA outcomes between our work and earlier assessments (Table 1) contrast with reports of the generality of WRA predictions across geographies with similar climates [39]. Although we sought to conservatively interpret the literature and answered questions only when we found specific evidence, our scores are consistently higher than those from other efforts. However, several recent publications (e.g., [4, 15, 16]) provided data not available to earlier studies. Additionally, the greater range of environments in the US versus a more regional scale effort increased the potential habitat suitability for some taxa. Improvements in both of the available data and guidance for application of the WRA and the secondary screen should reduce discrepancies in scoring and the probability of cognitive bias [11], increasing the reliability of the WRA results. We found only a weak correlation between WRA score and the number of questions answered when data for all conclusions were combined (Figure 2), indicating that risk prediction is largely independent of the amount of data available on taxa (see also [23]).

The hypothesis that taxa in the subgenus *Symphyomyrtus* are likely to be more invasive than taxa in other subgenera [15] is not supported by our data. While we had insufficient numbers of taxa in other subgenera to specifically test this hypothesis, taxa in this subgenus spanned the range of low to high risk results (Table 1). This suite of species suggests that *Symphyomyrtus* taxa are more likely to be cultivated than taxa from other subgenera (see also [15]).

Of the eucalypts that are currently most likely to be cultivated in the Southern USA [40], four (*E. amplifolia, E. benthamii, E. dunnii*, and *E. dorrigoensis*) are predicted to be low invasion risks, and two (*E. camaldulensis* and *E. viminalis*), high risks. The remaining two taxa likely to be cultivated (*E. macarthurii* and *E. urograndis*) need further evaluation. *Eucalyptus grandis, E. robusta*, and *E. saligna*, which have also received increasing attention

[14, 41], are all predicted to pose a high risk of invasion. We suggest a precautionary approach for using eucalypts in pulp, bioenergy, and other products by focusing on the taxa that have a low probability of becoming invasive [42] or by selection against traits likely to increase invasion risk (Figure 2).

Examination of the WRA results for high risk taxa may indicate specific traits that may be modified through plant breeding or genetic alteration that significantly reduce that risk [43, 44]. Selection, plant breeding, and genetic modifications have been used to reduce the probability of invasion in other taxa [3, 43, 44]. The biofuel grass hybrid *Miscanthus* × *giganteus* is sterile, unlike one of its highly invasive parents, *Miscanthus sinensis* [43]. Allelopathy, a trait that can support invasiveness, has been modified in species such as rice (*Oryza sativa*) [45]. Not surprisingly, our results suggest that reducing fecundity would reduce the probability that taxa will become harmful invaders. While reducing the time to maturity may negatively influence productivity, eliminating seed production appears feasible [46–48] and would effectively eliminate concerns about invasion.

If predicted high risk eucalypts are cultivated, those plantings should be treated as experimental testing of the predictions made by the WRA [49]. Maximizing the utility of that approach would require careful tracking of seedling establishment over multiple decades. These data are critical for evaluation of the actual invasiveness of taxa and refinement of weed risk assessment approaches.

## 5. Conclusions

Given the growing interest in identifying and cultivating bioenergy crop species, the WRA is increasingly being used to evaluate the invasion risk of those species (e.g., [24, 26, 42, 43]). An accompanying approach is to specify the best management practices that reduce invasion risk. These practices might span from taxon selection as described above, to cultivation and monitoring practices. Examples for eucalypts may be to avoid cultivation near waterways [15] and manage plantations to reduce seed production, including harvesting stems prior to seed maturation (see additional specific and limited uses for *E. grandis* cultivars identified by the University of Florida at http://plants.ifas.ufl.edu/assessment/conclusions.html). If the risk of invasion is outweighed by the likely benefits, creation of a fund designed to cover any necessary control costs for species with WRA scores > 6 would be advisable. If propagule pressure is the key to invasiveness in eucalypts [15, 38], one approach might be to restrict the extent of cultivation of any one taxon. However, the key to avoiding costly invasion impacts will likely be selection for sterility and vigilant control of even apparently slow spread from cultivation sites.

## Acknowledgments

This paper was developed for the Symposium on the Assessment and Management of Environmental Issues Related to *Eucalyptus* Culture in the Southern USA and benefited from discussion with meeting participants and comments from reviewers. D. R. Gordon was supported by MWV, Inc. and the Florida Chapter of The Nature Conservancy. S. K. Morris and A. L. Cooper were supported by a Tropical and Subtropical Agriculture Research (TSTAR) grant to the University of Florida.

## References

[1] P. E. Hulme, "Trade, transport and trouble: managing invasive species pathways in an era of globalization," *Journal of Applied Ecology*, vol. 46, no. 1, pp. 10–18, 2009.

[2] M. Springborn, C. M. Romagosa, and R. P. Keller, "The value of nonindigenous species risk assessment in international trade," *Ecological Economics*, vol. 70, no. 11, pp. 2145–2153, 2011.

[3] N. O. Anderson, S. M. Galatowitsch, and N. Gomez, "Selection strategies to reduce invasive potential in introduced plants," *Euphytica*, vol. 148, no. 1-2, pp. 203–216, 2006.

[4] D. M. Richardson and M. Rejmánek, "Trees and shrubs as invasive alien species—a global review," *Diversity and Distributions*, vol. 17, no. 5, pp. 788–809, 2011.

[5] R. N. Mack, "Cultivation fosters plant naturalization by reducing environmental stochasticity," *Biological Invasions*, vol. 2, no. 2, pp. 111–122, 2000.

[6] D. Pimentel, R. Zuniga, and D. Morrison, "Update on the environmental and economic costs associated with alien-invasive species in the United States," *Ecological Economics*, vol. 52, no. 3, pp. 273–288, 2005.

[7] M. Williamson and A. Fitter, "The varying success of invaders," *Ecology*, vol. 77, no. 6, pp. 1661–1666, 1996.

[8] P. C. Pheloung, P. A. Williams, and S. R. Halloy, "A weed risk assessment model for use as a biosecurity tool evaluating plant introductions," *Journal of Environmental Management*, vol. 57, no. 4, pp. 239–251, 1999.

[9] D. R. Gordon, D. A. Onderdonk, A. M. Fox, and R. K. Stocker, "Consistent accuracy of the Australian Weed Risk Assessment system across varied geographies," *Diversity and Distributions*, vol. 14, no. 2, pp. 234–242, 2008.

[10] P. Caley and P. M. Kuhnert, "Application and evaluation of classification trees for screening unwanted plants," *Austral Ecology*, vol. 31, no. 5, pp. 647–655, 2006.

[11] P. E. Hulme, "Weed risk assessment: a way forward or a waste of time?" *Journal of Applied Ecology*, vol. 49, no. 1, pp. 10–19, 2011.

[12] J. P. Schmidt, P. M. Springborn, and J. M. Drake, "Bioeconomic forecasting of invasive species by ecological syndrome," *Ecosphere*, vol. 3, no. 5, article 46, p. 12, 1890.

[13] R. P. Keller, D. M. Lodge, and D. C. Finnoff, "Risk assessment for invasive species produces net bioeconomic benefits," *Proceedings of the National Academy of Sciences of the United States of America*, vol. 104, no. 1, pp. 203–207, 2007.

[14] D. L. Rockwood, A. W. Rudie, S. A. Ralph, J. Y. Zhu, and J. E. Winandy, "Energy product options for *Eucalyptus* species grown as short rotation woody crops," *International Journal of Molecular Sciences*, vol. 9, no. 8, pp. 1361–1378, 2008.

[15] M. Rejmánek and D. M. Richardson, "Eucalypts," in *Encyclopedia of Biological Invasions*, D. Simberloff and M. Rejmánek, Eds., pp. 203–209, University of California Press, Berkeley, Calif, USA, 2011.

[16] P. H. M. da Silva, F. Poggiani, A. M. Sebbenn, and E. S. Mori, "Can Eucalyptus invade native forest fragments close to

commercial stands?" *Forest Ecology and Management*, vol. 261, no. 11, pp. 2075–2080, 2011.

[17] D. Dougherty and J. Wright, "Silviculture and economic evaluation of eucalypt plantations in the southern U.S.," *BioResources*, vol. 7, no. 2, pp. 1994–2001, 2012.

[18] T. H. Booth, "Eucalypts and their potential for invasiveness particularly in frost-prone regions," *International Journal of Forestry Research*, vol. 2012, Article ID 837165, 7 pages, 2012.

[19] P. I. Becerra and R. O. Bustamante, "The effect of herbivory on seedling survival of the invasive exotic species *Pinus radiata* and *Eucalyptus globulus* in a Mediterranean ecosystem of Central Chile," *Forest Ecology and Management*, vol. 256, no. 9, pp. 1573–1578, 2008.

[20] K. X. Ruthrof, "Invasion by *Eucalyptus megacornuta* of an urban bushland in Southwestern Australia," *Weed Technology*, vol. 18, no. 1, pp. 1376–1380, 2004.

[21] G. G. Forsyth, D. M. Richardson, P. J. Brown, and B. W. Van Wilgen, "A rapid assessment of the invasive status of *Eucalyptus* species in two South African provinces," *South African Journal of Science*, vol. 100, no. 1-2, pp. 75–77, 2004.

[22] L. Henderson, *SAPIA NEWS* No. 12, South Africa Agricultural Research Council-Plant Protection Research Institute, Southern African Plant Invaders Atlas, 2009, http://www.arc.agric.za/uploads/images/0_SAPIA_NEWS_No._12.pdf.

[23] C. C. Daehler, J. S. Denslow, S. Ansari, and H. C. Kuo, "A risk-assessment system for screening out invasive pest plants from Hawaii and other Pacific Islands," *Conservation Biology*, vol. 18, no. 2, pp. 360–368, 2004.

[24] C. E. Buddenhagen, C. Chimera, and P. Clifford, "Assessing biofuel crop invasiveness: a case study," *PLoS ONE*, vol. 4, no. 4, article e526, 2009.

[25] W. Dawson, D. F. R. P. Burslem, and P. E. Hulme, "The suitability of weed risk assessment as a conservation tool to identify invasive plant threats in East African rainforests," *Biological Conservation*, vol. 142, no. 5, pp. 1018–1024, 2009.

[26] D. R. Gordon, K. J. Tancig, D. A. Onderdonk, and C. A. Gantz, "Assessing the invasive potential of biofuel species proposed for Florida and the United States using the Australian weed risk assessment," *Biomass and Bioenergy*, vol. 35, no. 1, pp. 74–79, 2011.

[27] J. L. Lockwood, P. Cassey, and T. Blackburn, "The role of propagule pressure in explaining species invasions," *Trends in Ecology and Evolution*, vol. 20, no. 5, pp. 223–228, 2005.

[28] D. R. Gordon, B. Mitterdorfer, P. C. Pheloung et al., "Guidance for addressing the Australian weed risk assessment questions," *Plant Protection Quarterly*, vol. 25, no. 2, pp. 56–74, 2010.

[29] D. R. Gordon and C. A. Gantz, "Screening new plant introductions for potential invasiveness: a test of impacts for the United States," *Conservation Letters*, vol. 1, no. 5, pp. 227–235, 2008.

[30] U.S. Department of Agriculture, Agricultural Research Service, National Genetic Resources Program, Germplasm Resources Information Network—(GRIN) [Online Database], National Germplasm Resources Laboratory, Beltsville, Md, USA, 2012, http://www.ars-grin.gov/cgi-bin/npgs/html/taxgenform.pl?language=en.

[31] U.S. Department of Agriculture, Natural Resources Conservation Service, The PLANTS Database, National Plant Data Center, Baton Rouge, Louisiana, USA, 2012, http://plants.usda.gov.

[32] A. V. Slee, M. I. H. Brooker, S. M. Duffy, and J. G. West, EUCLID Eucalypts of Australia, 3rd Edition, Centre for Plant Biodiversity Research, Canberra, Australia, 2006, http://www.anbg.gov.au/cpbr/cd-keys/Euclid/sample/html/index.htm.

[33] The Council of Heads of Australasian Herbaria, Australia's Virtual Herbarium, 2012, http://avh.ala.org.au/.

[34] S. G. Southerton, P. Birt, J. Porter, and H. A. Ford, "Review of gene movement by bats and birds and its potential significance for eucalypt plantation forestry," *Australian Forestry*, vol. 67, no. 1, pp. 44–53, 2004.

[35] J. G. Virtue, S. J. Bennett, and R. P. Randall, "Plant introductions in : how can we resolve "weedy" conflicts of interest?" in *Proceedings of the 14th Australian Weeds Conference*, B. M. Sindel and S. B. Johnson, Eds., pp. 42–48, Weed Society of New South Wales, Sydney, Australia, 2004.

[36] J. A. Crooks, "Lag times and exotic species: the ecology and management of biological invasions in slow-motion," *Ecoscience*, vol. 12, no. 3, pp. 316–329, 2005.

[37] I. Kowarik, "Time lags in biological invasion with regard to the success and failure of alien species," in *Plant Invasions-General Aspects and Special Problems*, P. Pysek, M. Rejmánek, and M. Wade, Eds., pp. 15–38, SPB Academic, Amsterdam, The Netherlands, 1995.

[38] M. Rejmánek, D. M. Richardson, S. I. Higgins, M. J. Pitcairn, and E. Grotkopp, "Ecology of invasive plants—state of the art," in *Invasive Alien Species: A New Synthesis*, H. A. Mooney, R. N. Mack, J. A. McNeely, L. Neville, P. J. Schei, and J. Waage, Eds., pp. 104–161, Island Press, Washington, DC, USA, 2005.

[39] K. Y. Chong, R. T. Corlett, D. C. J. Yeo, and H. T. W. Tan, "Towards a global database of weed risk assessments: a test of transferability for the tropics," *Biological Invasions*, vol. 13, no. 7, pp. 1571–1577, 2011.

[40] D. W. Gerhardt, Director of Operations Support, MeadWestvaco, Corp., Personal Communications, 2012.

[41] R. Gonzalez, T. Treasure, J. Wright et al., "Exploring the potential of *Eucalyptus* for energy production in the Southern United States: financial analysis of delivered biomass. Part I," *Biomass and Bioenergy*, vol. 35, no. 2, pp. 755–766, 2011.

[42] A. S. Davis, R. D. Cousens, J. Hill, R. N. Mack, D. Simberloff, and S. Raghu, "Screening bioenergy feedstock crops to mitigate invasion risk," *Frontiers in Ecology and the Environment*, vol. 8, no. 10, pp. 533–539, 2010.

[43] J. N. Barney and J. M. DiTomaso, "Nonnative species and bioenergy: are we cultivating the next invader?" *BioScience*, vol. 58, no. 1, pp. 64–70, 2008.

[44] A. R. Jakubowski, M. D. Casler, and R. D. Jackson, "Has selection for improved agronomic traits made reed canarygrass invasive?" *PLoS ONE*, vol. 6, no. 10, Article ID e25757, 2011.

[45] M. Olofsdotter, "Getting closer to breeding for competitive ability and the role of allelopathy-an example from rice (*Oryza sativa*)," *Weed Technology*, vol. 15, no. 4, pp. 798–806, 2001.

[46] D. De Martinis and C. Mariani, "Silencing gene expression of the ethylene-forming enzyme results in a reversible inhibition of ovule development in transgenic tobacco plants," *Plant Cell*, vol. 11, no. 6, pp. 1061–1071, 1999.

[47] C. Zhang, K. H. Norris-Caneda, W. H. Rottmann, J. E. Gulledge, and J. E. S, "Control of pollen mediated gene flow in transgenic trees," *Plant Physiology*, vol. 159, no. 4, pp. 1319–1334, 2012.

[48] M. A. W. Hinchee, Chief Science Officer, ArborGen, Inc., Personal Communications, 2012.

[49] S. L. Flory, K. A. Lorentz, D. R. Gordon, and L. E. Sollenberger, "Experimental approaches for evaluating the invasion risk of biofuel crops," *Environmental Research Letters*, vol. 7, no. 4, Article ID 045904, 2012.

# 13

# Environmental and Socioeconomic Indicators for Bioenergy Sustainability as Applied to *Eucalyptus*

**Virginia H. Dale, Matthew H. Langholtz, Beau M. Wesh, and Laurence M. Eaton**

*Oak Ridge National Laboratory, Environmental Sciences Division, Center for BioEnergy Sustainability, Oak Ridge, TN 37831, USA*

Correspondence should be addressed to Virginia H. Dale; dalevh@ornl.gov

Academic Editor: John Stanturf

*Eucalyptus* is a fast-growing tree native to Australia and could be used to supply biomass for bioenergy and other purposes along the coastal regions of the southeastern United States (USA). At a farmgate price of 66 dry Mg , a potential supply of 27 to 41.3 million dry Mg year$^{-1}$ of *Eucalyptus* could be produced on about 1.75 million ha in the southeastern USA. A proposed suite of indicators provides a practical and consistent way to measure the sustainability of a particular situation where *Eucalyptus* might be grown as a feedstock for conversion to bioenergy. Applying this indicator suite to *Eucalyptus* culture in the southeastern USA provides a basis for the practical evaluation of socioeconomic and environmental sustainability in those systems. Sustainability issues associated with using *Eucalyptus* for bioenergy do not differ greatly from those of other feedstocks, for prior land-use practices are a dominant influence. Particular concerns focus on the potential for invasiveness, water use, and social acceptance. This paper discusses opportunities and constraints of sustainable production of *Eucalyptus* in the southeastern USA. For example, potential effects on sustainability that can occur in all five stages of the biofuel life cycle are depicted.

## 1. Introduction

As society moves forward toward considering energy options other than petroleum-based fuels, bioenergy is an important alternative to evaluate. In addition to developing the ability to provide energy, it is important to identify ways to do so in a sustainable manner. The concept of sustainability refers to activities that support long-term balance in environmental, social, and economic conditions in particular circumstances. Brundtland [1] defined it as the capacity of an activity to operate while maintaining options for future generations. Yet development and use of energy always has some environmental impacts, for example, on water and air quality and biodiversity. The challenge, therefore, is to develop means to address tradeoffs in the costs and benefits in energy choices while considering effects on both environmental and socioeconomic aspects of sustainability. The first step in determining these effects is developing a means to quantify and measure Brundtland's broad definition of sustainability. Building on prior efforts, this paper discusses proposed indicators of sustainability and attempts to apply them to evaluate the potential for using *Eucalyptus* for sustainable bioenergy in the southeastern United States (USA). However the application of sustainability indicators in this situation is limited by the paucity of pertinent information. Hence, this analysis also suggests key information that needs to be obtained in order to evaluate sustainability of using *Eucalyptus* for bioenergy in the southeastern USA.

Approaches to bioenergy options should consider a diversity of feedstock options that are suitable in different regions and contexts. Feedstocks being considered for bioenergy in the southeastern USA include forest and agriculture wastes as well as dedicated perennial energy crops such as herbaceous grasses and fast-growing trees [2]. There is no one feedstock type suitable for all places. The appropriate conditions for growing feedstocks in a region depend on prevailing climate and soils, past land-use practices, and existing equipment and experience of the growers. In addition, available forest, agriculture, and other residues are also bioenergy feedstocks.

*Eucalyptus*, a fast-growing tree native to Australia, is currently being grown in the southeastern USA for mulch and is being considered as a potential feedstock for future bioenergy production. The purpose of this paper is to discuss (1) the locations and amounts of feedstock that *Eucalyptus* could provide in the southeastern USA and (2) how environmental and socioeconomic indicators can be used to evaluate the sustainability of a bioenergy industry based on *Eucalyptus*. While this paper focuses on sustainability of *Eucalyptus* for bioenergy in the southeastern USA, we designed it to serve as a template for how sustainability implications of bioenergy crop options can be considered at the regional scale. However difficult challenges remain such as obtaining the data necessary for such quantitative evaluation and determining appropriate and useful methods for collective evaluation of the many components of sustainability.

## 2. Short-Rotation Woody Crops and *Eucalyptus* Potential as a Bioenergy Crop in the Southeastern USA

*Eucalyptus* spp. is the world's most widely planted hardwood genus. Its fast, uniform growth, self-pruning behavior, and ability to coppice make it desirable for timber, pulpwood, and bioenergy feedstocks. High yield is an important attribute for any short-rotation woody crop (SRWC), for it improves the economics and reduces the area needed for production. In Brazil, *Eucalyptus* hybrids such as *E. grandis* × *E. urophylla* produce 22 to 27 dry Mg ha$^{-1}$ yr$^{-1}$ [7]. In Florida *E. grandis* can achieve more than 34 dry Mg ha$^{-1}$ yr$^{-1}$ [8], rivaling yields of such potential feedstocks as *Sacchrum* spp. (energy cane) and *Pennisetum purpureum* (napier grass). Hence there is great interest in *Eucalyptus* as a bioenergy feedstock.

Estimating the location of where *Eucalyptus* might be planted to support the bioenergy industry is a prerequisite for considering its effects. *Eucalyptus* production for bioenergy in the southeastern USA is likely to occur along the southeastern Atlantic and Gulf of Mexico coastal regions of the USA where *Eucalyptus*'s lack of hardiness to frost entails a low to moderate risk (Figure 1) [3]. This area encompasses some places with existing production of *Eucalyptus*. For example, *Eucalyptus grandis* has been grown as a commercial crop in Florida primarily for mulchwood for the past five decades [9], largely relying on its ability to sprout prolifically subsequent to coppicing. ArborGen has developed a freeze-tolerant *Eucalyptus*, that has a tolerance down to −8.9°C while maintaining high productivity [7]. It is not certain how climate changes and associated changes in hardiness zones may affect the potential areas where *Eucalyptus* might grow. In any case, *Eucalyptus* will most likely be grown along the coastal areas of the southeastern USA where both frost hardiness and salt tolerance may be an issue.

We provide estimates of the potential supplies of *Eucalyptus* for bioenergy by utilizing projections generated from the Billion-Ton Update, which estimated the forest and agricultural resource potential for the expansion of bioenergy and bioproducts industries [2]. Estimates of biomass supplies were produced for a range of prices, and the amounts and locations were specified at the county/parish level with projections from 2012 to 2030. Feedstocks include all major primary and secondary forest and agricultural residues, major waste feedstocks, and energy crops grown specifically for bioenergy, including SRWCs. The models in the Billion-Ton Update incorporate yields and production budgets that represent commercial-scale production of various SRWC species, including willow (*Salix* spp.), loblolly pine (*Pinus taeda*), poplar (*Populus* spp.), and, of interest to this paper, *Eucalyptus*.

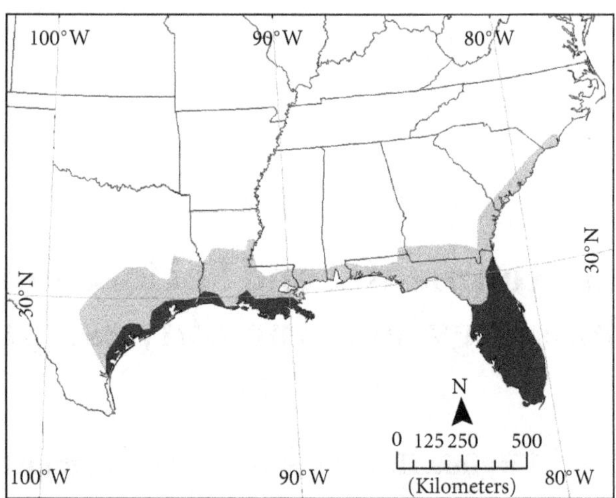


Figure 1: Map of locations for potential feedstock locations for *Eucalyptus* in the United States that could be used for bioenergy (as estimated by Kline and Coleman [3] based on the USDA Plant Hardiness Zones [6] and interviews with experts).

Projections of biomass production were made for the Billion-Ton Update using supply/cost curves generated by POLYSYS [10, 11] for each major feedstock group for a baseline and a high-yield case. The baseline case assumes a continuation of the USA Department of Agriculture's 10-year forecast of yields for major food and forage crops to 2018 and then extrapolates it to 2030. The high-yield scenario assumes increased yields and higher adoption of no-till cultivation for traditional crops. All energy crops are assumed to have annual yield increase of 1% for the baseline case, and three levels of increase (2%, 3%, and 4%) were considered for the high-yield scenario. In addition, the POLYSYS model assumes that, in order for energy crops to be grown in a county, the crops must provide a higher net return than the commodity crops or pastures that they displace, and there can only be limited impacts on food, feed, exports, and fiber production. Furthermore, pasture can only convert to energy crops if the displaced forage is made up through intensification. Energy crops are not allowed on irrigated cropland or pasture. Best Management Practices (BMPs) are assumed to be used for establishment, cultivation, maintenance, and harvesting of energy crops. Additionally, energy crops are allowed to compete against each other for land on a per-acre net return basis. Other assumptions of the POLYSYS analysis used by the Billion-Ton Update are

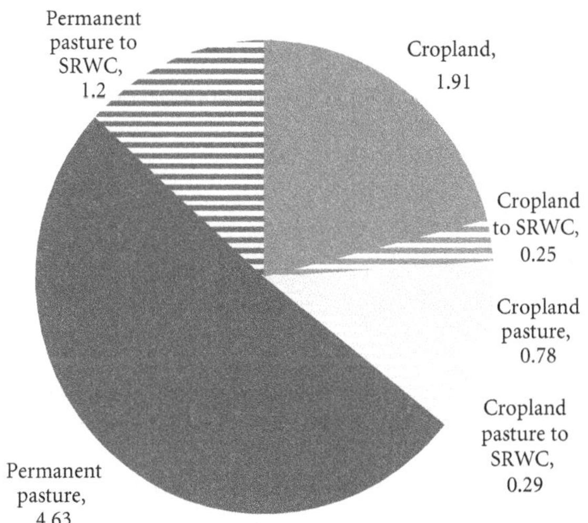


FIGURE 2: Current allocation of cropland, cropland pasture, and permanent pasture to SRWC within the potential geographic range of *Eucalyptus* (Billion-Ton Update Base Case Scenario assuming \$66/Mg$^{-1}$ farmgate price, results for year 2030).

detailed in the full report (see Appendix B of the report for general modeling assumptions) [2].

To quantify an upper limit of sustainable production of *Eucalyptus* in response to a bioenergy market as constrained by the POLYSYS assumptions summarized above, we disaggregated the SRWC production estimates from the Billion-Ton Update for the 192 counties in the *Eucalyptus* production ranges shown in Figure 2. Those 192 counties were identified as having centroids within the low- and moderate-risk *Eucalyptus* ranges shown in Figure 1. County-level POLYSYS results for SRWC production in these 192 counties were used to estimate potential *Eucalyptus* production (yield and land area) in the USA. POLYSYS simulates SRWCs in this range as any tree species that is managed as single-stem for eight-year rotations and yielding a mean annual increment of about 13 dry Mg ha$^{-1}$ yr$^{-1}$, with yields projected to increase with future improvements. Actual *Eucalyptus* production practices would deviate from these assumptions, for some of the simulated SRWC production will be met with pine, poplars, or other species.

We estimate a supply potential in year 2030 of 27 to 41 million dry Mg year$^{-1}$ of *Eucalyptus* production potential in the Southeast by assuming all SRWC production is realized by *Eucalyptus* within the baseline and a high-yield case estimated by the Billion-Ton Update and shown in Figure 2. This calculation derives from simulating a farmgate price of \$66 dry Mg$^{-1}$(\$60 dry ton$^{-1}$) under the baseline and high-yield (4% yield increase) scenarios. Under these assumptions, the Billion-Ton Update estimates that 1.0 to 1.5 billion dry Mg year$^{-1}$ of biomass are available from all sources in the conterminous USA by 2030 [2]. These projections include 114 to 285 million dry Mg year$^{-1}$ of SRWC, of which 27 to 41 million dry Mg year$^{-1}$ are produced in the 192 counties identified above in 2030.

To illustrate the scale of potential landscape change that might be attributable to future *Eucalyptus* production, land use and conversion from this same simulation is shown in Table 1 and Figure 2. Assuming a farmgate price of \$66 dry Mg$^{-1}$, these model results suggest that up to 0.25, 0.29, and 1.20 million hectares of cropland, cropland pasture, and permanent pasture within the geographic range of *Eucalyptus* production could be converted to SRWCs by the year 2030. This amount represents about 19% of the agricultural land and 4.5% of total land in these 192 counties.

POLYSYS is constrained to only allow SRWC production on non-forested land but also projects feedstock supplies to 2030 from logging residues, thinnings, and pulpwood from forest land. Depending on policy, economics, and landowner values, forestland might also be brought into *Eucalyptus* production. For example, this same POLYSYS simulation produces 200,900 Mg of softwood pulpwood in 2030 from the 192 selected counties. Assuming a mean annual increment of 11 Mg ha$^{-1}$ yr$^{-1}$, this material could be drawn from about 18 thousand hectares of forestland, some of which could be converted from pine to *Eucalyptus* or other SRWC plantations. Hence the potential aggregate change of the landscape of about 1.8 million ha warrants critical evaluation of possible effects.

## 3. Assessing Sustainability of the *Eucalyptus* Biofuel Supply Chain via Indicators

To assess sustainability, means of quantifying it have to be specified. Brundtland's broad definition of sustainability is useful but is nonspecific. Therefore, many groups have been working toward establishing a set of indicators that can be used to quantify bioenergy sustainability (e.g., the Roundtable on Sustainable Biofuels [12], Global Bioenergy Partnership [13], and Council on Sustainable Biomass Production [14]). However, implementation is hampered when indicators are too numerous, too costly, and too broad [15] as is the case for current efforts.

Thus, our team of researchers at Oak Ridge National Laboratory considered bioenergy sustainability indicators proposed by many groups and selected a small set of measureable indicators of bioenergy sustainability using the criteria of being practical, sensitive to stresses, unambiguous, anticipatory, predictive, calibrated with known variability, and sufficient when considered collectively [16]. These conditions are also prerequisites for energy security [17] as well as other aspects of sustainability. Furthermore, the selected indicators are less cumbersome than those proposed by other groups because we assume they only apply in situations that have basic legal, regulatory, and enforcement services and transparent, stable, and legitimate governance. This final assumption is critical, for it avoids situations where bioenergy has been called on to resolve major development challenges such as lack of land tenure or government corruption.

We hypothesize that the selected suite of 35 environmental and socioeconomic indicators provides a practical and consistent way to assess the sustainability of a particular situation where a feedstock might be grown and converted

Table 1: Area of cropland, cropland pasture, and permanent pasture (1) in the USA lower forty-eight states, (2) in *Eucalyptus* ranges in the Southeast, (3) potentially converted to *Eucalyptus* in a Base Case Scenario, and (4) potentially converted to *Eucalyptus* in a High-yield Scenario.

| | Cropland | Cropland Pasture (million hectares) | Permanent Pasture |
|---|---|---|---|
| (1) USA (lower 48 states) total[a] | 125.82 | 13.15 | 155.59 |
| (2) Total in *Eucalyptus* range[b] | 2.17 | 1.07 | 5.83 |
| (3) Converted from (2) to SRWC, Base Case[c] | 0.25 | 0.29 | 1.20 |
| (3) Converted from (2) to SRWC, High-yield[d] | 0.27 | 0.28 | 1.08 |

[a] Census of Agriculture, 2007.

[b] Includes counties with centroids contained by both low- and moderate-risk *Eucalyptus* ranges from Kline and Coleman [3] shown in Figure 1.

[c] Areas in (2) above that are converted to SRWC in the Billion-Ton Update (DOE 2011) [2], assuming \$66 dry $Mg^{-1}$ farmgate price, Base Case Scenario.

[d] Areas in (2) above that are converted to SRWC in the Billion-Ton Update (DOE 2011) [2], assuming \$66 dry $Mg^{-1}$ farmgate price, High-Yield Scenario.

Table 2: List of recommended environmental indicators for bioenergy sustainability (derived from [4]).

| Category | Indicator | Units |
|---|---|---|
| Soil quality | (1) Total organic carbon (TOC) | Mg/ha |
| | (2) Total nitrogen (N) | Mg/ha |
| | (3) Extractable phosphorus (P) | Mg/ha |
| | (4) Bulk density | $g/cm^3$ |
| Water quality and quantity | (5) Nitrate concentration in streams (and export) | concentration: mg/L; export: kg/ha/yr |
| | (6) Total phosphorus (P) concentration in streams (and export) | concentration: mg/L; export: kg/ha/yr |
| | (7) Suspended sediment concentration in streams (and export) | concentration: mg/L; export: kg/ha/yr |
| | (8) Herbicide concentration in streams (and export) | concentration: mg/L; export: kg/ha/yr |
| | (9) Peak storm flow | L/s |
| | (10) Minimum base flow | L/s |
| | (11) Consumptive water use (incorporates base flow) | feedstock production: $m^3$/ha/day; biorefinery: $m^3$/day |
| Greenhouse gases | (12) $CO_2$ equivalent emissions ($CO_2$ and $N_2O$) | $kgC_{eq}$/GJ |
| Biodiversity | (13) Presence of taxa of special concern | Presence |
| | (14) Habitat area of taxa of special concern | Ha |
| Air quality | (15) Tropospheric ozone | Ppb |
| | (16) Carbon monoxide | Ppm |
| | (17) Total particulate matter less than 2.5 $\mu$m diameter ($PM_{2.5}$) | $\mu g/m^3$ |
| | (18) Total particulate matter less than 10 $\mu$m diameter ($PM_{10}$) | $\mu g/m^3$ |
| Productivity | (19) Aboveground net primary productivity (ANPP)/Yield | $gC/m^2$/year |

to bioenergy. The 19 environmental indicators of bioenergy sustainability fall into the categories of soil quality, water quality and quantity, greenhouse gases, biodiversity, air quality, and productivity (Table 2) [4]. Socioeconomic aspects of bioenergy sustainability are defined by 16 indicators that fall into the categories of social wellbeing, energy security, trade, profitability, resource conservation, and social acceptability (Table 3) [5]. These indicators constitute a way to assess the capacity of bioenergy systems to advance toward the goal of sustainability. Here we consider how these 35 indicators can be applied to the use of *Eucalyptus* to produce bioenergy in the southeastern USA.

Indicators of bioenergy sustainability can be applied conceptually to a region, but actual application should be context specific [18]. For example, sustainability of *Eucalyptus* depends on a variety of factors, such as prevailing environmental conditions, ongoing management, previous land practices, and intended use of the product. While we discuss how these indicators might be applied to *Eucalyptus* deployment in the southeastern USA for bioenergy, actual evaluation of the sustainability of *Eucalyptus* depends on the specific situation and management, and much of that information is not yet known. Therefore, when appropriate and possible, we rely on information from other locations and uses of *Eucalyptus* other than for bioenergy.

To illustrate their application, we discuss how potential effects on sustainability of using *Eucalyptus* for bioenergy occur in all five stages of the biofuel life cycle (Table 4): feedstock production, feedstock logistics, conversion to biofuel, biofuel logistics, and biofuel end uses. Each is discussed below. All feedstock types have effects (e.g., on greenhouse gas emissions, air quality, profitability, social well being, trade, energy security, resource conservation and social acceptability) that are distributed throughout the supply

TABLE 3: List of recommended socioeconomic indicators for bioenergy sustainability (derived from Dale et al. (2013) [5]).

| Category | Indicator | Units |
|---|---|---|
| Social well being | Employment | Number of full time equivalent (FTE) jobs[1] |
| | Household income | Dollars per day |
| | Work days lost due to injury | Average number of work days lost per worker per year |
| | Food security | Percent change in food price volatility |
| Energy security | Energy security premium | Dollars/gallon biofuel |
| | Fuel supply volatility | Standard deviation of monthly percentage price changes over one year |
| External trade | Terms of trade | Ratio (price of exports/price of imports) |
| | Trade volume | Dollars (net exports or balance of payments) |
| Profitability | Return on investment (ROI)[1] | Percent (net investment/initial investment) |
| | Net present value (NPV)[2,3] | Dollars (present value of benefits minus present value of costs) |
| Resource conservation | Depletion of non-renewable energy resources | Amount of petroleum extracted per year (MT) |
| | Fossil Energy Return on Investment (fossil EROI) | Ratio of amount of fossil energy inputs to amount of useful energy output (MJ) (adjusted for energy quality) |
| Social acceptability | Public opinion | Percent favorable opinion |
| | Transparency | Percent of indicators for which timely and relevant performance data are reported[5] |
| | Effective stakeholder participation | Percent of documented responses to stakeholder concerns and suggestions reported on an annual basis |
| | Risk of catastrophe[4] | Annual probability of catastrophic event |

[1] FTE employment includes net new jobs created, plus jobs maintained that otherwise would have been lost, as a result of the system being assessed.
[2] Conventional economic models can address long-term sustainability issues by extending the planning horizon, projecting as an infinite geometric series, or calculating with a low discount rate.
[3] Can be expanded to include non-market externalities (e.g., water quality, GHG emissions).
[4] A catastrophic event can be defined as an event or accident that has more than 10 human fatalities, affects an area greater than 1000 ha, or leads to extinction or extirpation of a species.
[5] For example this measure could be the percent of all social, economic and environmental indicators identified via stakeholder consultation or the percent of the 35 indicators listed here and in McBride et al. [4] for which relevant baseline, target, and performance data are reported and made available to the public on a timely basis (at least annually).

chain; however much more is known about the feedstock production stage for *Eucalyptus*.

*3.1. Feedstock Production.* Feedstock production builds from the current condition of the land, soil, and water resources and encompasses propagation, site preparation, establishment, and management. Sustainability effects of bioenergy that are specific to *Eucalyptus* and other SRWC are largely concentrated in the feedstock production stage of the life cycle (Table 4). As with any dedicated biomass plantation, *Eucalyptus* plantations can affect all six categories of environmental indicators (soil quality, water quality and quantity, greenhouse gases, biodiversity, air quality, and productivity), and the effects are specific to each location, prior conditions, and management practice.

Resource conditions prior to establishment of plantations have significant implications on effects of these attributes. These conditions include the soil, water, and air quality, as well as biodiversity and habitat circumstances of the area prior to the establishment of the crop. The sign and degree of effects are different for each situation. The effects can be negative where clearing natural forests compromises biodiversity or soil conditions and depend on the spatial scale being considered [19]. Carbon sequestration of *Eucalyptus* plantations on prior pasture lands is influenced by precipitation patterns and intervals between harvests [20]. The effects can be positive in cases where plantations replace little or poorly managed vegetation, or negative if the plantations are poorly managed and replace well-established and productive stands. For example, when established on former pasture land in southern Europe, *E. nitens* and *E. globulus* enhance carbon sequestration in both biomass and soil [21]. And studies of *E. nitens* in Australia confirm that management via thinning, pruning, and nitrogen fertilization has interactive effects on above-ground biomass and biomass partitioning among crown, bole, and roots [22]. As another example, *Eucalyptus* has been demonstrated to provide beneficial impacts on soil quality, water quality and quantity, greenhouse gases, and biodiversity when *Eucalyptus* plantations are established for purposes of mine land reclamation or phytoremediation (e.g., [23–29]) and could be used on other degraded land. As with any other bioenergy crop, appropriate lands and management practices must be used if sustainability is to be achieved.

Water use by *Eucalyptus* grown for bioenergy is a concern where water is scarce, as is the case during droughts and

Table 4: Categories of sustainability indicators that experience environmental or socioeconomic effects within the *Eucalyptus*-to-biofuel supply chain. Major effects for *Eucalyptus* and other fast-growing non-native crops are depicted by *, and additional effects exhibited by all feedstocks are depicted by +. A blank means there is no effect in that category.

| Categories of indicators | | Environmental Effects | | | | | | Socioeconomic Effects | | | | | |
|---|---|---|---|---|---|---|---|---|---|---|---|---|---|
| Steps in biofuel supply chain | Components of each biofuel supply chain step | Soil quality | Water quality and quantity | Greenhouse gases | Biodiversity | Air quality | Productivity | Profitability | Social well being | External trade | Energy security | Resource conservation | Social acceptability |
| Feedstock Production | Resource Conditions | * | * | * | * | * | * | + | + | | | + | * |
| | Feedstock Type | * | * | * | * | * | * | * | + | + | + | + | * |
| | Management | * | * | * | * | * | * | + | + | | | + | * |
| Feedstock Logistics | Harvesting & Collection | + | * | + | + | + | + | * | + | | | | + |
| | Processing | | | + | | + | | + | + | | | | + |
| | Storage | | | + | | + | | + | + | | + | | + |
| | Transport | | | + | + | + | | + | + | + | | + | + |
| Conversion to Biofuel | Conversion Process | | + | + | | + | | + | + | + | + | + | + |
| | Fuel Type | | | + | | | | + | + | + | + | + | |
| | Coproducts | | + | + | | + | | + | + | + | + | + | |
| Biofuel Logistics | Transport | | | + | | + | | + | + | + | | | + |
| | Storage | | | + | | + | | + | + | + | + | | + |
| Biofuel End Uses | Engine Type and Efficiency | | | + | | + | | + | + | + | + | + | + |
| | Blend Conditions | | | + | | + | | + | + | + | + | + | + |

for selected areas of the southeastern USA, including parts of the 192 countries where *Eucalyptus* might be grown. The water scarcity issue is localized and relates more to population growth and demand than to inherent supply limits. Of most concern is groundwater recharge due to deep rooting in areas where the primary drinking water source is groundwater (as in peninsular Florida). As a fast-growing tree, *Eucalyptus* can use significant amounts of water. This trait may be a concern in areas where groundwater is scarce or may be an asset in applications such as phytoremediation or reclaiming saturated clay settling areas of mined lands [24]. The main question of water use is how tradeoffs in allocation are addressed. Once established, *Eucalyptus* can tolerate drought and water scarcity. For example, *E. occidentalis* was able to produce 22 tons/ha in the dry land Mediterranean climate of southwestern Australia [30]. Eucalypts are able to make use of soil water to depths of 8 to 10 m within 7 years of planting and are able to penetrate clay subsoils [30]. As with other categories of indicators, the interpretation of the values of water quality and quantity indicators is specific to each situation.

Similar to any agricultural or forest land use, mismanagement can result in negative environmental impacts, while appropriate management can enhance or at least maintain environmental quality. The question then becomes, "what are appropriate management practices for *Eucalyptus* in the southeastern USA?" For example, management practices of *Eucalyptus* plantations can serve to control soil erosion, with implications for soil and water quality, as well as yield. On many sites in the southeastern USA that are available for planting *Eucalpytus*, both competing vegetation and low fertility will need to be addressed.

Expansive monocultures managed with stringent control of competing vegetation are likely to reduce biodiversity. Conversely, a mosaic of *Eucalyptus* stands interspersed on the landscape that includes native vegetation and a diversity of stand structures may have less impact on biodiversity. Preplantation land-use conditions also have implications for biodiversity. For example, higher diversity can occur in pine plantations established on cutover forest land than planted on former agricultural land [31]. Hence, establishing *Eucalyptus* plantations on land previously cleared for rowcrops or pasture in the southeastern USA should be designed to not jeopardize existing biodiversity. Maintaining land in forest or increasing forest area can promote biodiversity via habitat provision services of forests and forest edges.

Areas with high native biodiversity should be excluded from plantation development. In the southeastern USA, high-diversity forest lands are often in federal ownership [32] and are excluded from providing bioenergy feedstocks by the Renewable Fuel Standard [33].

Based on concern about the invasiveness of *Eucalyptus* because it is a foreign plant to the USA, The Nature Conservancy evaluated it using the Australian Weed Risk Assessment system [34]. Some *Eucalyptus* species are considered by Florida to be naturalized in disturbed areas and not invasive [35]. Using a check list to evaluate invasiveness, *E. amplifolia* requires further evaluation, but *E. camaldulensis* and *E. grandis* are considered invasive [36]. Even in Brazil, where the amount of *E. grandis* plantations are the largest (4.2 m ha in 2010), *E. grandis* is not considered invasive for several reasons. The species has very few small seeds within a fire-protective capsule. These capsules help the seeds grow after a fire but prevent them from growing otherwise, for the seed must be on exposed soil to germinate, with survival requiring no surrounding vegetation and full sunlight [6]. These seeds also do not have any characteristics that facilitate dispersal by wind, water, or other means. Hence tree height and the wind conditions are the main factors influencing how far the seeds will travel, and seeds typically fall within a distance of 1.3 times the height of the tree [37]. Furthermore, in order to reduce invasiveness, ArborGen has successfully engineered a *Eucalyptus* hybrid that does not produce pollen [30]. Introduction of *Eucalyptus* species into new areas and large-scale plantations requires careful evaluation of their potential for invasiveness [34].

Furthermore, salt-affected soil usually does not support high productivity due to the degradation of the soil. To increase both soil quality and profits, salt-tolerant species such as *E. camaldulensis* can be grown and harvested on salt-affected soils [36]. *E. occidentalis* was able to produce 31 tons/ha on salinized soils in southwestern Australia that had previously been abandoned by agriculture [36].

*Eucalyptus* plantations can also affect all aspects of the socioeconomic components of sustainability: social well-being, energy security, trade, profitability, resource conservation, and social acceptability, as does any bioenergy crop. These effects can be positive if the bioenergy system is well managed and located in a place where benefits can accrue. For example, a refinery could be located where rural jobs are in decline, and the establishment of a new industry based on *Eucalyptus* could revitalize the community while providing a new energy source that might be competitive with fossil fuels. The biggest difference in social acceptability from most other SRWC being proposed for bioenergy in the southeastern USA is that *Eucalyptus* is not a native species and has high water demands and potential for invasiveness. Use of *Eucalyptus* has been initially challenged in many places where it is planted but is not native. However, as one example of the turnaround in its public acceptance, expansion of *Eucalyptus* forestry in Ethiopia resulted in 96% of growers and 90% of the district experts supporting that expansion largely for economic reasons and despite environmental concerns [38]. In the USA outside of Florida, there are no state or federal restrictions on planting non-native *Eucalyptus*, and Florida's restriction is based on invasiveness, not on non-native status. Furthermore, the frost-tolerant hybrid mentioned earlier is a genetically modified organism, which is regulated under federal laws.

As with other forest practices, the use of *Eucalyptus* for bioenergy provides an opportunity to retain land in forest versus succumbing to other land pressures such as development or urban expansion. The demand for bioenergy and value of the *Eucalyptus* for that purpose as compared to other activities on the land determine where and how *Eucalyptus*-based bioenergy will occur. Retaining land in productive forestry could also provide rural socioeconomic benefits such as jobs and profit from the land. While much focus now for bioenergy in the southeastern USA is on perennial grasses, cost projections for *Eucalyptus*-delivered feedstock may be more economical in some areas. For example, the estimated lowest cost based on simulations of switchgrass is \$67 $Mg^{-1}$, and for *Eucalyptus* is \$55 $Mg^{-1}$ for the southeastern USA [39].

Currently 9.6 percent of the land in seven Gulf South states where *Eucalyptus* might grow is in plantation forests [37]. With the forest industry downturn in the southeastern USA, both jobs and forest land are being lost [40, 41]. At the same time, more land is being developed for urban and suburban use, and bioenergy crops, such as *Eucalyptus*, may offer an opportunity to counteract these trends [42]. To this end, some developments are incorporating a landscape design that includes both forests and houses within the overall planning. For example, a housing development in the coastal region near Ravenal, South Carolina, allocates a portion of the total planned area to forestry where several *Eucalyptus* spp. are being grown in test trials.

*3.2. Feedstock Logistics.* Feedstock logistics include the harvesting, processing, storage and transport of the feedstock to the refinery. Of particular environmental concern in *Eucalyptus* feedstock logistics is effects on water quality and quantity during harvest and on biodiversity during transport. Biofuel cost is highly sensitive to the delivered cost of the *Eucalyptus* feedstock, which can constitute 35–50% of the total cost of ethanol production [43, 44].

*3.3. Conversion to Biofuel, Biofuel Logistics, and Biofuel End Uses.* Conversion is the process of changing the feedstock into biofuel and depends on the fuel type selected and any coproducts created. Sometimes the coproducts have more value than the fuel produced. Biofuel logistics is the step of moving fuel (often by truck, rail, or barge) to the end users and storing it. End use involves the engine type in which the fuel is used as well as how much of the biofuel is blended with other fuels. For example, second-generation bioethanol can be acquired from *Eucalyptus globulus* if refined by specialized autohydrolysis processing, which breaks down the lignocellulose into soluble fragments, followed by Simultaneous Saccharification and Fermentation (SSF) processing, which is the fermentation process [45]. However, because there are no feedstock conversion processes to date that use *Eucalyptus*, there is limited information on how *Eucalyptus* might differ from other feedstocks in its effects on the last three steps of the life cycle: conversion to biofuel, biofuel logistics, or biofuel end use.

## 4. Conclusion, Opportunities, and Constraints for *Eucalyptus*-Based Bioenergy

This paper discusses a suite of sustainability indicators that can be applied to *Eucalyptus*-based bioenergy production in the southeastern USA. While this bioenergy production system has the potential to be environmentally, economically, and socially sustainable, context-specific information is needed before these indicators can be applied to determine conditions under which a system is sustainable. For *Eucalyptus* growing in the southeastern USA, key concerns and hence critical data needs revolve around potential for invasiveness, water use and social acceptability. Sustainability indicators should be applied as specific projects are deployed.

There are several opportunities provided by using *Eucalyptus* and other SRWC as feedstocks in the southeastern USA. Most importantly they could provide a new source of bioenergy and associated social and environmental benefits. They may provide a means to retain or expand the area of land in a forest land use, versus having them become developed, and thereby improve biodiversity conditions and water quantity and quality. *Eucalyptus* and other SRWC plantations may also provide rural jobs.

However, constraints exist to the full deployment of *Eucalyptus*-based bioenergy in the southeastern USA. Current environmental, sociopolitical, economic, and conditions may limit the places where *Eucalyptus* might be planted. These limits include pressures for land development, the value of wood and its products, and soils conditions that result from past land use. Furthermore, not all requisite information is currently available at the temporal and spatial scales of resolution at which it is needed to estimate the potential for a successful bioenergy industry based on *Eucalyptus* or to validate this approach. Therefore, we encourage the collection of data on the indicators in Tables 2 and 3 so that a quantitative evaluation can be made. Necessary information includes current environmental and socioeconomic conditions as well as factors affecting energy choices and their impacts. Another constraint is lack of information on the best management techniques for establishing and growing *Eucalyptus* in the southeastern USA. The processes for converting *Eucalyptus* to bioenergy are in their infancy and require development as well. There is a need to develop the industry for producing and converting *Eucalyptus* to bioenergy. As the bioenergy system based on *Eucalyptus* is deployed, it will be necessary to identify and address public perceptions and risks. For example, there is widespread concern that *Eucalyptus* is an invasive species. Finally, genomes of *Eucalyptus* need to be developed that can deal with environmental stresses that occur in the southeastern USA (such as those that are resistant to frost).

Once (and if) these constraints are surmounted, the benefits of a *Eucalyptus*-based bioenergy system can possibly be achieved. The forest industry is well positioned to tackle these constraints to feedstock provision using *Eucalyptus*. Brazil has much experience in growing eucalypts where they constitute about 90% of the forest plantations. However, it is not clear how much of that knowledge and technology can be transferred to the southeastern USA. The deployment of the bioenergy industry is still in development, and it is unknown how much *Eucalyptus* will differ from the conversion of other feedstocks. This analysis demonstrates that the sustainability issues associated with using *Eucalyptus* for bioenergy do not differ greatly from those of other feedstocks. In all cases, it is the specifics of how the industry is developed and deployed that determine the effects on sustainability of current systems.

## Acknowledgments

Maggie Davis provided helpful comments on an earlier version of this paper. John Stanturf and three anonymous reviewers also provided useful suggestions. This research was supported by the USA Department of Energy (DOE) under the Biomass Technologies Office. Oak Ridge National Laboratory is managed by UT-Battelle, LLC, for DOE under Contract DE-AC05-00OR22725.

## References

[1] G. H. Brundtland, Ed., *Our Common Future: The World Commission on Environment and Development*, Oxford University, Oxford, UK, 1987.

[2] United States Department of Energy, *U.S. Billion-Ton Update: Biomass Supply for a Bioenergy and Bioproducts Industry*, Oak Ridge National Laboratory, Oak Ridge, Tenn, USA, 2011.

[3] K. L. Kline and M. D. Coleman, "Woody energy crops in the southeastern United States: two centuries of practitioner experience," *Biomass and Bioenergy*, vol. 34, no. 12, pp. 1655–1666, 2010.

[4] A. C. McBride, V. H. Dale, L. M. Baskaran et al., "Indicators to support environmental sustainability of bioenergy systems," *Ecological Indicators*, vol. 11, no. 5, pp. 1277–1289, 2011.

[5] V. H. Dale, R. A. Efroymson, K. L. Kline et al., "Indicators for assessing socioeconomic sustainability of bioenergy systems: a short list of practical measures," *Ecological Indicators*, vol. 26, pp. 87–102, 2013.

[6] United States National Arboretum, *USDA Plant Hardiness Zone Map, Based on USDA Miscellaneous*, US Department of Agriculture, Washington, DC, USA, 2003, http://www.usna.usda.gov/Hardzone/.

[7] M. Hinchee, W. Rottmann, L. Mullinax et al., "Short-rotation woody crops for bioenergy and biofuels applications," *In Vitro Cellular and Developmental Biology*, vol. 45, no. 6, pp. 619–629, 2009.

[8] J. A. Stricker, D. L. Rockwood, S. A. Segrest, G. R. Alker, G. M. Prine, and D. R. Carter, "Short Rotation Woody Crops for Florida, University of Florida," http://www.treepower.org/papers/strickerny.doc, 2000.

[9] D. L. Rockwood, "History and status of *Eucalyptus* improvement in Florida," *International Journal of Foresty Research*, vol. 2012, Article ID 607879, 10 pages, 2012.

[10] D. G. D. Ugarte and R. E. Ray, "Biomass and bioenergy applications of the POLYSYS modeling framework," *Biomass and Bioenergy*, vol. 18, pp. 291–308, 2000.

[11] M. Langholtz, R. Graham, L. Eaton, R. Perlack, C. Hellwinkel, and D. G. D. Ugarte, "Price projections of feedstocks for biofuels and biopower in the U.S.," *Energy Policy*, vol. 41, pp. 484–493, 2012.

[12] "Roundtable on Sustainable Biofuels (RSB). Roundtable on Sustainable Biofuels—Indicators of Compliance for the RSB Principles & Criteria. École Polytechnique Fédérale de Lausanne," RSB-IND-01-001 (Version 2. 0), http://rsb.epfl.ch/files/content/sites/rsb2/files/Biofuels/Version%202/Indicators/11-03-08%20RSB%20Indicators%202-0.pdf, 2011.

[13] Global Bioenergy Partnership (GBEP), *The Global Bioenergy Partnership Sustainability Indicators for Bioenergy*, GBEP Secretariat, FAO, Environment, Climate Change and Bioenergy Division, Rome, Italy, 1st edition, 2011, http://www.arb.ca.gov/fuels/lcfs/workgroups/lcfssustain/Report_21_December.pdf.

[14] Council on Sustainable Biomass Production (CSBP), *Draft Provisional Standard For Sustainable Production of Agricultural Biomass*, Council on Sustainable Biomass Production, 2011, http://www.csbp.org/.

[15] National Research Council (NRC), *Monitoring Climate Change Impacts: Metrics at the Intersection of the Human and Earth Systems*, The National Academies Press, Washington, DC, USA, 2008, http://www.nap.edu/catalog.php?record_id=12965.

[16] V. H. Dale and S. C. Beyeler, "Challenges in the development and use of ecological indicators," *Ecological Indicators*, vol. 1, no. 1, pp. 3–10, 2001.

[17] B. K. Sovacool and I. Mukherjee, "Conceptualizing and measuring energy security: a synthesized approach," *Energy*, vol. 36, no. 8, pp. 5343–5355, 2011.

[18] R. A. Efroymson, V. H. Dale, K. L. Kline et al., "Environmental indicators ofbiofuel sustainability: what about context?" *Environmental Management*. In press.

[19] J. Tassin, A. P. Missamba-Lola, and J. N. Marien, "Biodiversity of *Eucalyptus* plantations," *Bois et Forêts des Tropiques*, vol. 309, pp. 27–35, 2011.

[20] S. T. Berthrong, G. Pineiro, E. G. Jobbagy, and R. B. Jackson, "Soil C and N changes with afforestation of grasslands across gradients of precipitation and plantation age," *Ecological Applications*, vol. 22, pp. 76–86, 2012.

[21] C. Perez-Cruzado, P. Mansilla-Salinero, and R. Rodriguez-Soalleiro, "Influence of tree species on carbon sequestration in afforested pastures in a humid temperate region," *Plant and Soil*, vol. 353, pp. 333–353, 2012.

[22] D. I. Forrester, J. J. Collopy, C. L. Beadle, and T. G. Baker, "Interactive effects of simultaneously applied thinning, pruning and fertiliser application treatments on growth, biomass production and crown architecture in a young *Eucalyptus* nitens plantation," *Forest Ecology and Management*, vol. 267, pp. 104–116, 2012.

[23] M. Langholtz, D. R. Carter, D. L. Rockwood, and J. R. R. Alavalapati, "The influence of CO2 mitigation incentives on profitability of *Eucalyptus* production on clay settling areas in Florida," *Biomass and Bioenergy*, vol. 33, no. 5, pp. 785–792, 2009.

[24] M. Langholtz, D. R. Carter, D. L. Rockwood, and J. R. R. Alavalapati, "The economic feasibility of reclaiming phosphate mined lands with short-rotation woody crops in Florida," *Journal of Forest Economics*, vol. 12, no. 4, pp. 237–249, 2007.

[25] D. L. Rockwood, D. Carter, L. Ma, C. Tu, and G. R. Alker, *Phytoremediation of Contaminated Sites Using Wood Biomass*, Florida Center For Solid and Hazardous Waste Management, Gainesville, Fla, USA, 2001.

[26] D. L. Rockwood, C. Naidu, S. Segrest et al., "Short-rotation woody crops and phytoremediation: opportunities for agroforestry?" in *Proceedings of the New Vistas in Agroforestry—A Compendium for the 1st World Congress of Agroforestry*, pp. 51–63, Kluwer Academic, Dordrecht, The Netherlands, 2004.

[27] D. L. Rockwood, G. Peter, M. Langholtz, and B. Becker, "Genetically improved Eucalypts for novel applications and sites in Florida," in *Proceedings of the Southern Forest Tree Improvement Conference*, Raleigh, NC, USA, 2005.

[28] B. Tamang, D. L. Rockwood, M. Langholtz, E. Maehr, B. Becker, and S. Segrest, "Vegetation and soil quality changes associated with reclaiming phosphate-mine clay settling areas with fast growing trees," in *Proceedings of the 32nd Annual Conference on Ecosystem Restoration and Creation*, Tampa, Fla, USA, 2005.

[29] S. Wullschleger, S. Segrest, D. L. Rockwood, and C. Garten Jr, "Enhancing soil carbon sequestration on phosphate mine lands in Florida by planting short-rotation bioenergy crops," in *Proceedings of the 3rd Annual Conference on Carbon Capture and Sequestration*, Washington, DC, USA, 2004.

[30] R. J. Harper, S. J. Sochacki, K. R. J. Smettem, and N. Robinson, "Bioenergy feedstock potential from short-rotation woody crops in a dryland environment," *Energy and Fuels*, vol. 24, no. 1, pp. 225–231, 2010.

[31] C. W. Hedman, S. L. Grace, and S. E. King, "Vegetation composition and structure of southern coastal plain pine forests: an ecological comparison," *Forest Ecology and Management*, vol. 134, no. 1–3, pp. 233–247, 2000.

[32] M. Leslie, G. K. Meffe, J. L. Hardesty, and D. L. Adams, *Conserving Biodiversity on Military Lands: A Handbook for Natural Resource Managers*, The Nature Conservancy, Virginia, Va, USA, 1996.

[33] United States Government, *The Energy Independence and Security Act of 2007 (H.R. 6)*, 2007.

[34] D. R. Gordon, K. J. Tancig, D. A. Onderdonk, and C. A. Gantz, "Assessing the invasive potential of biofuel species proposed for Florida and the United States using the Australian Weed Risk Assessment," *Biomass and Bioenergy*, vol. 35, no. 1, pp. 74–79, 2011.

[35] H. Trevor Booth, "Eucalypts and Their Potential for Invasiveness Particularly in Frost-Prone Regions," *International Journal of Forestry Research*, vol. 2012, Article ID 837165, 7 pages, 2012.

[36] B. Wicke, E. Smeets, V. Dornburg et al., "The global technical and economic potential of bioenergy from salt-affected soils," *The Royal Society of Chemistry*, vol. 4, pp. 2669–2681, 2011.

[37] D. Zhang and M. Polyakov, "The geographical distribution of plantation forests and land resources potentially available for pine plantations in the U.S. South," *Biomass and Bioenergy*, vol. 34, no. 12, pp. 1643–1654, 2010.

[38] D. Jenbere, M. Lemenih, and H. Kassa, "Expansion of eucalypt farm forestry and its determinants in Arsi Negelle District, South Central Ethiopia," *Small-Scale Forestry*, vol. 11, pp. 389–405, 2012.

[39] R. Gonzalez, T. Treasure, J. Wright et al., "Exploring the potential of *Eucalyptus* for energy production in the Southern United States: financial analysis of delivered biomass. Part I," *Biomass and Bioenergy*, vol. 35, no. 2, pp. 755–766, 2011.

[40] D. G. Hodges, A. J. Hartsell, C. Brandeis, T. J. Brandeis, and J. W. Bentley, "Recession effects on the forest and forest products

industries of the South," *Forest Products Journal*, vol. 61, pp. 614–624, 2011.

[41] C. W. Woodall, C. E. Keegan, C. B. Sorenson et al., "An overview of the forest products sector downturn in the United States," *Forest Products Journal*, vol. 61, pp. 595–603, 2011.

[42] V. H. Dale, K. L. Kline, L. L. Wright, R. D. Perlack, M. Downing, and R. L. Graham, "Interactions among bioenergy feedstock choices, landscape dynamics, and land use," *Ecological Applications*, vol. 21, no. 4, pp. 1039–1054, 2011.

[43] S. Gonzalez-Garcia, M. T. Moreira, G. Feijoo, and R. J. Murphy, "Comparative life cycle assessment of ethanol production from fast-growing wood crops (black locust, *Eucalyptus* and poplar)," *Biomass & Bioenergy*, vol. 39, pp. 378–388, 2012.

[44] R. Gonzalez, T. Treasure, R. Phillips et al., "Converting *Eucalyptus* biomass into ethanol: Financial and sensitivity analysis in a co-current dilute acid process. Part II," *Biomass and Bioenergy*, vol. 35, no. 2, pp. 767–772, 2011.

[45] A. Romani, G. Garrote, and J. C. Parajo, "Bioethanol production from autohydrolyzed *Eucalyptus globulus* by Simultaneous Saccharification and Fermentation operating at high solids loading," *Fuel*, vol. 94, pp. 305–312, 2012.

# 14

# The High Input of Soil Organic Matter from Dead Tree Fine Roots into the Forest Soil

**Hans Å. Persson**

*Department of Ecology, Swedish University of Agricultural Sciences, P.O. Box 7044, 750 07 Uppsala, Sweden*

Correspondence should be addressed to Hans Å. Persson, hans.persson@ekol.slu.se

Academic Editor: Jingxin Wang

The spatial and temporal dynamics of tree fine roots were investigated in six boreal forests types in Eastern Sweden, close to the Swedish Forsmark and Laxemar nuclear power plants. Four dry and two wet forest types were included in the study. The amount of live and dead fine roots in terms of dry weight was estimated in soil cores. The live/dead ratios of fine roots (<1 mm in diameter) decreased with depth; very low ratios were observed in two wet forest sites. The proportions of dead fine roots to the total amounts of fine roots in the mineral soil horizons of those wet sites were 63 and 86%. The corresponding proportions in the mineral soil in dry forest sites were 45 and 45% and 49 and 48% at Forsmark and Laxemar, respectively. Sequential soil core sampling demonstrated a high variation in live and dead amounts of fine roots during the growth period. A high accumulation of carbon from dead tree fine root was found in all six forest types, in particular in the wet forest sites, but also in deeper soil horizons. Consequently, substantial amounts of organic matter from dead fine roots are continuously accumulated in the soil in boreal forests.

## 1. Introduction

Tree fine roots of forest trees are for their function forced to penetrate dry soil volumes often against mechanical resistance in densely packed soil layers. In spite of those difficulties, tree root systems explore the uppermost parts of the soil profile with a network of growing root tips. For an example, the root system of a 13-year-old Scots pine (*Pinus sylvestris*) reached an area of about 5 m from the tree stem and expanded about 0.4 m $yr^{-1}$ [1, 2]. The high tree density in this forest stand (1095 $ha^{-1}$) made it clear that the uppermost parts of the soil profile must have been completely interwoven by tree fine roots.

The lateral roots in a nearby 120-year-old Scots pine stand (tree density = 393 $ha^{-1}$) reached 15–20 m from the tree trunks [3]. The root system of those mature Scots pine trees had expanded about 0.1 m $yr^{-1}$during the 120-year period leaving an area of about 300 $m^{-2}$ penetrated by tree roots. The wide-spread network of fine roots on the structural roots increases the total surface area and length of the root system [4].

The white coloured area of the root tips behind the region of cell elongation is most active in the uptake processes [5, 6]. This absorption zone is often covered by mycorrhizal fungi or root hairs [7]. The seasonal change in the number of root tips seems to be very well correlated with the changes in the amount and length and surface area of living fine roots [8]. Fine roots are short lived in their effectiveness; death and decay of dead fine roots often take place within a few weeks.

The energy cost of forming and maintaining the fine root system of a single tree is very high [9–13].The amount of carbon allocated to the root systems is substantial [2, 8, 9, 12, 14–17]. The starch content in living fine roots (<1 mm in diameter) may reach as much as 30% of the dry weight [18]. Live fine root ramifications are constantly replaced and a substantial amount of dead fine-root tissues are formed.

Fine root production and litter formation of fine roots are important components of nutrient and carbon cycles in forest ecosystems [2, 8, 12, 14, 16, 17, 19–25]. Calculating carbon allocation to tree fine roots and turnover time is therefore essential in order to understand the patterns of carbon cycling in forest ecosystems.

The most commonly used methods of estimating fine-root production in forest ecosystems involve the measurements of live and dead amounts of fine roots in terms of dry weight from sequential soil cores. The contribution of different diameter classes to root biomass and turnover rate may vary [4, 15]. The most substantial turnover rate is to be found in fine roots. Some fine roots are transferred to long-living large-diameter roots [13, 15].

The live/dead ratio of fine roots is a valuable vitality criterion in the soil profile [26]. Persson and Ahlström [27] used the live/dead ratio of fine roots to quantify changes in the "vitality" of fine roots in nitrogen manipulated Norway spruce catchment areas. The fluctuations in the amount of live and dead fine roots during the growth period in these forest areas suggested that climatic alterations were the most important underlying sources of variation. Soil temperature and soil water conditions seem to be the most important factors influencing fine-root growth. The most substantial decomposition takes place in the upper soil layers where high temperature, supplies of air, water, and food allow microorganisms to thrive [28, 29].

The effects of anoxic conditions on the spatial and temporal distribution on the tree fine roots have recently been identified [15, 26, 30]. The live/dead ratio of fine roots may be reduced in water saturated forest soils as a result of oxygen deficiency, a low rate of decomposition, and a substantial accumulation of decomposing organic matter from dead fine roots [31–33].

The boreal forests are in many ways able to keep plant remains from decomposing, thus preventing the release of carbon into the air. Most carbon is accumulated in poorly and very poorly drained soils. A high share of forested wetlands is to be found in the boreal forest region [34–36]. In boreal forests tree fine roots are exposed to large seasonal variations of soil moisture, nutrient availability, and soil temperature [37, 38]. In water-saturated forest stands the accumulation of dead fine roots is expected to be very high.

We hypothesized that soil water saturation plays an important role in regulating the amount of deposed organic matter from dead fine roots. A low live/dead ratio of fine roots is expected in water-saturated forest stands as a result of the high accumulation of dead root tissues. Soil carbon from dead tree roots plays a key role in the global carbon cycle and is an important component in climate models. Humans have, and will likely continue to have, significant impacts on the size of this pool by forest management practices forest such as forest harvesting, clear-cutting, ploughing, and drainage.

## 2. Material and Methods

Investigations were carried out in six different forest types in areas surrounding two Swedish nuclear power plants at Forsmark and Laxemar [15, 26]. The forest sites at Forsmark were of coniferous *Calluna-Empetrum* type, coniferous fern type, and *Alnus* swamp of herb type [39]. A distinct hummock and hollow microtopography was developed at the *Alnus* forest of swamp herb type at Forsmark.

The sites at Laxemar were of herb rich oak forest type, coniferous *Vaccinium myrtillus* type, and *Alnus* shore forest type. The soil type varied between leptosol/regosols) gleysols at Forsmark and histosols/gleysol at Laxemar (see Table 1). Stone/boulder volumetric content (%) and soil moisture conditions differed considerably between the different sites [26].

The tree density (the number of trees $ha^{-1}$) was 1340, 780, and 3340 at the Forsmark and 200, 400, and 1600 at the Laxemar sites, respectively. The mean tree height was 16.3, 19.8, and 18.5 m at the Forsmark sites and 17.1, 21.0, and 11.6 m at the Laxemar sites, respectively. The mean tree height of the trees was 16.3, 19.8 and 18.5 m at the Forsmark and 17.1, 21.0, and 11.6 m at the Laxemar sites, respectively (Table 1). The basal area ($m^2$/ha) of the trees was 22.5, 27.0, and 17.9 at the Forsmark sites and 15.0, 15.5, and 17.5 at the Laxemar sites, respectively. The specific basal area was occupied by different tree species.

The *Alnus* swamp forest site at Forsmark consisted besides *Alnus glutinosa* (Table 1; basal area = 7.3 $m^2$ $ha^{-1}$) of *Betula verrucosa*, *Picea abies*, and *Pinus sylvestris* (total basal area = 10.3 $m^2$ $ha^{-1}$). The *Alnus* shore forest at Laxemar consisted of *Alnus glutinosa* (basal area = 17.5 $m^2$ $ha^{-1}$). The coniferous *Calluna-Empetrum* site, the coniferous fern site at Forsmark and the coniferous *Vaccinium* site at Laxemar were dominated by *Picea abies* (basal area = 22.5 and 15.5 $m^2$ $ha^{-1}$) and the herb rich oak forest at Laxemar by *Quercus robur* (basal area = 15.0 $m^2$ $ha^{-1}$). The average thickness of the humus layer was 15.3, 5.2, and 15.3 cm at the Forsmark sites and 11.5, 5.5, and 5.3 cm at the Laxemar sites.

In total 32 soil cores were taken in each forest site from the four corners of a quadrate covering 200 $m^2$, 8 taken randomly in each corner (north, east, south, and west) [15, 26]. Each soil sample was taken as deep as possible, namely, to a depth where stones and larger blocks prevented further penetration by the soil corer.

Soil cores were sampled in the middle of October 2004 for the Forsmark sites and in the end of April 2005 for the Laxemar sites. During the winter months the upper part of the soil profile was deep frozen, and only limited growth of the fine roots could take place. The soil cores were randomly distributed within the hummock and hollow microtopography pattern at the Alnus forest of swamp herb type at Forsmark. Half of the soil core samples were taken in the hummocks and the other half in the hollows. In all other forest sites no stratification was applied.

Additional sequential soil core sampling was carried at the coniferous fern forest type at Forsmark on in total 4 sampling occasions [26]. Besides the first sampling in the middle of October in 2004, samplings were carried out in the middle of April, in the beginning of August and in the end of October in 2005 [15]. The depth distribution of fine roots was measured, at intervals of 0-2.5 (H1), 2.5-5 (H2), and 5–10 cm (H3) of the LFH horizon and in 10 cm segments (M1-M3) for the mineral soil profile down to 30 cm.

A cylindrical steel corer, with an inner diameter of 4.5 cm, was used for soil core sampling. In total 32 soil cores were taken, each soil core sample was taken to a depth where stones and larger blocks prevented further penetration by the soil corer. Only few root fragments were found in the deepest soil layer. The uppermost 0–2.5 cm layer consisted of humus in all investigated sites.

The soil samples were transferred into plastic bags and transported to our laboratory and stored in a cold storage

Table 1: Site and stand characteristics at Forsmark and Laxemar. *Picea abies* = *P. a.*, *Pinus sylvestris* = *P. s.*, *Betula verrucosa* = *B. v.*, *Alnus glutinosa* = *A. g.*, *Quercus robur* = *Q. r.*

| Stand characteristics | Coniferous, *Calluna empetrum* | Coniferous fern | *Alnus* swamp herb | Herb rich oak forest | Coniferous *Vaccinium myrtillus* | *Alnus* shore forest |
|---|---|---|---|---|---|---|
| Soil moisture class[1] | Fresh | Fresh/moist | Moist | Fresh | Fresh | Moist |
| Soil type[1] | Leptosol | Regosol/Gleysol | Gleysol | Histosol/Gleysol | Histosol | Histosol |
| Tree age | 59–60 | 80–88 | 85–95 | 112 | 55 | 34 |
| Number of trees /ha | 1340 | 780 | 3340 | 200 | 400 | 1600 |
| Tree height (m) | 16.3 | 19.8 | 18.5 | 17.1 | 21.0 | 11.6 |
| Diameter at breast height (dbh in m) | 0.21 | 0.26 (*P. a.*) | 0.31 (*P. a*) | 0.36 | 0.32 | 0.14 |
| Basal area ($m^2$/ha) | 22.5 (*P. a.*) | 20.5 (*P. a.*)<br>6.5 (*B. v.*) | 5.3 (*B. v.*)<br>7.3 (*A. g.*)<br>3.0 (*P. a.*)<br>2.3 (*P. s.*) | 15.0 (*Q. r.*) | 15.5 (*P. a.*) | 17.5 (*A. g.*) |

[1] [40, 41].

at −4°C; a temperature that did not damage the live tissue and caused no change in ion concentrations [42]. The roots were sorted out from the soil cores immediately after thawing. In order to distinguish live roots from dead roots distinct morphological characteristics were used [43].

Live fine roots (<1 mm in diameter) were defined as roots with a varying degree of brownish/suberized tissues, often well branched, with the main part of the root tips light and turgid or changed into mycorrhizal root tips [43, 44]. In cases when there was a difficulty to judge if a root fragment was live or dead, it was cut lengthwise with a sharp dissection knife and the judgement was based on the colour between cortex and periderm. The stele of live roots was white to slightly brown and elastic. In roots considered as dead, the stele was brownish and easily broken, and the elasticity was reduced. Dead root fragments with a length <1 cm were regarded as soil organic matter. The dry weight was estimated for all root fractions after drying in an oven at 65°C to constant dry weight (at least for 24 hours).

The fine-root production and turnover rates were calculated from significant (Student's *t*-test) increments of live, dead, and live + dead fine roots at the coniferous fern forest at Forsmark. Comprehensive descriptions of calculation methods used are to be found in [8]. By turnover rate, in this context, was meant the annual fluxes in the live, dead, or in the live + dead fine-root compartment. Root turnover rate ($yr^{-1}$) was calculated from the annual increments in live, dead and, live + dead fine roots divided by the average amounts in those categories during the year of sampling.

From these calculations only minimum estimates of turnover rate can be obtained since the sampling frequency (only four sampling occasions) covered only some of the major fluctuations, but certainly not all increases. The risk for overestimation due to the random variations in the means is low with a low number of sampling occasions and a high number of samples on each sampling occasion [8].

## 3. Results

A substantial variation with depth in the total amount of live and dead tree fine roots was observed at the different forest sites (Figure 1). High amounts of dead fine roots were indicated in the mineral soil horizons. The proportion of dead fine roots varied at all sites and in all horizons from 6 to 88% (Table 2). The highest proportions of dead fine roots were found in the two *Alnus* forest sites at Forsmark and Laxemar, which were both classified as "moist," with a high topographic wetness index (cf. Table 1). The live/dead ratio of fine roots (<1 mm in diameter) varied in all sites from 14.1 in the uppermost 2.5 cm of the humus layer to 0.2 in the deepest parts of the mineral soil horizon (Table 2). The live/dead ratio is calculated only for soil cores with both live and dead fine roots.

Low amounts of live fine roots were found in the uppermost 2.5 cm of soil of the moist *Alnus* shore forest at Laxenar compared with the same soil segments at other forest sites (Figure 1 and Table 2). The total amounts of fine roots were very low at the latter site and substantial amounts of dead fine roots were found in the total soil profile. At the moist *Alnus* swamp forest at Forsmark, a more substantial proportion of live fine roots were observed in the total soil profile, resulting in a comparatively high live/dead ratio, in particular in the humus layers (Table 2).

The rooting density ($g\,dm^{-3}$) was low at all sites in the mineral soil horizons compared with in the humus layers except for in the coniferous fern forest at Forsmark (Figure 1). A low live/dead ratio of fine roots was found in the deepest soil horizons, where the rooting density was very low (Table 2). Extremely low rooting density was observed in the mineral soil horizons at the *Alnus* swamp forest at Forsmark, at the herb rich oak forest at Laxemar and at the *Alnus* shore forest. The soil corer was driven in all sites to depths in the mineral soil horizon (at least 2 dm) where only

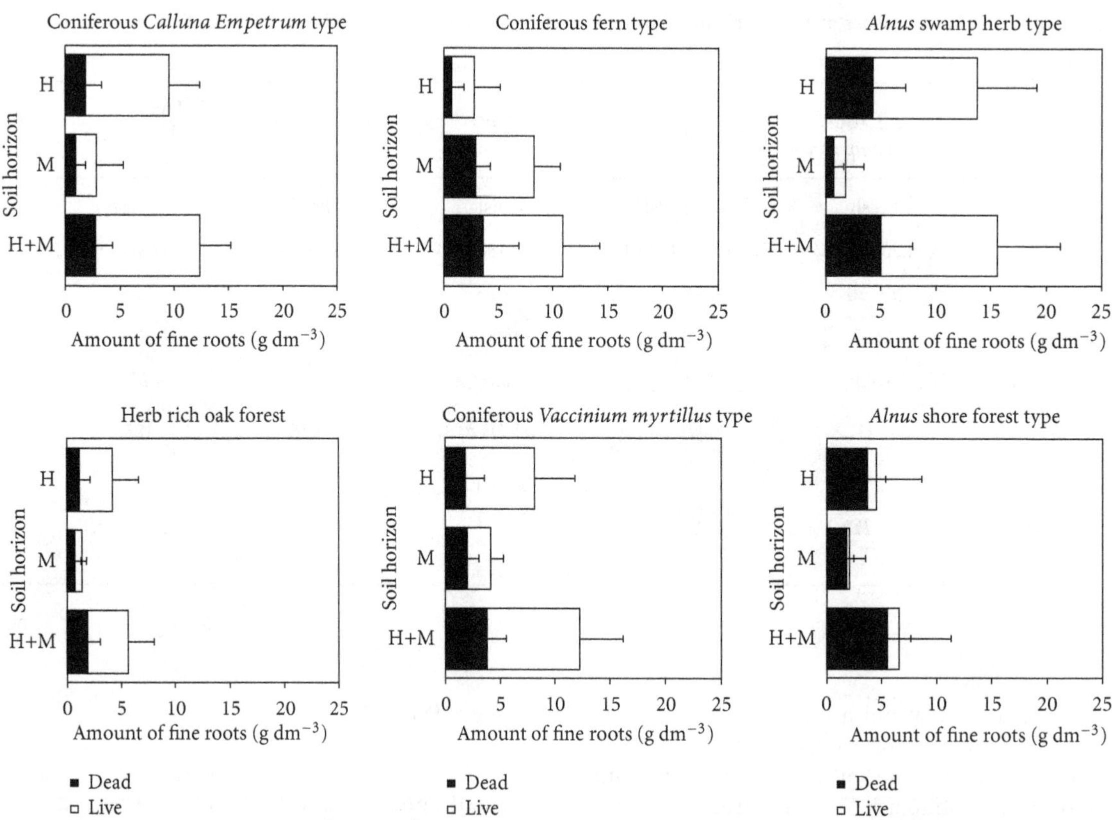


Figure 1: The amount of live and dead tree fine roots (g dm$^{-3}$; <1 mm in diameter) in the humus layer (H) and in the mineral soil horizon (M) compared with in the total soil profile (H+M) at different Swedish forest sites.

limited amounts of live or dead fine-root fragments were observed (Table 2).

The mean amount of live tree fine roots in the humus layer in relation to the total amount of live + dead fine roots in the soil profile was 48, 7, and 48% for the three Forsmark sites and 35, 25, and 6% for the three Laxemar sites, respectively (Figure 1, Table 2). The related amount of dead fine roots in the humus layer was 13, 2, and 30% for the Forsmark and 13, 2, and 28% for the Laxemar sites, respectively. Proportionally more live than dead fine roots were observed at all sites in the humus compared with in the mineral soil horizon, except for at the *Alnus* shore forest site at Laxemar (Figures 1 and 2). The density estimates (Figure 1; g dm$^{-3}$) give a more correct picture of the extensive distribution of live fine roots in the humus layer than simply the distribution per unit area in different soil layers (Figure 2 and Table 2; g m$^{-2}$).

The proportion of dead fine roots (<1 mm in diameter) of the total amount of fine roots in the humus layer (Table 2) was high in both moist *Alnus* forest at Forsmark and Laxemar sites (38 and 82%, resp.). A distinct hummock and hollow microtopography was developed at the *Alnus* forest of swamp herb type at Forsmark and tree fine roots were more frequently found in the uppermost aerated parts of hummocks. The tree layer consisted of a mixture of *Betula verrucosa*, *Picea abies*, and *Pinus sylvestris* (Table 1). At the water-saturated *Alnus glutinosa* shore forest at Laxemar, no

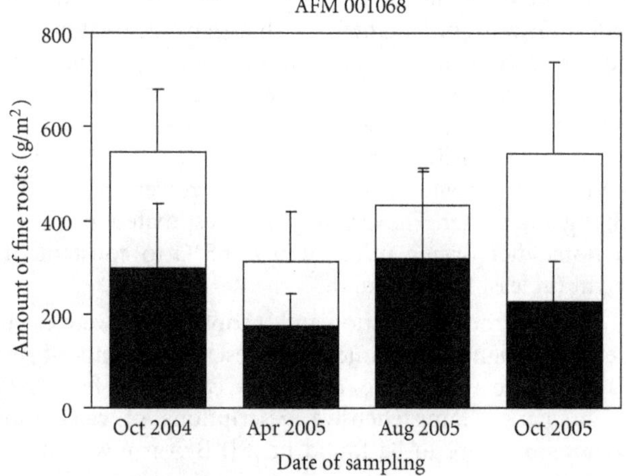


Figure 2: The amount of live (unfilled bar) and dead (black bar) fine roots (<1 mm in diameter) on the four sampling occasions in the coniferous fern site at Forsmark during 2004-2005. Mean values ±SD.

hummocks were found and anaerobic conditions reached even the uppermost parts of the soil horizon.

Substantial variations in live and dead amounts of fine roots and live/dead ratios were observed in the coniferous fern site at Forsmark during the sampling period confirming

Table 2: The distribution of live and dead amounts of tree fine roots (<1 mm in diameter), live/dead ratios, and dead (%) at different depths (H = humus; M = mineral soil) at the forest sites: Forsmark-1 (Coniferous, *Calluna Empetrum type*), Forsmark-2 (Coniferous fern type) and Forsmark-3 (*Alnus* swamp herb type) and Laxemar-1 (herb-rich oak forest), Laxemar-2 (Coniferous *Vaccinium myrtillus* type), and Laxemar-3 (*Alnus* shore forest type) forest sites. The live/dead ratio is calculated from the mean amounts of live and dead tree fine roots. Estimates are given as mean values ±SD ($n$ = 31-32).

| Site | Horizon | Tree roots ($g\,m^{-2}$) | | | |
|---|---|---|---|---|---|
| | | Live | Dead | Live/dead ratio | Dead (%) |
| Forsmark-1 | H 0–2.5 | 48 ± 42 | 3 ± 9 | 13.3 | 6 |
| | H 2.5–5 | 64 ± 29 | 15 ± 22 | 4.2 | 19 |
| | H 5–10 | 60 ± 63 | 21 ± 25 | 2.9 | 26 |
| | H 10–15 | 9 ± 20 | 9 ± 21 | 0.9 | 50 |
| | H 15–20 | 2 ± 9 | 2 ± 12 | 1.0 | 50 |
| | M 0–10 | 71 ± 96 | 58 ± 63 | 1.2 | 45 |
| | M 10–20 | 13 ± 59 | 10 ± 21 | 1.3 | 43 |
| Forsmark-2 | H 0–2.5 | 30 ± 35 | 4 ± 5 | 7.5 | 12 |
| | H 2.5–5 | 4 ± 12 | 4 ± 11 | 1.1 | 50 |
| | H 5–10 | 2 ± 6 | 3 ± 10 | 0.6 | 60 |
| | M 0–10 | 228 ± 138 | 136 ± 70 | 1.7 | 37 |
| | M 10–20 | 51 ± 67 | 73 ± 60 | 0.7 | 59 |
| | M 20–30 | 2 ± 8 | 5 ± 15 | 0.4 | 71 |
| Forsmark-3 | H 0–2.5 | 56 ± 58 | 11 ± 11 | 5.1 | 16 |
| | H 2.5–5 | 46 ± 36 | 23 ± 21 | 2.0 | 33 |
| | H 5–10 | 55 ± 52 | 51 ± 40 | 1.1 | 48 |
| | H 10–15 | 28 ± 35 | 25 ± 33 | 1.1 | 47 |
| | H 15–20 | 13 ± 30 | 9 ± 16 | 1.5 | 41 |
| | H 20–25 | 3 ± 9 | 6 ± 22 | 0.5 | 67 |
| | M 0–10 | 29 ± 73 | 45 ± 55 | 0.6 | 61 |
| | M 10–20 | 6 ± 11 | 14 ± 21 | 0.4 | 70 |
| Laxemar-1 | H 0–2.5 | 24 ± 28 | 7 ± 10 | 3.4 | 23 |
| | H 2.5–5 | 34 ± 37 | 15 ± 16 | 2.3 | 31 |
| | H 5–10 | 21 ± 28 | 8 ± 12 | 2.6 | 28 |
| | M 0–10 | 57 ± 42 | 48 ± 34 | 1.2 | 46 |
| | M 10–20 | 3 ± 7 | 9 ± 23 | 0.3 | 82 |
| Laxemar-2 | H 0–2.5 | 102 ± 61 | 18 ± 23 | 5.6 | 15 |
| | H 2.5–5 | 48 ± 52 | 17 ± 21 | 3.0 | 26 |
| | H 5–10 | 10 ± 29 | 11 ± 30 | 0.9 | 52 |
| | M 0–10 | 132 ± 69 | 71 ± 48 | 1.8 | 35 |
| | M 10–20 | 51 ± 49 | 71 ± 37 | 0.7 | 58 |
| | M 20–30 | 25 ± 49 | 31 ± 32 | 0.8 | 38 |
| | M 30–40 | 4 ± 10 | 26 ± 34 | 0.2 | 87 |
| Laxemar-3 | H 0–2.5 | 14 ± 17 | 47 ± 79 | 0.3 | 77 |
| | H 2.5–5 | 5 ± 9 | 38 ± 57 | 0.1 | 88 |
| | H 5–10 | 1 ± 4 | 7 ± 19 | 0.1 | 33 |
| | M 0–10 | 17 ± 25 | 115 ± 104 | 0.1 | 87 |
| | M 10–20 | 7 ± 16 | 35 ± 40 | 0.2 | 85 |
| | M 20–30 | 3 ± 8 | 21 ± 33 | 0.1 | 84 |
| | M 30–40 | 2 ± 8 | 10 ± 14 | 0.2 | 83 |

TABLE 3: The amount of live, dead, and live + dead fine roots (<1 mm in diameter) and differences on different sampling occasions, annual means, Σ annual increases of fine roots and turnover rate of live, dead, and live + dead fine roots at a fresh/moist coniferous fern forest site at Forsmark.

| Sampling number | Live | Dead | Live + dead |
|---|---|---|---|
| (1) | 317 ± 196 | 226 ±88[a] | 543 ± 205 |
| (2) | 113 ±79[a] | 321 ±184[a] | 434 ± 212 |
| (3) | 150 ± 112 | 180 ±62[b] | 330 ±136[a] |
| (4) | 248 ±134[a] | 299 ±136[b] | 546 ±212[a] |
| Annual means | 207 | 257 | 463 |
| Σ Annual increases | 135 | 214 | 216 |
| Turnover rate | 0.7 | 0.8 | 0.5 |

Sampling took place on 4 sampling occasions: October 20th, 2004 (1), April 18th, 2005 (2), August 2nd, 2005 (3), and October 28th, 2005 (4). Estimates are given as mean values ±SD ($n = 32$). Significant increases are marked by [a] and [b]. Differences are significant at $P = 0.05$ (Student's $t$-test).

a high turnover rate in live, dead, and live + dead fine roots (Table 3). Low amounts of dead fine roots were found at the remaining sites at Forsmark and Laxemar, in particular in the upper well-oxidized parts of the humus layer. Although, no sequential soil coring was carried out at the two moist *Alnus* forest sites, high turnover rates of live fine roots must explain the high accumulation rate of dead fine roots. Thus, anoxia may play an important role in regulating the rate of organic matter accumulation in forest soils.

## 4. Discussion

The methods used of estimating fine-root production and mortality should involve the measurements of live and dead dry weight of fine roots from soil cores in undisturbed soil horizons. The often reported discrepancy in the estimates of root litter formation in data from sequential coring may partly be due to imprecise definition of vitality and size classes of the fine roots. In our case distinct morphological characteristics were used in order to separate live and dead fine roots [8, 43]. In roots considered as dead, the stele was brownish and easily broken, and the elasticity was reduced [43]. Dark coloured tissues are frequently found in live fine roots and the colour in itself is not a reliable criterion of vitality. Depending on the pattern of cell death, several root functions can cease even before cell dies [45].

Substantial variations in live and dead fine roots and live/dead ratios usually occur in tree stands confirming our calculations of turnover rates [4, 17, 23, 27, 39, 46–50]. The dead amounts of fine roots normally do not persist for long in well-oxidized conditions, decay or complete disappearance is accomplished in a few days only [8, 49, 51]. Under anoxic conditions there is an accumulation of dead fine roots, since the decomposition will be reduced [33, 52, 53]. Under long-term anoxic conditions, for example, the case, at the *Alnus* swamp forest of herb type at Forsmark and at the *Alnus* shore forest at Laxemar, an accumulation of dead fine roots was observed.

Fine roots respond quickly to environmental changes and are rapidly penetrating favourable the soil horizons. We know that fine roots are sensitive to drought and that their live/dead ratios are decreasing with less water availability in the soil [54–57]. In water-saturated forest ecosystems the primary production of fine roots may exceed the decomposition of dead roots, also leading to a decreased live/dead ratio [32, 58, 59]. In many cases death takes place as a result of ageing, reduced carbohydrate supply, the influence of different climatic stress factors such temperature changes and frost [49, 60]. Increased fine-root herbivory especially in nutrient-rich patches is furthermore expected to significantly influence the carbon cycling patterns [61].

Although our sampling occasions were few, our data from the coniferous fern site at Forsmark suggest that substantial increases/decreases in live, dead, and live + dead fine roots will take place during the growth period (Table 3). The annual above-ground litter fall (mainly *Picea abies* needles) was 135 g m$^2$ compared with the annual below-ground fine-root litter supply of 257 g m$^2$ [62] (Table 3). Our study suggests that the annual above-ground tree litter fall from leaves/needles is less important in terms of dry weight than the annual belowground formation of fine-root litter.

Available information in the literature suggests a fine-root production with a seasonal pattern different from needle or leaf production [26, 63, 64]. Fine roots respond quickly to environmental changes and their life span is relatively short. Our investigation confirms that the growth pattern of the fine roots depends on where in the soil profile they are developed and that the live/dead ratio is decreasing with depth (Table 2).

"Vitality" in terms of live/dead ratios of fine roots should be expected to be high in the humus layer, since the extensive mycorrhizal infection in that layer makes the fine roots functional over a prolonged period of time [65–68]. Although no distinct seasonal pattern is reported in the literature a high growth and death rate of fine roots should be expected during the summer months [8, 69]. A high death rate was observed during the winter month and early spring (Table 3). Dead fine roots are decomposed quickly in well-oxidized soil layers ensuring a high live/dead ratio of fine roots.

Detritus from above-ground and below-ground plant tissues constitutes the primary source of carbon for soil organic matter [28]. The presence of soil organic matter improves the nutrient availability and reduces soil strength. Tree fine roots may play a more important role for the formation of soil organic matter than the needles/leaves [8, 22, 26, 70–73]. In strongly seasonal climates, the length of the growing season often increases the lifespan of needles/leaves, but fine roots may stay alive less long.

Sequential coring data at the ecosystem scale suggest an annual production of fine roots, frequently higher than the average amount of live fine-roots [8, 17, 22, 26, 74–78]. Most observations on fine-root turnover are underestimates and the costs and benefits of exudation, root hairs and the mycorrhizal fungi is not yet sufficiently clarified [43]. Techniques for obtaining root data are still in a formative stage. As more research is conducted on root methods, techniques will become more refined and standardized. Obtaining root data is essential for all kind of long-term field experiments, because plant responses may occur in the shifts in carbon allocation between above- and below-ground plant components.

Decomposition of fine roots is determined mainly by the interactions between soil temperature and oxygen accessibility [79]. Other factors regulating the fate of dead fine roots in the soil are the soil pH, the availability of decomposer organisms, and litter quality [22, 80, 81]. Only few studies have considered the interactions between all these factors [32, 33, 35, 52, 53, 71, 79, 82–84]. Estimates of root respiration to the total $CO_2$ efflux range from 10% to 90%, with considerable methodological uncertainties [35, 84].

The high amount of dead fine roots accumulated in our two anoxic forest sites unveils a high turnover rate of live fine roots. The changes between anoxic and well-oxidized soil conditions in a forest soil may cause death of both fine roots and mycorrhiza [33, 52]. The effects of anoxia on the metabolic cost on the plant root system and the mycorrhizal infection have so far received limited attention [32, 59, 85]. Uncertainties in the belowground carbon balances limit the establishment and improvement of policies regulating the atmospheric $CO_2$ concentrations.

Trees constitute major reservoirs of carbon in terrestrial ecosystems; large amounts of carbohydrates are annually transported from the shoots to the roots and stored in the root systems [4, 5, 9, 11, 13, 21, 25, 48, 64, 72]. High amounts of live and dead fine roots are found in forest ecosystems [8, 12, 15, 55, 80]. The vitality of the fine roots seem to depend on where in the soil profile they are developed; it is therefore essential, while studying the distribution of fine roots in forest ecosystems, to relate to the natural soil-horizons [15, 46]. The often well-developed organic-rich humus layer in the forest soil most effectively buffers the root system against drought and nutrient deficiencies [15, 69]. Few studies have so far examined patterns in live/dead ratios of fine roots in relation to soil water and mineral nutrient availability during the growth period [15, 24, 74]. Fine roots of trees in many ways are indicators of environmental change, soil nutrient status, and forest health [14, 16, 47].

## Acknowledgments

The author is grateful for financial support from the Swedish Nuclear Fuel and Waste Management Co. (SKB).

## References

[1] L. Kutschera, E. Lichtenegger, M. Sobotik, and D. Haas, "Wurzeln. Bewurzelung von Pflanzen in verschiedenen Lebensräumen. 5. Band der Wurzelatlas-Reihe," *Stapfia*, vol. 49, p. 331, 1997.

[2] H. Persson, "Fine-root production, mortality and decomposition in forest ecosystems," *Vegetatio*, vol. 41, no. 2, pp. 101–109, 1980.

[3] T. Persson, "Structure and function of northern coniferous forests," *Ecological Bulletin*, vol. 32, pp. 1–609, 1979.

[4] H. Persson, "Root system in arboreal plants," in *Plant Roots—the Hidden Half*, Y. Waisel, A. Eshel, and U. Kafksfi, Eds., pp. 187–204, 3rd edition, 2002.

[5] M. M. Caldwell, "Root structure: the considerable cost of belowground function," in *Topics in Plant Population Biology*, O. T. Solbrig, S. Jain, G. B. Johnson, and P. H. Raven, Eds., pp. 408–427, Columbia University Press, New York, NY, USA, 1979.

[6] P. Jensén and S. Pettersson, "Nutrient uptake in roots of Scots pine," *Ecological Bulletin*, vol. 32, pp. 229–237, 1977.

[7] A. H. Fitter, "Magnolioid roots—hairs, architecture and mycorrhizal dependency," *New Phytologist*, vol. 164, no. 1, pp. 15–16, 2004.

[8] H. Persson, "Root dynamics in a young Scots pine stand in Central Sweden," *Oikos*, vol. 30, pp. 508–519, 1978.

[9] G. Ågren, B. Axelsson, J. G. K. Flower-Ellis et al., "Annual carbon budget for a young Scots pine," *Ecological Bulletins*, vol. 32, pp. 307–313, 1980.

[10] R. Fogel, "Root turnover and productivity of coniferous forests," *Plant and Soil*, vol. 71, no. 1–3, pp. 75–85, 1983.

[11] R. B. Jackson, J. Canadell, J. R. Ehleringer, H. A. Mooney, O. E. Sala, and E. D. Schulze, "A global analysis of root distributions for terrestrial biomes," *Oecologia*, vol. 108, no. 3, pp. 389–411, 1996.

[12] I. A. Janssens, D. A. Sampson, J. Curiel-Yuste, A. Carrara, and R. Ceulemans, "The carbon cost of fine root turnover in a Scots pine forest," *Forest Ecology and Management*, vol. 168, no. 1–3, pp. 231–240, 2002.

[13] K. A. Vogt, D. J. Vogt, and J. Bloomfield, "Input of organic matter to the soil by tree roots," in *Plant Roots and Their Environment*, B. L. McMichael and H. Persson, Eds., pp. 171–190, Elsevier Science, Amsterdam, The Netherlands, 1991.

[14] L. Finér, H.-S. Helmisaari, K. Lõhmus et al., "Variation in fine root biomass of three European tree species: beech (*Fagus sylvatica* L.), Norway spruce (*Picea abies* L. Karst.), and Scots pine (*Pinus sylvestris* L.)," *Plant Biosystems*, vol. 141, no. 3, pp. 394–405, 2007.

[15] H. Persson and I. Stadenberg, "Fine root dynamics in a Norway spruce forest (*Picea abies* (L.) Karst) in eastern Sweden," *Plant and Soil*, vol. 330, no. 1, pp. 329–344, 2010.

[16] H. S. Helmisaari, K. Makkonen, S. Kellomäki, E. Valtonen, and E. Mälkönen, "Below- and above-ground biomass, production and nitrogen use in Scots pine stands in eastern Finland," *Forest Ecology and Management*, vol. 165, no. 1–3, pp. 317–326, 2002.

[17] I. Ostonen, K. Lõhmus, and R. Lasn, "The role of soil conditions in fine root ecomorphology in Norway spruce (*Picea*

*abies* (L.) Karst.)," *Plant and Soil*, vol. 208, no. 2, pp. 283–292, 1999.

[18] A. Ericsson and H. Persson, "Seasonal changes in starch reserves and growth of fine roots of 20-year old Scots pines," *Ecological Bulletin*, vol. 32, pp. 239–250, 1980.

[19] M. R. Bakker, *Effect des amendements calciques sur les racines fines de chêne (*Quercus petrea et robur*): conséquences des changements dans la rhizosphère*, Doctoral thesis, l'Université Henri Poincaré, Nancy, France, 1998.

[20] M. R. Bakker, J. Garbaye, and C. Nys, "Effect of liming on the ectomycorrhizal status of oak," *Forest Ecology and Management*, vol. 126, no. 2, pp. 121–131, 2000.

[21] R. J. Norby, J. Ledford, C. D. Reilly, N. E. Miller, and E. G. O'Neill, "Fine-root production dominates response of a deciduous forest to atmospheric $CO_2$ enrichment," *Proceedings of the National Academy of Sciences of the United States of America*, vol. 101, no. 26, pp. 9689–9693, 2004.

[22] H. Helmisaari and H.-S. Helmisaari, *Long-Term Forest Fertilization Experiments in Finland and Sweden*, Swedish Environmental Protection Agency, 1992.

[23] H. S. Helmisaari and L. Hallbäcken, "Fine-root biomass and necromass in limed and fertilized Norway spruce (*Picea abies* (L.) Karst.) stands," *Forest Ecology and Management*, vol. 119, no. 1–3, pp. 99–110, 1999.

[24] C. Stober, G. A. Eckart, and H. Persson, "Root growth and response to nitrogen," in *Carbon and Nitrogen Cycling in European Forest Ecosystems, Ecological Studies*, E. D. Schulze, Ed., vol. 142, pp. 99–121, Springer, Berlin, Germany, 2000.

[25] K. A. Vogt, C. C. Grier, and D. J. Vogt, "Production, turnover and nutrient dynamics of above- and below-ground detritus of world forest," *Advances in Ecological Research*, vol. 15, pp. 303–377, 1986.

[26] H. Persson and I. Stadenberg, "Spatial distribution of fine-roots in boreal forests in eastern Sweden," *Plant and Soil*, vol. 318, no. 1-2, pp. 1–14, 2009.

[27] H. Persson and K. Ahlström, "Fine-root response to nitrogen supply in nitrogen manipulated Norway spruce catchment areas," *Forest Ecology and Management*, vol. 168, no. 1–3, pp. 29–41, 2002.

[28] S. Fontaine, A. Mariotti, and L. Abbadie, "The priming effect of organic matter: a question of microbial competition?" *Soil Biology and Biochemistry*, vol. 35, no. 6, pp. 837–843, 2003.

[29] S. Fontaine, G. Bardoux, L. Abbadie, and A. Mariotti, "Carbon input to soil may decrease soil carbon content," *Ecology Letters*, vol. 7, no. 4, pp. 314–320, 2004.

[30] A. Löfgren, S. Miliander, J. Truvé, and T. Lindborg, "Carbon budgets for catchments across a managed landscape mosaic in southeast Sweden: contributing to the safety assessment of a nuclear waste repository," *Ambio*, vol. 35, no. 8, pp. 459–468, 2006.

[31] M. R. Bakker, L. Augusto, and D. L. Achat, "Fine root distribution of trees and understory in mature stands of maritime pine (*Pinus pinaster*) on dry and humid sites," *Plant and Soil*, vol. 286, no. 1-2, pp. 37–51, 2006.

[32] A. J. Cantelmo Jr. and J. G. Ehrenfeld, "Effects of microtopography on mycorrhizal infection in Atlantic white cedar (*Chamaecyparis thyoides* (L.) Mills.)," *Mycorrhiza*, vol. 8, no. 4, pp. 175–180, 1999.

[33] R. H. Jones, B. G. Lockaby, and G. L. Somers, "Effects of microtopography and disturbance on fine-root dynamics in wetland forests of low-order stream floodplains," *American Midland Naturalist*, vol. 136, no. 1, pp. 57–71, 1996.

[34] Y. Abernethy and R. E. Turner, "US Forested Wetlands: 1940–1980. Field-data surveys document changes and can guide national resource management," *BioScience*, vol. 37, pp. 721–727, 1987.

[35] G. Rapalee, S. E. Trumbore, E. A. Davidson, J. W. Harden, and H. Veldhuis, "Soil carbon stocks and their rates of accumulation and loss in a boreal forest landscape," *Global Biogeochemical Cycles*, vol. 12, no. 4, pp. 687–701, 1998.

[36] H. Sjörs, "Amphi-atlantic zonation, nemoral to arctic," in *North Atlantic Biota and Their History*, A. Löve and D. Löve, Eds., pp. 109–125, Pergamon Press, Oxford, UK, 1963.

[37] S. C. Chang and E. Matzner, "Soil nitrogen turnover in proximal and distal stem areas of European beech trees," *Plant and Soil*, vol. 218, no. 1-2, pp. 117–125, 2000.

[38] G. L. Tierney, T. J. Fahey, P. M. Groffman et al., "Environmental control of fine root dynamics in a northern hardwood forest," *Global Change Biology*, vol. 9, no. 5, pp. 670–679, 2003.

[39] Anon. Nordiska Ministerrådet. Vegetationstyper. Representativa naturtyper och hotade biotoper I Norden. Remissupplaga, 1978.

[40] L. Lundin, E. Lode, J. Stendahl, P.-A. Melkerud, L. Björkvall, and A. Thorstensson, "Soil and site types in the Forsmark area," SKB R-04-08, Sv. Kärnbränslehantering AB, 2004.

[41] L. Lundin, E. Lode, J. Stendahl, L. Björkvall, and J. Hansson, "Soil and site types in the Oskarshamn area," SKB R-05-15, Sv. Kärnbränslehantering AB, 2005.

[42] A. Clemensson-Lindell and H. Persson, "Effects of freezing on rhizosphere and root nutrient content using two soil sampling methods," *Plant and Soil*, vol. 139, no. 1, pp. 39–45, 1992.

[43] K. A. Vogt and H. Persson, "Measuring growth and development of roots," in *Techniques and Approaches in Forest Tree Ecophysiology*, J. P. Lassoie and T. M. Hinckley, Eds., pp. 477–501, CRC Press, 1991.

[44] R. Agerer, Ed., *Colour Atlas of Ectomycorrhizae*, Eduard Dietenberger Gmbh, München, Germany, 1987–1993.

[45] L. H. Comas, D. M. Eissenstat, and A. N. Lakso, "Assessing root death and root system dynamics in a study of grape canopy pruning," *New Phytologist*, vol. 147, no. 1, pp. 171–178, 2000.

[46] W. Borken, G. Kossmann, and E. Matzner, "Biomass, morphology and nutrient contents of fine roots in four Norway spruce stands," *Plant and Soil*, vol. 292, no. 1-2, pp. 79–93, 2007.

[47] H. S. Helmisaari, J. Derome, P. Nöjd, and M. Kukkola, "Fine root biomass in relation to site and stand characteristics in Norway spruce and Scots pine stands," *Tree Physiology*, vol. 27, no. 10, pp. 1493–1504, 2007.

[48] J. W. Raich and K. J. Nadelhoffer, "Belowground carbon allocation in forest ecosystems: global trends," *Ecology*, vol. 70, no. 5, pp. 1346–1354, 1989.

[49] E. R. C. Reynolds, "Root distribution and the cause of its spatial variability in *Pseudotsuga taxifolia* (Poir.) Britt," *Plant and Soil*, vol. 32, no. 1, pp. 501–517, 1970.

[50] H. J. van Praag, S. Sougnez-Remy, F. Weissen, and G. Carletti, "Root turnover in a beech and a spruce stand of the Belgian Ardennes," *Plant and Soil*, vol. 105, no. 1, pp. 87–103, 1988.

[51] W. H. Lyford, "Rhizography of non-woody roots of trees in forest floor," in *The Development and Function of Roots*, J. G. Torrey and D. T. Clarkson, Eds., pp. 179–196, Academic Press, London, UK, 1975.

[52] F. P. Day, J. P. Megonigal, and L. C. Lee, "Cypress root decomposition in experimental wetland mesocosms," *Wetlands*, vol. 9, no. 2, pp. 263–282, 1989.

[53] V. J. Lieffers and R. L. Rothwell, "Rooting of peatland black spruce and tamarack in relation to depth of water table," *Canadian Journal of Botany*, vol. 65, no. 5, pp. 817–821, 1987.

[54] A. F. M. Olsthoorn, "Fine root density and root biomass of two Douglas-fir stands on sandy soils in the Netherlands. 1. Root biomass in early summer," *Netherlands Journal of Agricultural Science*, vol. 39, no. 1, pp. 49–60, 1991.

[55] H. Persson, H. Majdi, and A. Clemensson-Lindell, "Effects of acid deposition on tree roots," *Effects of Acid Deposition and Tropospheric Ozone on Forest Ecosystems in Sweden*, pp. 158–167, 1995.

[56] J. Puhe, H. Persson, and I. Börjesson, "Wurzelwachtum und wurzelshäden in skandinavischen nadelwäldern," *AFZ*, vol. 20, pp. 488–492, 1986.

[57] D. Santantonio and R. K. Hermann, "Standing crop, production, and turnover of fine roots on dry, moderate, and wet sites of mature Douglas-fir in western Oregon," *Annals of Forest Science*, vol. 42, pp. 113–142, 1985.

[58] D. Santantonio, R. K. Hermann, and W. S. Overton, "Root biomass studies in forest ecosystems," *Pedobiologia*, vol. 17, pp. 1–31, 1977.

[59] B. A. D. Hetrick, "Mycorrhizas and root architecture," *Cellular and Molecular Life Sciences*, vol. 47, pp. 355–362, 2005.

[60] J. D. Marshall and R. H. Waring, "Predicting fine root production and turnover by monitoring root starch and soil temperature," *Canadian Journal of Forest Research*, vol. 15, no. 5, pp. 791–800, 1985.

[61] G. N. Stevens and R. H. Jones, "Patterns in soil fertility and root herbivory interact to influence fine-root dynamics," *Ecology*, vol. 87, no. 3, pp. 616–624, 2006.

[62] K. Mjöfors, M.-B. Johansson, Å. Nilsson, and R. Hyvönen, "Input and turnover of forest tree litter in the Forsmark and Oskarshamn areas," SKB R-07-23, Sv Kärnbränslehantering AB, 2007.

[63] K. Makkonen and H. S. Helmisaari, "Assessing fine-root biomass and production in a Scots pine stand—comparison of soil core and root ingrowth core methods," *Plant and Soil*, vol. 210, no. 1, pp. 43–50, 1999.

[64] H. Persson, "The distribution and productivity of fine roots in boreal forests," *Plant and Soil*, vol. 71, no. 1–3, pp. 87–101, 1983.

[65] M. C. Brundrett, "Coevolution of roots and mycorrhizas of land plants," *New Phytologist*, vol. 154, no. 2, pp. 275–304, 2002.

[66] D. M. Durall, J. D. Marshall, M. D. Jones, R. Crawford, and J. M. Trappe, "Morphological changes and photosynthate allocation in ageing *Hebeloma crustuliniforme* (Bull.) Quel. and *Laccaria bicolor* (Maire) Orton mycorrhizas of *Pinus ponderosa* Dougl. ex Laws," *New Phytologist*, vol. 127, no. 4, pp. 719–724, 1994.

[67] J. E. Hooker, K. E. Black, R. L. Perry, and D. Atkinson, "Arbuscular mycorrhizal fungi induced alteration to root longevity of poplar," *Plant and Soil*, vol. 172, no. 2, pp. 327–329, 1995.

[68] H. Marschner, *Mineral Nutrition of Higher Plants*, Academic Press, 2nd edition, 2002.

[69] D. Gaul, D. Hertel, W. Borken, E. Matzner, and C. Leuschner, "Effects of experimental drought on the fine root system of mature Norway spruce," *Forest Ecology and Management*, vol. 256, no. 5, pp. 1151–1159, 2008.

[70] R. W. Ruess, K. Van Cleve, J. Yarie, and L. A. Viereck, "Contributions of fine root production and turnover to the carbon and nitrogen cycling in taiga forests of the alaskan interior," *Canadian Journal of Forest Research*, vol. 26, no. 8, pp. 1326–1336, 1996.

[71] R. A. Scheffer and R. Aerts, "Root decomposition and soil nutrient and carbon cycling in two temperate fen ecosystems," *Oikos*, vol. 91, no. 3, pp. 541–549, 2000.

[72] G. Scarascia-Mugnozza, G. A. Bauer, H. Persson, G. Matteucci, and A. Masci, "Tree biomass, growth and nutrient pools," in *Carbon and Nitrogen Cycling in European Forest Ecosystems. Ecological Studies*, E. D. Schulze, Ed., vol. 142, pp. 49–62, Springer, Berlin, Germany, 2000.

[73] P. Vanninen and A. Mäkelä, "Fine root biomass of scots pine stands differing in age and soil fertility in southern Finland," *Tree Physiology*, vol. 19, no. 12, pp. 823–830, 1999.

[74] B. John, H. N. Pandey, and R. S. Tripathi, "Vertical distribution and seasonal changes of fine and coarse root mass in *Pinus kesiya* royle ex.Gordon forest of three different ages," *Acta Oecologica*, vol. 22, no. 5-6, pp. 293–300, 2001.

[75] Y. Son and J. H. Hwang, "Fine root biomass, production and turnover in a fertilized Larix leptolepis plantation in central Korea," *Ecological Research*, vol. 18, no. 3, pp. 339–346, 2003.

[76] S. Usman, S. P. Singh, and Y. S. Rawat, "Fine root productivity and turnover in two evergreen central Himalayan forests," *Annals of Botany*, vol. 84, no. 1, pp. 87–94, 1999.

[77] Y. S. Yang, G. S. Chen, P. Lin, J. S. Xie, and J. F. Guo, "Fine root distribution, seasonal pattern and production in four plantations compared with a natural forest in Subtropical China," *Annals of Forest Science*, vol. 61, no. 7, pp. 617–627, 2004.

[78] S. Zewdie, M. Fetene, and M. Olsson, "Fine root vertical distribution and temporal dynamics in mature stands of two enset (*Enset ventricosum* Welw Cheesman) clones," *Plant and Soil*, vol. 305, no. 1-2, pp. 227–236, 2008.

[79] E. A. Davidson, E. Belk, and R. D. Boone, "Soil water content and temperature as independent or confounded factors controlling soil respiration in a temperate mixed hardwood forest," *Global Change Biology*, vol. 4, no. 2, pp. 217–227, 1998.

[80] D. L. Godbold, H. W. Fritz, G. Jentschke, H. Meesenburg, and P. Rademacher, "Root turnover and root necromass accumulation of Norway spruce (*Picea abies*) are affected by soil acidity," *Tree Physiology*, vol. 23, no. 13, pp. 915–921, 2003.

[81] H. Eijsackers and A. J. B. Zehnder, "Litter decomposition: a Russian matriochka doll," *Biogeochemistry*, vol. 11, no. 3, pp. 153–174, 1990.

[82] R. D. Newsom, T. T. Kozlowski, and Z. C. Tang, "Responses of Ulmus americana seedlings to flooding of soil," *Canadian Journal of Botany*, vol. 60, no. 9, pp. 1688–1695, 1982.

[83] E. J. Sayer, E. V. J. Tanner, and A. W. Cheesman, "Increased litterfall changes fine root distribution in a moist tropical forest," *Plant and Soil*, vol. 281, no. 1-2, pp. 5–13, 2006.

[84] R. J. Norby and R. B. Jackson, "Root dynamics and global change: seeking an ecosystem perspective," *New Phytologist*, vol. 147, no. 1, pp. 3–12, 2000.

[85] D. L. Godbold, M. R. Hoosbeek, M. Lukac et al., "Mycorrhizal hyphal turnover as a dominant process for carbon input into soil organic matter," *Plant and Soil*, vol. 281, no. 1-2, pp. 15–24, 2006.

# 15

# Functioning of South Moravian Floodplain Forests (Czech Republic) in Forest Environment Subject to Natural and Anthropogenic Change

**Emil Klimo,[1] Jiří Kulhavý,[1] Alois Prax,[1] Ladislav Menšík,[1] Pavel Hadaš,[2] and Oldřich Mauer[3]**

[1] *Department of Forest Ecology, Mendel University in Brno, Zemedelska 3, 613 00 Brno, Czech Republic*
[2] *Data Processing, Ořechovka 1727, 696 62 Strážnice, Czech Republic*
[3] *Department of Silviculture, Mendel University in Brno, Zemedelska 3, 613 00 Brno, Czech Republic*

Correspondence should be addressed to Ladislav Menšík; ladislav.mensik@mendelu.cz

Academic Editor: Leon Bren

South Moravian floodplain forests at the confluence of the Morava and Dyje Rivers, which are related to the floodplain forests of Austria and Slovakia to a considerable degree, have been strongly affected by changes in forest environment caused by natural and anthropogenic impacts. The dominant change factors encompassed changes in the 12–14th centuries resulting in the formation of a flooded alluvium and a significant transition of hardwood floodplain to softwood floodplain. Their further development was affected particularly by forestry activities, and they saw a gradual transformation into hardwood floodplain forests with dominant species of oak, ash, hornbeam, and others. The primary impact in the 20th century was stream regulation and the construction of three water reservoirs, which resulted predominantly in changes in the groundwater table. Response to these changes was registered particularly in the herb layer. The contemporary forest management adjusts to environmental changes and makes efforts to alleviate the negative impacts of previously implemented changes through restoration projects.

## 1. Introduction

The Czech Republic houses approximately 33,000 ha floodplain forests concentrated in four main sites (Figure 1) [1].

(1) South Moravia (Morava and Dyje Rivers)—15,800 ha,

(2) Litovelské Pomoraví—10,400 ha,

(3) the Odra River alluvium—600 ha,

(4) the Labe River alluvium—6,300 ha.

Over the course of time, South Moravian floodplain forests, situated in the alluvium of the lower reach of the Morava River and primarily at the confluence of the Morava and Dyje, have been subject to a complex evolution related to changes in landscape development and human impact (site no 1 in Figure 1.) The prevailing vegetation type is "*Fraxino pannonical ulmetum*" [2]. According to [3], several cold and dry glacial periods have alternated with warm and humid interglacial periods in the Quaternary over the past two million years. This is characterized by alternating gully erosion and accumulation stages and a gradual formation of today's landscape morphology. During these periods the landscape was gradually covered with vegetation, as the buried humus horizons attest to this. River gravels deposited in the conditions of the so-called rampaging river, under which the riverbed frequently changed its course within the entire floodplain. Later, river sands and sandy gravels were deposited in the streams of meandering rivers. These processes conditioned the resulting heterogeneity of alluvial cover. The youngest fine-grained deposits come in the form of alluvial silts and clays which level out river floodplains to their existing shape. These processes were interrupted, as humus horizons at different depths of the soil profile attest to. At the time of a relative stabilization of the natural conditions, first human settlements appeared in the Morava and Dyje alluvia. Poláček [4] states that in the period 8,000–6,000 BC sandy

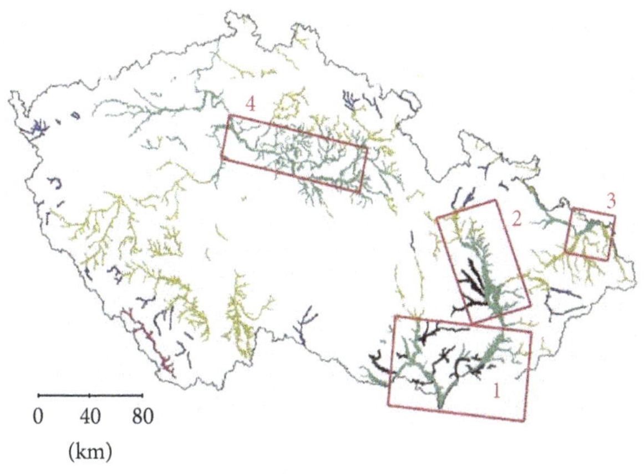


FIGURE 1: Distribution of floodplain forests in the Czech Republic [1].

islands in the river floodplain represented sought-after settlement areas for Mesolithic hunters and fishermen. Owing to massive colonization and forest felling, particularly in the early and late Bronze Ages (2000–750 BC), erosion intensified and the areas of herb growths consequently increased. An important milestone in the Morava and Dyje Rivers floodplain colonization was the arrival of Slavs in Moravia in the first third of the sixth century AD. Naturally, these tribes were attracted to the fertile areas along rivers. The river constituted a source of life, and the surrounding oak-hornbeam forests provided plenty of timber, game, and forest crops. The High Middle Ages (1250–1492 AD) saw a major change in the hydrological and geomorphological relationships in rivers which was accompanied by massive floods and a gradual levelling of the river floodplain by alluvial silts [4].

## 2. Changes in Land Use

As a result of changes in natural conditions, the species composition of floodplain forests gradually shifted towards the so-called softwood floodplain. Colonization of the submontane regions, massive deforestation, changes in land use, and a general devastation of the landscape all contributed to floods and the accompanying erosion. The origins of changes in land use in the Morava and Dyje River floodplains lie both in natural and anthropogenic factors, the latter eventually becoming dominant [5]. Provided a complex study of the changes in land use in her dissertation. Between 1836 and 1999 the total length of the Morava River decreased from 344.86 km to 268.02 km. The most pronounced changes were recorded along the river's lower reach, in the area of the Lower Morava River Valley, where the course was shortened by 48 km. This shortening accelerated the stream flow and decreased water seepage into the surrounding soil environment. Table 1 reveals that meadows were the most frequently represented form of land use in the 19th century. This changed radically in the 1950s, however, when meadows were gradually ploughed up and grazing in the floodplain was put to an end. Forests appear to be a highly stable component of the floodplain throughout the entire period. In this respect, however, the replacement of 1,400 ha of floodplain forests by water bodies (the Nové Mlýny Water Reservoirs) in the 1980s must be noted.

Generally it may be said, as [5] observes, that meadows and forests, that is, ecosystems of a high retention capacity, prevailed in the Morava River floodplain in the 19th century. Arable land was found on sites further from the river. Particularly the massive floods of 1997 showed that forest geobiocenoses and meadows are capable of retaining and slowing down the extreme water runoff to a certain degree.

The Early Middle Ages were characterized by prevailing communities of hardwood floodplain forests with oak, elm, and ash. These forests were considerably open and served as pastures as well. After pasture farming disappeared, individual oak stems often remained, under which stands of ash originated through natural regeneration, which later contributed to the establishment of nature reserves.

The evolution of floodplain forests on the lower reaches of the Morava and Dyje Rivers was influenced by land ownership as well. From this perspective, the Liechtenstein family played a historically significant role. Starting from 1249 they had gradually become the biggest landowners of both farm and forest land until their estate was confiscated by the state after World War 2. The Liechtensteins introduced amelioration measures which drained excessive water and resulted in a gradual transformation of the softwood floodplain forests into a floodplain with prevailing hardwood species, such as oak (*Quercus robur L.*), ash (*Fraxinus excelsior L.*; *Fraximus angustifolia Vahl*), or hornbeam (*Carpinus betulus L.*).

## 3. Water Management Measures

Apart from a high groundwater table, the basic characteristic which holds for all floodplain ecosystems is regular natural inundation by flooded rivers, usually in times of spring floods. South Moravian forests used to be thus affected by water from the Dyje, Kyjovka, and Morava Rivers, but regular floods in these rivers' lower reaches have long been a thing of the past. The last natural flood in the Dyje River took place in 1972 and in the Morava and Kyjovka Rivers in 1977. The elimination of natural flooding is a result of the construction of the Nové Mlýny Water Reservoirs on the Dyje River (1968–1988) and completed regulation of the rivers' lower reaches in the same period. On the one hand, this measure improved the living conditions of the local inhabitants and increased the safety of their activities, while on the other hand it had a highly adverse effect on the floodplain ecosystem. Floods occur at present as well, albeit in the form of disastrous events in the vegetation season, and always have a significantly negative impact on the floodplain forests, namely, their regeneration.

The completed regulation of the lower reaches of the Dyje and Morava Rivers in the 1970s and 1980s threatened the very existence of the floodplain forest complex. It was partly due to recessing of riverbeds in the channelized streams, followed by a drop in the groundwater table levels in forests by up to 1 m (Figure 2).

However, the biggest problem in the floodplain ecosystems was caused by the absence of regular inundation.

TABLE 1: Land use in the Morava River floodplain [5].

| | 1836 | | 1877 | | 1953 | | 1999* | |
|---|---|---|---|---|---|---|---|---|
| | km$^2$ | % | km$^2$ | % | km$^2$ | % | km$^2$ | % |
| Forests | **177.2** | **27.89** | **168.97** | **26.58** | **159.92** | **25.16** | **162.23** | **25.52** |
| Meadows | **273.52** | **43.03** | **222.61** | **35.02** | **179.26** | **28.20** | **53.86** | **8.47** |
| Pastures | **28.7** | **4.51** | **21.92** | **3.45** | **6.95** | **1.1** | **0** | **0.0** |
| Arable land | **136.65** | **21.5** | **196.78** | **30.95** | **235.94** | **37.11** | **329.28** | **51.8** |
| Gardens and orchards | 0.85 | 0.13 | 4.4 | 0.69 | 12.53 | 1.97 | 0.47 | 0.07 |
| Settlements | 16.3 | 2.56 | 19.36 | 3.05 | 38.24 | 6.01 | 66.16 | 10.41 |
| Transport routes | 0.39 | 0.06 | 0.85 | 0.13 | 2.06 | 0.321 | 2.24 | 0.35 |
| Water areas | 2.02 | 0.32 | 0.81 | 0.13 | 0.8 | 0.13 | 21.46 | 3.38 |
| Total | 635.7 | 100 | 635.7 | 100 | 635.7 | 100 | 635.7 | 100 |

*The data are adopted from project "Restoration of the Ecological Continuum of the Morava River."

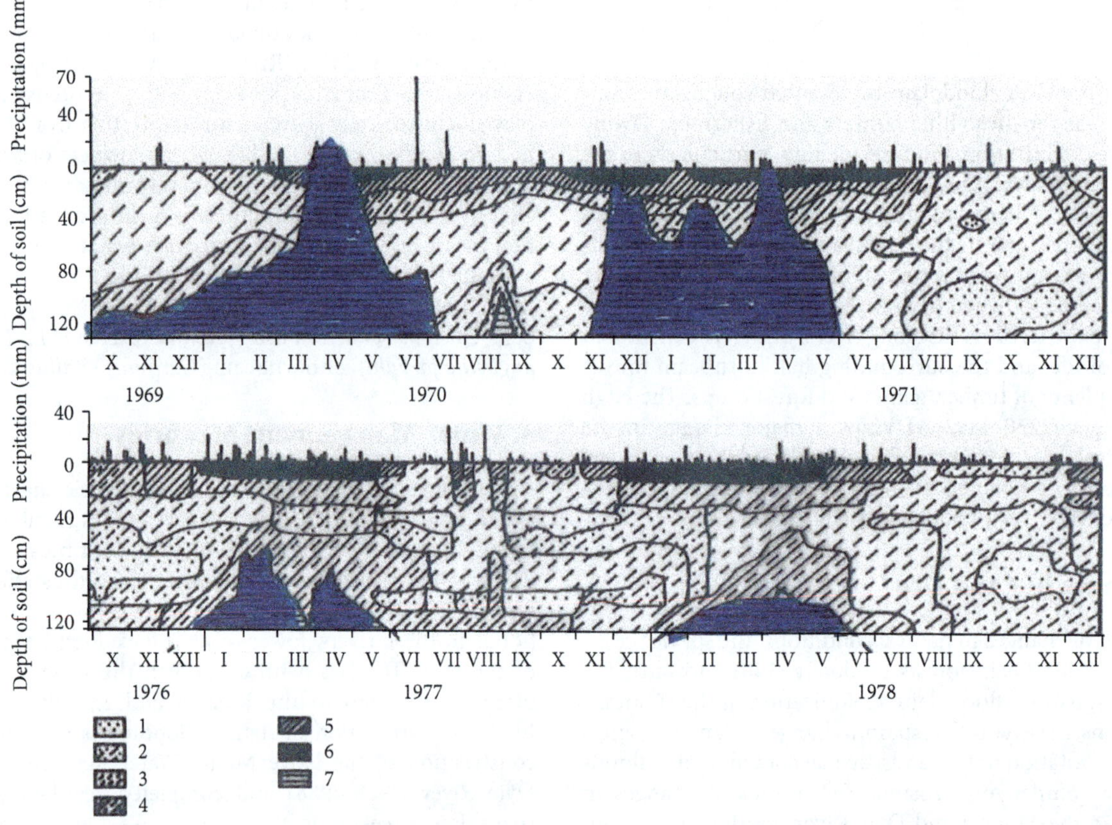


FIGURE 2: The dynamics of soil moisture (chronoisopleths) measured on research site in a period of floods (1969–1971) and in a period without floods (1976–1978). Soil moisture in percent by volume (1) 25–30, (2) 30–35, (3) 35–40, (4) 40–45, (5) 45–50, (6) 50–55, and (7) groundwater [7].

The floodplain forests began a slow transformation into a different, drier type of forests.

*3.1. Development of Moisture Balance.* An objective assessment of the development of abiotic conditions (temperatures and precipitation) in the floodplain forest ecosystem with respect to the assurance of sustainable management of these floodplain forests also involves the moisture balance. Calculation of moisture balance (MB) for a period 1851–2011 is based on a basic relation [8]. The basic relation was adjusted for the value of horizontal precipitation as follows:

$$\mathrm{MB} = \left[\left(R_a + \mathrm{HP}\right) - R_n\right] - \left(\mathrm{ETP}_a - \mathrm{ETP}_n\right), \quad (1)$$

where $R_a$ is monthly precipitation total in the given year in mm; $R_n$ is long-term precipitation total in the given month; $\mathrm{ETP}_a$ is monthly total potential evapotranspiration in the given year in mm; $\mathrm{ETP}_n$ is long-term total potential evapotranspiration in the given month in mm. The sum of moisture balance is calculated for the months of April through to September. Potential evapotranspiration (ETP) is derived by indirect method according to Thornthwaite [9].

The long-term monthly precipitation totals, average monthly air temperatures, and potential evapotranspiration in the area of floodplain forests are presented in Table 1. The moisture balance calculation also took into account fixed values of horizontal precipitation (HP) totals for stand microclimate, derived as an average of localities under study. The precipitation improves the total moisture balance. Development of moisture balance (MB) in the area of floodplain forests in 1851–2011 is illustrated in Figure 3 and Table 2.

It follows from Figure 3 that there are great changes occurring in the development of moisture balance between individual years. The changes reflect very markedly the temperature and precipitation conditions both at a level of mesoclimate and stand microclimate. In years with minus water balance, vegetation growth and distribution as well as development of pests and troublesome forms of insects in the ecosystem of floodplain forests are apparently limited by soil water deficit and by groundwater table under soil surface more than by any other environmental factor. In these situations, the significance of stand microclimate is further increasing, particularly its capacity to partly mitigate stresses induced by drought via temperature and relative air humidity, which play an important role in the development of horizontal precipitation (dew, and etc.).

The results of moisture balance suggest clearly that precipitation cannot be the only source of water for the existence of the floodplain forest ecosystem. Water balance specifies, for example, moisture available for the natural habitat of oak, which according to [10] ranges around 250 mm. Days with water balance of 250 mm and higher were observed to occur only 5 times during the period of 150 years (1851–2011).

Figure 3 further indicates that a moisture balance forecast may be compared with its actual development. The forecast reveals that the trend of actual (measured) MB is reaching the values of expected moisture balance of the 2020 time horizon (average of the period 2010–2030). A significant deterioration of MB may be expected in future, its levels falling to deficit values of −56 to −88 mm, along with a further increase in climate extremities (alternating of surplus and deficit MB values in consequent years). This trend became apparent approximately in 1965.

The expected changes in average monthly air temperatures and monthly precipitation totals were adopted from global circulation models (as a so-called multimodel median of 8 global circulation AR4 models: RCGCM, CNRM, CSIRO, IPSL, MIROC, ECHAM5, MRI, and HADCM3 [11]).

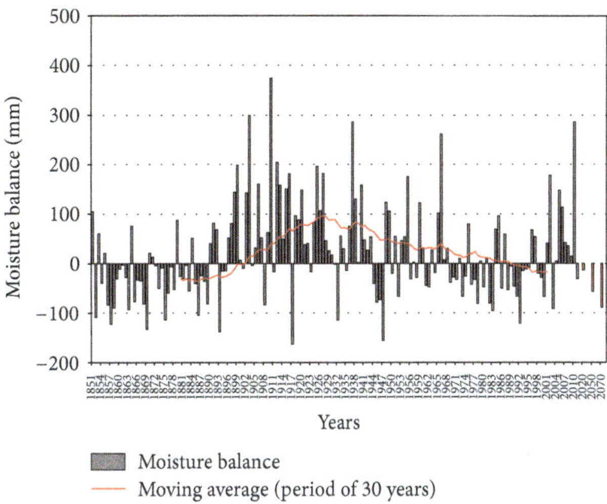


FIGURE 3: Development of moving averages (30 years) of moisture balance in the area of South Moravian floodplain forests in the period of 1851–2011 and prediction of moisture balance for time horizons of 2020, 2050, and 2070 which also represent a 30-year period.

## 4. Development of the Groundwater Table in the Confluence Floodplain of the Morava, Dyje, and Kyjovka Rivers

Detailed measurements of groundwater table (GWT) in the floodplain of the Morava and Dyje Rivers have been taken for 18 years. During this time, extreme states were monitored in the development of groundwater table, caused by flood discharges in the core streams of the Morava and Dyje and by the occurrence of a prolonged period of a dry hot climate.

Figure 4 shows the development of average daily groundwater table (GWT) values below the terrain level at the Ranšpurk station standing for the water regime in the region of the Soutok forest district. Development of GWT in this locality significantly differs from the development in the Dyje River floodplain. A different method of hydrotechnical regulations performed in the Morava River (discharge regime is closer to the natural state) and reciprocal mixing of the discharge regimes of the Morava, Dyje, and Kyjovka Rivers entering the confluence floodplain is reflected in the GWT development in the Ranšpurk locality. The water regime is also affected by artificial flooding which has been taking place in this locality since 1996. The lowest GWT level values, reaching the depth of 2,7 m below the surface (2003), occur in the drier period of the year (from the start of summer until the onset of winter). Maximum GWT level values (apart from the overflowings) at the level of 20 to 60 cm below the surface occur at the beginning of the growing season.

A characteristic feature of GWT of the confluence floodplain locality is the preservation of typical GWT dynamics. This means that the maximum level of GWT is time dependent on the beginning of the growing season and that the minimum level occurs in the autumn. This feature of the GWT dynamics was attained also at the beginning of the measurements in 1995 (Figure 4). However, conditions of moisture stress (for the longer part of the growing season, the GWT level is below the level of −1,60 m) form in the alluvial plain ecosystem due to GWT "critical level." Owing to increased discharges as a result of a gradual increase in precipitation amounts in the upper reach of the Morava River basin and partly also due to managed artificial flooding, the GWT level is progressively rising to the lower horizons of flood loams of heavier texture. GWT level between 1996 and

Table 2: Long-term monthly precipitation totals ($R_n$), average monthly air temperatures ($T_n$), and potential evapotranspiration ($EVP_n$) in the area of floodplain forests for the period 1851–2011 (all values are given in mm).

| Parameter | (I) | (II) | (III) | (IV) | (V) | (VI) | (VII) | (VIII) | (IX) | (X) | (XI) | (XII) | (I)–(XII) |
|---|---|---|---|---|---|---|---|---|---|---|---|---|---|
| $EVP_n$ | 0.8 | 2.6 | 16.9 | 47.9 | 83.0 | 103.2 | 116.1 | 110.9 | 68.6 | 39.8 | 10.2 | 1.8 | 601.8 |
| $R_n$ | 27.6 | 26.0 | 30.3 | 36.7 | 55.5 | 62.8 | 66.1 | 57.7 | 40.6 | 37.9 | 36.9 | 33.1 | 511.4 |
| $T_n$ | −1.4 | 0.2 | 4.3 | 9.6 | 14.5 | 17.8 | 19.6 | 18.9 | 15.0 | 9.6 | 4.1 | 0.1 | 9.35 |
| $HP^+$ | 3.3 | 2.1 | 1.1 | 2.6 | 2.6 | 3.8 | 3.2 | 5.8 | 5.8 | 3.9 | 4.9 | 3.1 | 42.2 |

$^+$Average values of horizontal precipitation are derived from dew and rime for the period of 1998–2002 from three studied localities of stand microclimate [6].

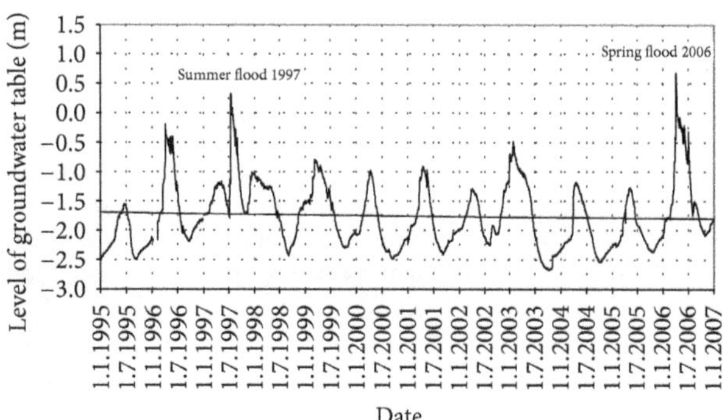


Figure 4: Development of daily averages of the groundwater table level 1995–2006 in the monitoring site Ranšpurk in the locality of the forest district Soutok in the confluence floodplain of the Morava and Dyje Rivers. The groundwater table level is expressed by the depth of the water level below the terrain. The trend of groundwater table is fitted with a polynomial of the 2nd degree.

1997 was quickly rising not only in the spring period but the growing trend could be observed also in the course of the whole year. The reason for this was the predominantly increasing total precipitation in the upper reach of the Morava River basin which first provoked local floods in the Bruntál Region and then, in July 1997, disastrous floods in the entire Morava River catchment area. The July 1997 flood in the locality Ranšpurk culminated in the overflowing of water at the height of 33 cm above the terrain. Between 1998 and 2002, the GWT level dynamics in the spring period became stabilized around the levels of −0,50 to −1,00 m; in the summer, the GWT level dropped to −2,20 to −2,50 m below the terrain. The year 2003 presents a certain exception because the Morava R. first showed a more distinct discharge growth from the melting snow in the upper reach of the river's catchment area. In the dry and extremely hot climate during the summer, the favourable GWT level (over −1,60 m) could be maintained by means of artificial flooding only until June. In August and September, the GWT level rapidly dropped to the absolute minimum levels of −2,70 m. In 2004 and 2005, the GWT dynamics started improving slightly thanks to the growing spring discharges from the thawing snow pack. This process culminated in the second largest flood registered in the period of 1995–2006 in the Morava River basin. The spring flood of 2006 reached its peak in the locality of Ranšpurk at the turn of March and April, with flood water overflowing at the maximum level of 67 cm above the terrain.

The specific character of soil moisture regime in the natural conditions of floodplain forest ecosystems is conditioned by the interrelations and effect of five crucial factors. It primarily encompasses the diverse soil conditions of alluvial sediments consisting of a variety of soil textures, ranging from light sandy soils to heavy clay soils, all in various, usually, stratified combinations. The subsoil of these soils usually contains gravel depositions of varying depth. The second factor represents the hydrological conditions of rivers, such as the flow intensity and dynamics throughout the year. The third factor includes local climatic conditions in general, particularly atmospheric precipitation, its yields, and intensity. The fourth factor of no lesser importance with respect to assessing the soil moisture regime of a given place or site is the terrain microrelief, which significantly affects the soil moisture regime. The fifth factor encompasses the vegetation cover, mostly in the form of floodplain forests and meadows, sometimes fields.

As for the forest floor heterogeneity, we can remark that the main factor is the species composition of forest stands, which is furthermore emphasized by the creation of pure stands as a result of clear-felling forest regeneration. The importance of differences, particularly between stands of oak and ash, is also demonstrated by the dynamics of decomposition processes during the year, as determined by the method of litter bags. It has been found that the forest decomposition of litter in a floodplain forest occurs, generally, in ash and hornbeam stands during the winter season [12].

Following the above-mentioned facts it is evident that in the course of assessing the hydrological and soil moisture regimes in floodplain ecosystems various situations may arise based on the listed factors.

Prax et al. [13] state that the primary source of groundwater in the Morava River floodplain is river water seeping through the highly permeable gravel-sandy subsoil. Precipitation also plays a certain role, mostly by positively affecting the moisture balance of soil profiles, namely, their surface horizons. This is closely related to the influence of water brought in by the restoration channels. Growth-ring analyses of the narrow-leaved ash radial increment revealed that every sudden and long-term change in the groundwater table regime results in a decreased radial increment. The period which ash stands require to adjust to changed natural conditions takes approximately from seven to ten years [12].

## 5. Ecological Response to Changes in Water Regime and Restoration of the Floodplain Forest Environment

A team of the Department of Forest Ecology of Mendel University in Brno studied this issue systematically in the 1990s. The obtained results were summarized in a publication [14, 15]. We present the following conclusions reached by the team.

The primary dominant species of the tree layer, such as the pedunculate oak (*Quercus robur L.*) and ash (*Fraximus excelsior L.*; *Fraximus angustifolia Vahl*), did not show a significant response to changes in water regime.

The layer of shrubs and young trees with prevailing *Cornus sanguinea* L. responded to a change in the water regime by a decrease in the leaf area index.

The herb layer responded to changes in water regime significantly, displaying a 60–70% decrease in biomass within ten years. Apart from this, a decline in the species diversity of the herb layer was monitored at some localities.

The trend of changes in the composition of herb layer communities is shifting towards drier communities.

During the times of regular floods, native grassland ecosystems worked as important producers and fulfilled important ecological and economic functions. After floods had been eliminated, vast areas of native grasslands were ploughed up. The rich supplies of humus and nitrogen generated by the grassland ecosystems had a positive effect on farm production at the initial stages following the ploughing-up.

Apart from changes in the water regime caused by the elimination of floods, climate change must be taken into consideration as well. The climate in the South Moravian region in the period following the flood control measures is going through a warm and dry period compared with the long-term means for annual temperature (9.0°C; 9.3°C in 1973–1982) and annual precipitation (524 mm; 453 in 1973–1982). In the period following the flood control measures there have been annual precipitation deficits, most frequently in August and September [15].

In the late 1980s, foresters were the first to realize the potentially disastrous effects of the flood control measures on floodplain forests. In the 1990s, in collaboration with hydrologists and other experts they prepared projects aimed at bringing water back inside the floodplain forests. In 1991–1999 a complex network of water channels, sluice gates, culverts, and small dams was built, allowing the control of water in vernal and permanent pools, as well as the groundwater table and its natural fluctuation in the vegetation season. After many years, wetland biotopes with permanent or periodic inundation reappeared in the forests. Forest stand vitality improved significantly as well. This influence is spatially limited to the distance of approximately 10–20 m, where it combines with groundwater table affected by the primary recipient, while in the spring time the channels function as drainage elements (Figures 5, 6, and 7).

However, it was generally noted that the restoration channel network displays a relatively low functionality. The preparation and construction of three Nové Mlýny Water Reservoirs provoked numerous discussions. The aim of this project was to protect the surrounding landscape from floods and to allow utilization of water supplies in dry periods of the year for irrigation of farm crops. Foresters took advantage of this possibility in the course of the implemented restoration projects of large-scale artificial inundation which, however, was gradually abandoned.

## 6. South Moravian Floodplain Forest Management

It is rather difficult to set an exact time which may separate a mere utilization of floodplain forests and their systematic management. Nožička [16] notes that planned forest management was introduced at the Liechtenstein estates in the 1760s. The rotation period for coppice forests was set at 20 to 40 years. Introduction of exotic tree species from North America dates back to 1799. According to Vybíral [17], it is safe to say that intensive cultivation of floodplain forests starts at the beginning of the 19th century. Last remnants of native natural forests disappeared, followed by a period of transformation of the originally predominantly softwood floodplain forests into cultivated forests with dominant long-lived, broad-leaf species, the so-called hardwood floodplain forests. The Liechtenstein foresters were aware of the high production capacity of floodplain sites in Slavonia, where original complexes of floodplain oak forests with growing stock exceeding 1,000 $m^3$ per hectare could still be found. In the late 19th and early 20th centuries, up to 50 tons of Slavonic acorns was therefore imported to South Moravia every year.

Softwood floodplain forests with low representation of oak began to be restored on large clear-cut areas, and between the rows of regenerated trees farm crops were grown [17]. This has resulted in the following species composition of most stands of South Moravian floodplain forests: pedunculate oak (*Quercus robur L.*) 41%, narrow-leaved ash (*Fraximus excelsior L.*; *Fraximus angustifolia Vahl*) 29%, poplars (usually cultivars) 13%, willow (*Salix alba L.*) 4%, maple (*Acer campestre L.*) 3%, hornbeam (*Carpinus betulus L.*) 2%, linden (*Tilia cordata Mill*) 2%, and other deciduous species and pine 3%.

The present species composition of dominant tree species in floodplain forests does not correspond with the natural site composition. Floodplain forests with dominant hardwood tree species (pedunculate oak/*Quercus robur L.*/, narrow-leaved ash/*Fraximus excelsior L.*; *Fraximus angustifolia Vahl*) evolved as a result of long-term, systematic management measures. Traditionally, floodplain forests regenerated on areas of up to several dozen hectares, with the area of exploited clearcuts being gradually limited to 5 ha. At present the Forest Act allows floodplain forest clearcuts of up to 2 ha. Utilization of pedunculate oak regeneration appears to be optimal, but it faces several problems. The crucial issue preventing the dominant use of the said method is a shortage of oak forests with suitable lower storey as well as prepared stands (without a high representation of ash). Another limitation is the fact that floodplain oak stands have very low production. The year 1999 proved to be an exception, as oaks had not produced virtually any mast for

(a)

(b)

FIGURE 5: Wetland restoration: tree vegetation formed during the period of 1982–1990 (South Moravia).

FIGURE 6: Large-scale artificial inundation.

FIGURE 8: Site prepared for artificial regeneration.

FIGURE 7: Water pools restored by means of channels.

FIGURE 9: Artificial regeneration of oak (*Quercus robur L.*).

over 30 years. The question why it was the case is subject to further study. For the time being, the only tested and so far proved hypothesis states that pedunculate oaks growing in the existing stands in the Czech Republic have very small crowns, masting very little or not at all, while stands of oak trees with large crowns growing in two-storey stands mast significantly more. This can be observed mainly abroad (Croatia, France), where natural regeneration of this tree species is commonly adopted. Low seed yields and massive damage caused by rodents also pose threats. Nevertheless, should a strong mast year of pedunculate oak (*Quercus robur L.*) come, it is advantageous to subject suitable stands to natural regeneration (the best method being scarification of the soil surface prior to regeneration), whereby game, rodent, and weed control are required. A successfully naturally regenerated stand encompasses a minimum of 3 seedlings per 1 $m^2$ after the first year of regeneration (Figures 8 and 9).

Starting from 1998, artificial stand regeneration is done on sites prepared by wood chippers.

(i) In winter or spring, wood chippers processing brushwood into coarse chips are deployed in clearcuts with brushwood left after harvesting.

(ii) The chips are left on the site where they work as a mulch layer and partly prevent weed growth.

(iii) At the end of the vegetation season (August, September) chemical weed control is applied from tractor-mounted sprayer.

(iv) Mechanized sowing or planting is done ideally in autumn months (November, December) or in early spring.

Apart from artificial regeneration of oak (*Quercus robur L.*), a similar method is applied, for example, in regeneration of the black walnut (*Juglans nigra L.*) which at present covers over 350 ha. Additionally, artificial regeneration of "classical" forest clones of hybrid poplars and cherry trees (*Prunus avium L.*) is underway as well.

## 7. Conclusion

It may be concluded that South Moravian floodplain forests have deservedly become the focus of keen interest of foresters, research institutions, nature conservationists, and international organizations alike. July 2003 marked the designation of the Lower Morava Biosphere Reserve by UNESCO in Paris, which encompassed virtually all important floodplain forests along the lower reaches of the Morava and Dyje Rivers and made them part of the World Network of Biosphere Reserves. South Moravian floodplain forests underwent major changes of their natural conditions in Early Middle Ages as well as changes brought about by foresters' efforts to cultivate high-quality production forests in the area. Regulation of the Morava and Dyje Rivers in the 20th century and the consequent changes in groundwater table also made a major impact on the condition of the floodplain forests. The implementation of restoration projects is therefore driven by efforts to enhance the forest environment and thus sustain both the production and nonproduction functions of the South Moravian floodplain forests.

## Acknowledgments

The paper was prepared under the financial support of the NAZV QJ1220033 Research Project "Optimization of Water Regime in the Model Area of the Morava River Alluvium" and within the Joint Project "The Functioning of Floodplain Forest Ecosystem in the Sava and Drava (Croatia) and Morava and Dyje (Czech Republic) Rivers Watershed under Changed Environment, Catastrophic Flooding and other Anthropogenic Impact."

## References

[1] E. Klimo, H. Hager, S. Matič, I. Anič, and J. Kulhavý, *Floodplain Forests of the Temperate Zone of Europe*, Lesnická Práce, Kostelec nad Černými lesy, 2008.

[2] Z. Neuhauslova, *Map of Potential Vegetation of the Czech Republic*, Academia Praha, 1998.

[3] P. Havlíček, "Geology of the dyje and morava confluence," in *Floodplain Forests of the Morava and Dyje Floodplain*, H. Kordiovský, Ed., pp. 12–19, Moraviapress Břeclav, 2011.

[4] L. Poláček, "The prehistory and history of the river valley landscape," in *Morava River Floodplain Meadows—Importance, Restoration and Management*, J. Šeffer and V. Stanová, Eds., pp. 25–36, DAPHNE, Bratislava, Slovakia, 1999.

[5] H. Kilianová, *Assessment of changes in forest geobiocenoses in the Morava River floodplain during the 19$^{th}$ and 20$^{th}$ centuries [Ph.D. dissertation]*, Olomouc, Czech Republic, 2001.

[6] P. Hadaš, "Temperature and humidity conditions of the floodplain forest with respect to stand microclimate and mesoclimate," *Ekológia*, vol. 22, supplement 3, pp. 19–46, 2003.

[7] P. Prax, "The hydrophysiological properties of the soil and changes in them," in *Floodplain Forest Ecosystem II. After Water Management Meausures*, M. Penka, M. Vyskot, E. Klimo, and F. Vašíček, Eds., pp. 145–168, Amsterdam coed. Academia, Prague, Czech Republic, 1991.

[8] M. Možný, "Potential evapotranspiration as an important agroclimatic characteristic," *Meteorologické Zprávy*, vol. 46, no. 5, pp. 152–156, 1993 (Czech).

[9] W. Mottl, "Abschätzung der potentiellen Evapotranspiration aus Klimadaten und Vergleich verschiedener Berechnungsmethoden," *Österreichische Wasserwirtschaft*, vol. 35, no. 9/10, pp. 247–254, 1983.

[10] B. Vinš et al., "Impacts of a possible climatic change on forests in the Czech Republic—Territorial study of climate change" (Czech), National Climatic Programme of Czech Republic. Element, 19, 2, ČHMÚ Praha, pages 135, 1996.

[11] J. Kalvová, E. Holtanová, M. Motl, J. Mikšovský, P. Pišoft, and A. Raidl, "Odhad rozsahu změn klimatu České republiky pro tři časová období 21. Století na základě výstupů AR4 modelů (Estimated Scope of Climate Change in the Czech Republic for Three Time Periods in the 21$^{st}$ Century Based on AR4 Model Outputs)," *Meteorologické Zprávy*, vol. 2, no. 63, pp. 57–66, 2010.

[12] E. Klimo, V. Hybler, H. Lorencová, and J. Štykar, "The heterogeneity of soil properties and biodiversity in floodplain forests of southern moravia in natural conditions and under antropogenic impacts," *Ekologia Bratislava*, vol. 30, no. 3, pp. 296–314, 2011.

[13] P. Prax, A. Prax, M. Kloupar, J. Heteša, and I. Sukop, "Optimization of the floodplain ecosystem hydrological regime following anthropogenic interference and its inclusion in the tvrdonice forest district management plan," Grantová služba LČR, Teplice. Závěrečná zpráva, pages 27, 2005.

[14] M. Penka, M. Vyskot, E. Klimo, and F. Vašíček, *Floodplain Forets Ecosystem II. After Water Management Meausures*, Elsevier, Prague, Czech Republic, 1991.

[15] F. Vašíček, "Changes in the structure and biomass of the herb layer under the condition of a medium moisture gradient," in *Floodplain Forest Ecosystem II. After Water Management Meausures*, M. Penka, M. Vyskot, E. Klimo, and F. Vašíček, Eds., pp. 197–242, Amsterdam coed. Academia, Prague, Czech Republic, 1991.

[16] J. Nožička, *An Outline of the Evolution of our Forests*, Praha, Czech Republic, 1957.

[17] J. Vybíral, "Role of foresters in the floodplain landscape," in *Floodplain Forests of the Morava and Dyje Floodplain*, H. Kordiovský, Ed., pp. 163–172, Moraviapress, Břeclav, Czech Republic, 2004.

16

# Opinions of Forest Managers, Loggers, and Forest Landowners in North Carolina regarding Biomass Harvesting Guidelines

**Diane Fielding, Frederick Cubbage, M. Nils Peterson, Dennis Hazel, Brunell Gugelmann, and Christopher Moorman**

*Department of Forestry and Environmental Resources, North Carolina State University, Raleigh, NC 27695-8008, USA*

Correspondence should be addressed to Diane Fielding, dianefielding68@gmail.com

Academic Editor: Thomas V. Gallagher

Woody biomass has been identified as an important renewable energy source capable of offsetting fossil fuel use. The potential environmental impacts associated with using woody biomass for energy have spurred development of biomass harvesting guidelines (BHGs) in some states and proposals for BHGs in others. We examined stakeholder opinions about BHGs through 60 semistructured interviews with key participants in the North Carolina, USA, forest business sector—forest managers, loggers, and forest landowners. Respondents generally opposed requirements for new BHGs because guidelines added to best management practices (BMPs). Most respondents believed North Carolina's current BMPs have been successful and sufficient in protecting forest health; biomass harvesting is only an additional component to harvesting with little or no modification to conventional harvesting operations; and scientific research does not support claims that biomass harvesting negatively impacts soil, water quality, timber productivity, or wildlife habitat. Some respondents recognized possible benefits from the implementation of BHGs, which included reduced site preparation costs and increases in proactive forest management, soil quality, and wildlife habitat. Some scientific literature suggests that biomass harvests may have adverse site impacts that require amelioration. The results suggest BHGs will need to be better justified for practitioners based on the scientific literature or linked to demand from new profitable uses or subsidies to offset stakeholder perceptions that they create unnecessary costs.

## 1. Introduction

Woody biomass from southeastern US forests is expected to play an important role in meeting the nation's future energy needs [1, 2]. Woody biomass includes small diameter trees, tops, limbs, or otherwise nonmerchantable forest products which are used for energy production and have not been utilized previously. The expansion of woody biomass-based energy raises concerns of forest sustainability, environmental degradation, and negative impacts on biodiversity [3–5]. North Carolina has mandated forest practice guidelines (FPGs), which are linked to recommended forestry best management practices (BMPs)—voluntary guidelines that were developed to protect water quality. However, these guidelines do not specifically address the harvesting of woody biomass, and removing this material has the potential to affect soil productivity, water quality, and forest biodiversity and could negatively affect forest health [3, 6]. These concerns led to the development of woody biomass harvesting guidelines (BHGs) in five states between 2007 and 2011, and several other states currently have BHGs under consideration [7]. The increasing demand for energy from biomass in proliferation of BHGs makes understanding how stakeholders perceive the suggested guidelines a critical need.

## 2. Study Objectives

The objective of this study was to assess the perspectives of key players in the forestry sector in North Carolina regarding

BHGs, compare these opinions with the scientific literature on BHGs, and draw conclusions about the implications for biomass markets and policy in North Carolina and the Southeastern US. The forest managers, loggers, and forest landowners interviewed during our study represent groups that will be directly affected by biomass harvesting policies and markets. Their views and willingness to adopt such standards will be crucial in determining the development and success of proposed BHGs.

## 3. Biomass Harvesting Guidelines

To mitigate the impact of biomass removal, BHGs generally recommend leaving 15 to 30 percent of harvestable coarse woody debris (CWD) [7]—which is dead wood material with a small end diameter of at least three inches and a length of at least three feet [8]—on the site following a harvest.

BHGs typically provide guidance on the form of debris retention, either spreading out or piling the woody debris that remains on a harvesting site. Minnesota established BHGs in 2007 which recommended that fine woody debris should be spread out relatively evenly across the site, rather than left in piles. Minnesota BHGs also recommend retaining and scattering 20 percent of the residual tops and limbs following a harvest. Wisconsin recommends retaining and scattering the tops and limbs from 10 percent of trees harvested on the site (one tree of every 10 trees harvested). Maine biomass retention guidelines recommend leaving as much dead wood on site as possible [9].

## 4. Literature

Some literature has addressed opinions about woody biomass either directly or indirectly. We summarize it here briefly and then compare our results with that literature more extensively in the discussion.

*4.1. Forest Managers.* Few studies have analyzed foresters' perspectives of harvesting woody biomass. Schulte et al. [10] conducted interviews with foresters in the US Midwest to learn more about the market for woody biomass. Forester respondents suggested leaving 33 to 55 percent of residues on site following a harvest for the improvement of soil quality and wildlife habitat. Enrich et al. [11] conducted surveys of foresters across the USA to gather perspectives of opportunities and challenges of harvesting wood for energy. According to respondents, the primary method of harvesting biomass was in conjunction with a conventional harvest. Aguilar and Garrett [12] surveyed foresters and discovered that the main opportunities from utilizing woody biomass were considered to be increased business for loggers and harvesters and an increase in commercial thinnings.

*4.2. Loggers.* Literature addressing loggers' opinions regarding harvesting biomass is uncommon. Abbas et al. [13] conducted semistructured interviews with Minnesota loggers to better understand the logistics of harvesting woody biomass compared with a conventional harvest. Abbas et al. discuss operational and financial challenges that loggers faced when harvesting small diameter trees which include the use of harvesting equipment that is not designed to handle small woody material and the cost burden of harvesting low-value wood products with expensive machinery. A few studies have addressed logger perceptions of BMPs—Milauskas and Wang [14] surveyed loggers in West Virginia and found 89 percent of loggers "always" comply with BMPs. More recently, Bolding et al. [15] surveyed Virginia logging business owners regarding BMPs and discovered that BMPs take significantly more time to be implemented in mountainous regions compared to the coastal plain. Becker et al. [16] reported that forestry stakeholders in the USA including loggers, believed a guaranteed supply of woody biomass was necessary before investments made in harvesting equipment. Communication analyses of the forestry community, particularly the perspectives of loggers regarding environmental policy, have been rare [17].

*4.3. Landowners.* Many studies have addressed landowner perspectives regarding forest management [17–20], but few have addressed landowner perspectives on BHGs or biomass harvesting. Londo [21] examined Mississippi forest landowners and reported that NIPF landowners had a low level of knowledge regarding BMPs. Many studies have been conducted to determine the effectiveness of forest landowner financial incentive programs [22–24]. Paula et al. [25] surveyed forest landowners in Alabama and found the majority willing to supply timber residues for energy purposes.

Most previous studies regarding woody biomass focused on the costs of harvesting biomass, rather than the experiences and attitudes of the individuals directly involved in the market. A small but growing number of studies explore the perspectives of the forestry community related to the harvesting of woody biomass, but no published research has examined their perceptions of BHGs. This study investigates the opinions of forestry stakeholders who would be directly affected by woody BHGs, if implemented in North Carolina, since they are the ones who would be putting the guidelines into practice.

## 5. Methods

We interviewed forest owners, foresters, and loggers to assess their opinions about BHGs. In addition, we examined the BHGs that have been developed in other states and reviewed the scientific literature about potential environmental problems with biomass harvests that could prompt the need for BHGs.

We sought to obtain and understand the perspectives of North Carolina forest landowners, professional loggers, and forest managers about BHGs using a qualitative research approach. We interviewed forest owners, foresters, and loggers to assess their opinions about BHGs. In addition, we examined the BHGs that have been developed in other states and reviewed the scientific literature about potential environmental problems with biomass harvests that could prompt the need for BHGs. We then compared the opinions of respondents in the forest business sector with the literature and approaches taken in other states to assess reasonable policy approaches to BHGs that could be taken in North Carolina or similar southern states.

The individual participants in the study were identified based on key initial contacts and subsequent followups. Five key informants were recommended through sources including the North Carolina Association of Professional Loggers, the State Board of Registration for Foresters, the Association of Consulting Foresters of America, the North Carolina Tree Farm Program, and the North Carolina Forestry Association. The group of key informants included one professional logger, three professional forest managers, and one landowner. The remaining respondents were identified through the snowball sampling method and interviewed to gain insight into the forest industry [26]. Rapport was established with interview subjects while meeting at their homes, places of employment, various logging sites, and public places.

From May 2010 to April 2011, semistructured interviews were conducted with 20 forest managers, 20 loggers, and 20 landowners. We used an interview guide, but the interviews were informant directed, and as a result the respondent was allowed to control the trajectory of the interview and decide which topics of conversation were of importance to them. Consistent with recommended qualitative research practice, the interview guides were allowed to evolve continually throughout the interview process. The questions addressed respondents experience with woody biomass harvesting and their opinions of the operational and financial feasibility of BHGs.

We continued interviewing new informants until saturation, the point where additional interviews did not contribute new information, was reached [26, 27]. Interviews were recorded and audio files were transcribed into QSR Internationals' NVivo 8 qualitative data analysis software. After the transcription of an interview was completed, a pseudonym was assigned to the speaker to ensure confidentiality. For example, a quotation identified as (Peter) was spoken by an interviewee identified with a pseudonym of Peter.

Qualitative data were analyzed using an inductive approach which allowed common themes to emerge from data gathered from interviews. Transcripts were reviewed and analyzed for emerging themes or patterns [27, 28]. Major themes in each interview were identified through finding repetitions of particular subjects. Transcripts of each interview were coded into different categories using the qualitative data analysis software.

Table 1: Highest education of North Carolina forest manager, logger, and forest landowner respondents ($n = 60$).

| | $n$ | % |
|---|---|---|
| Less than high school | 1 | 1.7% |
| High school | 13 | 21.7% |
| Some college/technical college/trade school | 6 | 10.0% |
| 4-year college degree | 25 | 41.7% |
| Graduate degree | 15 | 25.0% |

## 6. Interview Results

The 60 respondents (Appendix) were predominately male (93%) with an average age of 55. Most respondents had at least a 4-year college degree (Table 1).

The 20 forest managers were consulting foresters (45%), employees of forestry government agencies (15%), and forest managers for private companies (40%). A majority of logger respondents (65%) were currently participating in the woody biomass market, meaning they operated a wood chipper or grinder and sold dirty chips (chipped wood containing bark) to a facility that utilized the material for energy production. The delivered price that logger respondents received for dirty chips was in the range of \$15 to \$30 per ton with an average of \$21 per ton ($n = 9$). According to the loggers interviewed, one to three loads (approximately 25 to 75 tons) of woody biomass are typically harvested per acre, although this amount varies greatly depending on stand condition before it is cut. If loggers were not currently participating in the woody biomass market, they were asked to provide reasons for this decision. Loggers blamed the high costs of entry into the chipping business and the lack of market opportunities to sell the material. However, many of the loggers who were not currently harvesting expressed that they would probably add a chipping component in the future.

Forest landowners interviewed owned an average of 1286 acres of North Carolina forestland ($n = 20$) across 17 different North Carolina counties. Nineteen of the 20 forest landowners interviewed had a management plan, and all 19 were currently implementing that plan. Thus this sample seems representative of the most active forest landowners, not the general population of landowners as a whole, who often lack forest management plans. The primary objective landowners held in forest ownership was earning income from timber harvests. Additional dominant objectives included providing wildlife habitat, recreation, and having a natural resource to pass down to their children.

Our sample was drawn from the more involved and progressive members of each of these professional groups. These respondents represent active participants in the forestry sector of the North Carolina Piedmont and Coastal Plain regions, where forests are typically managed for timber rather than aesthetic purposes. Thus, the results seem to be a reasonable case in representing the southern timber producers' perspective on additional forest practice regulations.

*6.1. Consensus Themes*

*6.1.1. Current Forest Guidelines Are Sufficient.* A majority of respondents (96.6%) believed that North Carolina BMPs had been successful and most commented that the guidelines were "performance based" with "results-driven" standards and high compliance rates and thereby reasonable and had high compliance rates and were reasonable and effective. The majority of the participants believed that harvesting woody biomass could negatively impact water quality; however, they viewed current BMPs as a sufficient protective measure with few exceptions. Lucas, a forest manager, was asked if he believed current BMPs were a success, and he responded, "Yes, I do. Our BMPs are results driven. If you have good results, why change what you are doing? If you look at it, (we are) 97 to 98 percent in compliance of BMPs across the state. To me, that is very successful." Jason, a logger, described why the present positive response of those in the forestry community towards the BMPs can be attributed to the mode of establishment, stating

> I think one of the reasons they have been successful is when it started, it was basically as a volunteer implementation program and they figured out real quick the best way to implement these things and the state and everybody else kind of went along with it, working hand in hand with the forestry associations and various groups to come up with plans that would work. It was not forced on everybody, it was kind of volunteered; therefore, they have come up with a solution that has worked.

*6.1.2. Biomass Harvesting: Few Modifications to Operations.* Respondents expressed that woody biomass harvesting is only an additional component to harvesting with little or no modification to logging operations. During a typical timber harvesting operation in the Southeastern US, material is felled and brought to the logging deck to be sorted into the different products for processing. Respondents explained that when a biomass harvesting component is introduced to a logging operation, there is limited modification to the flow of operations, only the addition of a product at the logging deck. Henry, a landowner, expressed that biomass harvests were not different from current harvest operations, saying "Biomass harvesting is nothing more than a clearcut. We are doing a good job of that now. There should not be any additional harvesting guidelines for biomass." Connor, a logger, said

> Biomass harvesting is no different than regular clearcut harvesting. You are going to have the same machinery running on the ground, you will have to observe the same buffer zones and forest practice guidelines—there is no difference—same type of equipment doing the same thing.

*6.1.3. Lack of Scientific Research Supporting BHGs.* Forest manager, logger, and landowner respondents believed the proposed BHGs were being established by policy makers without the support of objective scientific research. Respondents agreed there was a lack of scientific research supporting claims that harvesting biomass leads to adverse effects on soil, water quality, timber productivity, and wildlife habitat. For example, Felix, a forest manager, stated "There is no document in the literature that I've seen that says that nutrients are diminished by biomass harvesting." Adam reiterated this point saying, "There is no research of (biomass harvesting) causing depletion to soil." Shaun, a logger, was asked about the adoption of BHGs, and he responded, "I do not know what benefits would be...whoever sets those guidelines must have documentation saying why this has got to be done. What is it doing for the land? What are the benefits?" The prevailing opinion of forest landowners was that forest policies are created by political forces or emotions rather than science. For example, Elijah, a forest landowner stated, "A lot of our regulations come from people that do not understand." Another landowner, Stephen, agreed saying, "A lot of people make regulations who do not know anything about forests."

*6.1.4. Threats of BHGs to Viability of Biomass Harvesting.* Forest managers and loggers regarded the woody biomass market as having very low profit margins and worried that BHGs posed a threat to the viability of biomass harvesting. Forest managers and loggers continually declared how they were constrained by small profit margins due to the undeveloped biomass markets in North Carolina. High transportation costs paired with high fuel costs in harvesting operations were the primary reasons given for why it was not feasible to add a chipping component to a harvesting operation. There were also high costs of entry in the logging business. Adding a woody biomass component to a harvesting operation required extra equipment including a fuel chipper and a chip van. The principal difficulty in adding a fuel chipping component to a timber harvesting operation was the increase in amount of fuel used during the time it takes to harvest the woody biomass material.

Forest managers and loggers expressed concern of the introduction of BHGs worsening the already volatile market for biomass. They communicated that if restrictions were placed on a product that had such a low market value in comparison to other forest products, biomass harvesting would become unattractive and not economically possible. A small number of loggers believed that the adoption of BHGs would not impact their logging operations financially, particularly if they were currently working for a large timber company rather than operating independently since these loggers do not purchase their own timber.

*6.1.5. Accurate Estimation of Debris Is Not Possible.* Most forest managers and loggers believed that estimating a particular percent of woody debris to be left after a harvest would be highly problematic from an operational standpoint. Foresters expressed the view that estimating a percentage would not only be logistically challenging but would also increase enforcement costs. Foresters noted that hardwood stands typically have much more debris left on site compared to pine stands, which, if previously thinned and mature, may have very little debris left. Requiring standardized rules of specific percentages of CWD to be left on a site may be difficult with the high degree of variability among harvests. The debris left on a site may depend on characteristics of the stand rather than whether there has been a biomass harvest or not. In addition, forest manager participants believed that violations of BHGs would be a challenge to remediate, particularly if debris was removed without leaving the recommended amount on site.

*6.1.6. BMPs Reflect Public Distrust of Forest Industry.* One additional theme that emerged from the interviews with forest managers was that they believed BHGs reflected public distrust of forest industry. This idea of "us versus them" describes the polarization between those who work in the woods and those who influence legislation. Forest managers described how they are skeptical of forest guidelines that are politicized and are a result of polemical lobbying by environmental groups. The reason many foresters gave for the lack of collaboration between environmental policy makers and foresters is because individuals working in the forest industry are seen as the "bad people" who cannot be trusted. Phillip explained this notion stating, "Contrary to what some in the environmental community believe, the mission of a forester is not to cut the last tree."

Forest managers also held the belief that forest guidelines often are established by policy makers without communicating with or considering the perspectives of those working in the forest industry. For example, one consulting forester when speaking about forest guidelines stated, "The new buffer rules were put into place without ever talking to foresters" (Patrick). Brad, a governmental agency forester stated, "It is unfortunate that a lot of folks do not want to believe what the foresters say anymore." Lucas, a forest manager in the industry sector, stated

> I would say a very high percentage of those people, their opinion of biomass harvesting is not based upon their opinion of biomass harvesting, it is based upon their opinion of logging in general. That is just my opinion but it can be verified by a lot of surveys and things that the general public says and does out there.

The public's view of forestry was not so apparent in logger responses although there were mentions of the disconnect between environmentalists and loggers. One logger, Kevin, explained how he hoped successful BMPs would make a difference in the public's perception of logging saying, "(BMPs) have definitely made a difference and I hope it makes a difference in what the public thinks about loggers."

*6.1.7. BMPs Reflect Public Fear of a Desolate Site.* The polarization between the public and those in the forest industry is illustrated by one particular vivid image of a desolate site, conjured by "outsiders" to the forest industry. Forest managers described the perceptions of the public visualizing timber harvests as "a vacuum cleaner that sucks out everything," a "concrete floor," or a "pool table." The desolate site rhetoric illustrated a perceived disconnect between the perceptions of the two groups. Our informants may have believed biomass harvesting would evoke this bleak image among the general public or environmental groups because of the fear of environmental degradation and loss of biodiversity of forestlands. However, these types of harvests are seen as only a temporary disturbance in the view of forest managers.

*6.1.8. Difficulties of Logging Business.* The topic of the current state of the economy and the effect on logging businesses permeated throughout most of the logger participants' interviews. One logger who declined to be interviewed only said, "If something does not change, you are going to see loggers go out of business." Walter stated, "The logger needs some help. That's something you need to put in your write up." He continued, discussing the difficulties of the business,

> I am 60 years old and I do not plan to stay in logging. It's too hard; it's nothing I would recommend a young man to do because it's too many rules and regulations right now. The logger do not get paid like he should...I have been in logging 14 years and I have not made a cent in the last two years. It's just getting harder and harder, fuel is high and everything you buy is high...and the logging rates for what you can get for wood is not going up.

Loggers expressed the need for governmental financial and legislative support. Most loggers interviewed owned or worked for small, family-owned logging businesses, and this often left them with small profit margins and little negotiating power.

*6.1.9. Definition of Biomass.* Forest landowners voiced concern over conflicting definitions of woody biomass in current political discussions. "I am not hearing anything that the BMPs need to be changed to address biomass concerns. Things I am hearing is that we need a definition of what biomass is," remarked Blaine. Duke stated, "I am all for biomass and developing the appropriate definition of it but I am not so sure there should be a limiting definition of what biomass is." Evelyn described how she was not in favor of the definition of woody biomass including whole trees for chipping. Verl, a landowner, also expressed how the lack of a clear definition of biomass was a limitation, saying

> Everyone is scared to death about what they are going to do on our definitions here. I talked to a fellow yesterday and he was wondering "Why do not we develop the (biomass) market?" And it's simple, who would make the economic investment, not even knowing if standing trees qualifies in the market. Everything is held hostage right now by definitional issues and debate over standards.

*6.1.10. Interest in Woody Biomass as an Additional Forest Product.* Only two of the twenty landowner interview respondents indicated they had previously had woody debris removed from their property to be used for energy purposes. Landowners who had not previously participated in the woody biomass market communicated interest in the expansion of the market in hopes that previously unmerchantable material could be used and provide additional income. In addition, landowners discussed the benefit of a cleaner site following a woody biomass harvest, which subsequently reduces fire hazards and improves forest health.

The primary reason given by landowners for not participating in the woody biomass market was the lack of market opportunities. Landowners who were aware of woody biomass market developments and recent legislation were eager to participate and be given the opportunity to use previously unmerchantable woody debris to reduce fire hazards and improve forest health.

*6.1.11. Government Support of Woody Biomass Market.* The majority of landowners agreed that the government should provide financial support for the biomass market and the positive externalities that landowners are providing to the public. Evelyn stated, "For the successful gathering of woody biomass, you're going to have to have tax incentives for loggers, lumber companies and for some landowners." Landowners were interested in federal woody biomass opportunities; however, there was discussion of the inadequacy of previous financial assistance programs such as the federal Biomass Crop Assistance Program (BCAP).

*6.1.12. Private Property Rights.* Many forest landowners were reluctant to accept any additional guidelines for a variety of reasons including the belief that forest guidelines are "unconstitutional" or are unnecessary since the private landowner should have the freedom in the responsibility over his or her own forest without an increase in environmental protections. Duke captured the overall attitudes of forest landowners toward forest regulations in his comment; "I do not think I need someone to tell me and my forester how to harvest the trees."

*6.1.13. Benefits of Biomass Harvesting and BHGs.* Two forest managers, three loggers and two of the forest landowners interviewed were not opposed to the adoption of BHGs and instead welcomed these guidelines, contrary to the beliefs of their counterparts. Although these respondents were in the minority, they did express their belief that removing more material from a site as woody biomass could be problematic and that there would be benefits to implementing BHGs. Spencer, a logger, explained "I would not be against (BHGs) because I think everything brought to the forestry field is for a good reason and a good cause. That is what we need, more guidance, guiding us the right way."

*6.1.14. Reduced Site Preparation Costs.* Forest manager respondents believed that the utilization of biomass can decrease the costs of site preparation for the landowner, which includes planting and bedding to enhance regeneration. Phillip stated, "(Harvesting of woody biomass) provides a cleaner site that is likely to require minimal site prep activity and less investment from a landowner for reforestation." Andrew stated, "The added benefit we get is by getting our site cleaned up better than it otherwise would" which decreases costs of future planting.

*6.1.15. BHGs Provide Increased Business for Consulting Foresters.* "The more rules that are in place for any forestry, the more confused the landowner will be, the more likely they will be to hire me," Jackson, a consulting forester, stated. A few consulting foresters noted that if BHGs are adopted in North Carolina, their business is likely to benefit since more people will seek the services of a registered forester to clarify and monitor the new guidelines. However, despite the possible personal advantage of increased guidelines, these foresters still opposed potential BHGs.

*6.1.16. Proactive Forest Management.* Many forest manager respondents thought that better markets for wood products would encourage landowners to be more proactive in forest management. Forest managers placed a great deal of emphasis on forestry decisions and practices being reliant on "the market." Adam, a forest manager, stated, "What causes good forestry practices? The market. We are not dummies and we know there has to be good markets. The better markets we have, the better we will practice good forestry." Michael stated, "If the logging cost made it uneconomical to harvest that landowner's tract, then that closes a market opportunity for that landowner, so tracts do not get harvested and forest management practices do not get implemented."

*6.1.17. Reduced Fire Hazard.* Forest managers and forest landowner respondents mentioned the reduction in fire hazard as a benefit to the harvest of woody biomass. In the past, prescribed burning was performed after a harvest. However, with the increase of liability and urbanization, less burning has been done so there is more debris left on harvesting sites, which can act as a fire hazard, particularly if left in large piles. Removing this debris for energy production can reduce a possible fire and smoke hazard.

*6.1.18. Soil Stabilization and Erosion Control.* Forest managers, loggers, and landowners expressed concern about biomass harvesting affecting soil quality; however, as mentioned previously, most demanded that sound research document the effects before operations are to be modified. Patricia, a forest manager, stated, "I think (the effect of biomass removal) would happen over time, a consistent removal of biomass in a particular area, over time, sure enough could affect nutrient properties of soil." Max, a logger, was asked about the recommendation of leaving 15 to 30 percent of CWD on site following a harvest, and he responded, "It seems like a fairly reasonable amount to be left, at least there is something being left to stabilize the soil and help the tree growing process. I do not think taking everything off the tract is a good idea." Max continued saying

> A lot of the time we are using all of that debris to stabilize the ground, to stop the run off, so if all that debris was getting chipped up and hauled off, then it could create an issue to where the guidelines may have to be changed, it really would—if we started to grind all that stuff up.

Several landowners expressed concern about soil damage and depletion of nutrients from harvesting woody biomass. "If you continue to take the trees off of there and do not leave some of the mass there to regenerate the soil, you are going to deplete the soil" (Brandon). "We do not pay much attention to the soils that take hundreds of years to develop that we can destroy with a single logging operation…I think there needs to be more attention to our soils by BMPs but I would not like to see another set of guidelines imposed on ones that are already there" (Duke). Cory, stated, "You cannot continue taking (woody debris) away and not adding something back to it." He continued saying

> Whenever I have had a harvest, I always encouraged my logger to make sure that (debris) was scattered about… I like to see that stuff scattered back on those sites so I do not get an erosion problem. (The debris) will rot and turn back to organic matter. I think that is an excellent guideline.

*6.2. Recommendations for BHGs.* When asked for suggestions for BHGs, most forest managers took this time to explain why guidelines specifically for biomass harvesting were not necessary. However, there were several recommendations for forestry guidelines to increase focus on soil quality in addition to water quality

> Instead of requiring a percentage left on site, I think it should be more related to soil impacts. In other words, like the FPGs…they do not really tell you how to do it, but the end result is to prevent water pollution and sediment and the streams. So I would think it would make sense for the BHGs to move towards preventing soil impacts. So if the logger has tracked equipment, maybe he can harvest all of the woody debris. If he does not, he has to leave 50 percent of the woody debris, if it is in a wet site (Patrick, forest manager).

Jackson, a consulting forester, recommended leaving a percentage of the ground surface covered with woody debris rather than a percentage of debris left on the ground. Brad, a government agency forest manager, suggested, "The BHG itself should be clear enough that it could be easily interpreted and easily measured, easily monitored." Richard, a consulting forester, believed that a five-year look back strategy would be the best way to determine if biomass harvesting was doing any damage on wildlife and soil before placing restrictions on the operations. He stated, "I think (a five year look back) would be much better than making the regulations onerous on the landowners and the buyers and everybody else."

Loggers who thought leaving 15 to 30 percent of CWD was reasonable believed that it should be done on problem areas. For example, Jack was asked about retaining 15 to 30 percent of CWD on site, and he responded, "It would be possible but only necessary in what we call the main skid trails." Clayton stated, "I would think if there was a slope, you would try to leave (debris) there. If it was more of a flat place, it may not take as much."

A small number of loggers suggested that guidelines be tailored to the region. For example, Jason suggested that guidelines should be different for the coast where the productivity of timber growth is higher than areas such as the Piedmont. He stated, "It just depends on what part of North Carolina you're talking about. One law will not fit across the whole state." Kevin was asked if North Carolina's BMPs should be adapted to address biomass harvesting and he responded, "I think in certain counties, maybe not in Neuse River or the Roanoke River area, but when you get in the western part of the state, I think the guidelines do need to change. Because we do not have a problem with erosion in this area like you do in Halifax and Warren County."

Jason, a logger, also offered advice for the formulation of guidelines saying

> (Policy makers) need to visit everybody who is in the biomass business now and get some comments from every single one of them before they come out with a set of rules. And they need to be loggers, they need to have had experience logging, they have to be able to show criteria. As long as you do not just give somebody a briefcase and say, here now you're the expert

> on biomass, and he do not know his damn way from Raleigh to the coast—I'm against that. But if the guy is legitimately trained and knows what he is doing, I am for it.

Landowners were asked if they had any recommendations for BHGs, and a variety of responses and suggestions were given. Some landowners took this opportunity to insist that additional forest guidelines were unnecessary. For example, Duke commented, "I am not sure you can regulate good forestry." A few landowners discussed enforcement and how without it guidelines were futile. Thomas stated

> If you do not have enforcement capabilities of guidelines or rules or regulations with teeth in them that are regularly used, somebody will do whatever they feel is in their best interest to do. So if you are going to have guidelines, have the gumption to have monitoring and enforcement, somewhere or another.

A few landowners suggested that for the woody biomass market to expand, woody material must be gathered for energy at the time of harvest rather than later. For example, Evelyn stated, "There does not need to be a delay, they don't need to come in a year later and gather the woody biomass. . .Come on in right behind, get it done at one time." Elijah stated, "I think it would be much more economical to be there on the spot with the logger doing it or a contractor doing it while the logger is there, as far as going in to the guideline."

*6.3. Forestry Organizations.* Both the North Carolina Forestry Association (NCFA) and the North Carolina Association of Professional Loggers (NCAPL) are important trade associations in the state, and several respondents were confident that guidelines regarding harvesting woody biomass would not be adopted because of these organizations' influence on the forestry sector and legislature. Jackson, a forest manager, stated

> Legislation will never pass in North Carolina. It is not going to. I think if you go back and look at the teeth we are trying to put in the board of registration laws for registered foresters, the forestry association has fought it all the way. Anything that requires a logger to learn about something, or spend more money, the forestry association fights you all the way.

Max, a logger, explained the NCAPL saying

> We have those guys as watchdogs and they are trying to get wood pellet companies into the state. Another thing too, with having that organization watching out for us, we don't see anything too intense as far as rules and regulations being passed. I do not think they would let it happen; they would do something to put a stop to it. So it's nice to have those guys and organizations in our corner as far as rules and regulations.

## 7. North Carolina Forestry Sector Opinions versus Biomass Harvesting Impacts Research

Our research interviews did not provide any background of biomass harvesting literature and science before or during the interviews. We were seeking the opinions of North Carolina forest landowners and professionals based on their existing knowledge and awareness of the issue. However, our review of literature on biomass harvesting impacts, the adoption of BHGs in other states, and the perceived fears of adverse environmental impacts by the public and environmental groups suggests discrepancies between the opinions of the forest business sector informants and perceptions of other groups concerned about biomass harvesting and its environmental impacts. These differences and some similarities are examined in this section.

*7.1. Current Guidelines Are Sufficient for Woody Biomass Harvests.* Our interview findings suggest broad support for BMPs among forestry stakeholder groups, which did not translate into support for BHGs. As in other studies, we found that respondents considered BMPs to have been successful for their effectiveness [29] and had high compliance rates [14, 30, 31]. Respondents believed that current guidelines were not only successful, but also sufficient in the protection of forest resources during harvesting operations. The widespread acceptance of BMPs may explain the perceived benefits of the current forest guidelines [32–34] and opposition to any additional guidelines since they are deemed unnecessary. There will likely be resistance from the North Carolina forestry community if BHGs are implemented. Perspectives of forestry stakeholders in states which have adopted BHGs would be beneficial to understanding the willingness to accept novel forest guidelines.

The perceived similarity between traditional operations and biomass operations also may explain why all groups believed that biomass harvesting should not be treated separately from a conventional harvest in terms of the guidelines applied. During a typical logging operation in the Southeast, after the tree is felled, the whole tree is usually skidded to the log deck where the limbs and tops are removed and the wood is sorted for each market. Adding a woody biomass harvesting component is the addition of a chipper and a chip van which are brought to the logging deck. The material which will be used for woody biomass is put through the chipper at the logging deck, and the operational process of harvesting trees in the woods is not altered.

Therefore, in the southeastern US, harvesting biomass does not significantly change operations in terms of harvesting material; it only supplies an additional product. However, harvesting operations differ across regions. For example, in the northeastern US, harvesters frequently leave the limbs and tops where the tree is felled, rather than bringing the material to the logging deck. These variations must be considered when adopting guidelines that may have been created for a region with different harvesting methods.

*7.2. Lack of Scientific Research Supporting BHGs.* Our findings indicate that the North Carolina forestry community does not believe the available science related to woody biomass harvesting is sufficient for the adoption of additional forest guidelines. All three of the groups agreed on the importance of the scientific method to achieve results that supported the creation of BHGs; however, they expressed concern that forest policy may be based on emotions and political expediency and ignore science and the practical constraints of harvesting operations. Similar studies found the same general attitudes towards new forest policy. For example, Holt [35, page 31] discovered that forest stakeholders in Oregon hoped that the biomass industry was "not politicized and driven by politics" and instead based upon funding and informed by science. Eliason et al. [36] found that natural resource professionals were wary of new guidelines and wanted to "understand the reasons" and "understand the end result…the end goal" before new forest guidelines were to be adopted. Dietz et al. [37] documented that forest professionals believed the public had little knowledge of forests. Similarly, Dirkswager et al. [38] interviewed landowners and found them to believe that policy makers who formulated biomass harvesting policies did not understand the economics of harvesting timber.

There is a large body of research that has explored the science behind the dynamics of woody biomass and the detrimental effects that harvesting the material could have on forest ecosystems. Harmon et al. [39] documented snags and downed coarse woody debris to be important to forest ecosystems and critical to wildlife habitat. Downed CWD has been shown to benefit a variety of organisms, including invertebrates, vertebrates, fungi, and plants [40, 41]. Downed CWD may also be important for nutrient retention [39]. It has been shown to limit erosion by reducing overland flow [42].

Despite the ample evidence for the importance of specific habitat elements (e.g., CWD, snags, cavity trees, etc.) to forest biodiversity, there is still reluctance among the forestry community to accept additional regulations to protect forest ecosystems in the event of increased biomass harvests. Conflict between practitioners, policy makers, and researchers likely emerges from efforts to set a specific number or amount of each habitat elements that should remain on an acre of forest land. This is comparable to the results of both Botkin [43] and Peterson [44] who found that the scientific method often exacerbates, rather than lessens, environmental conflict.

The growth in woody biomass markets may significantly increase the amount of CWD that is harvested [45], and Moorman et al. [46] suggest that the extraction of woody biomass may alter the dynamics of CWD and lead to lower amounts of the material due to shorter harvest rotations. This reduction in downed CWD could have detrimental effects on many species of wildlife including mice, small snakes, lizards, and salamanders [39, 47]. The Pennsylvania BHGs state that "good biomass practices can enhance and improve forestland; poor practices can damage and devalue it" [48, page 30]. Similarly, Hess and Zimmerman [45, page 6] discovered there to be a consensus among experts that the "absence of downed woody debris would be detrimental to biodiversity and ecological processes."

Some ambiguity in the science addressing the effects of woody biomass harvesting and the high level of conflict between interest groups suggests that it will be difficult to implement BHGs, since there is neither a top-down nor bottom-up pressure and little support from the interest groups interviewed here [49]. Per Matland [49], the coalition of local level actors will determine the policy implementation depending on their influence and strength. The resistance of loggers, landowners, and foresters to more BHGs will impede their adoption. Environmental interest groups and agencies may support the rules, but it appears they will face fairly pervasive opposition. The new conservative legislature in North Carolina, and much of the South, will also make development of new laws or even new BHG guidelines difficult.

On the other hand, a recent court ruling in North Carolina that all woody materials, even whole trees, qualify as biomass, along with the rapid expansion of more wood pellet plants, may cause enough pressure that BHGs will be developed. Also, all large energy and forestry companies already have corporate environmental policies, staff, and programs. Large companies also are ranked on their sustainability efforts by many sources, such as the Dow Jones Sustainability Index, which evaluates the sustainability performance of the companies listed on the Dow Jones and will surely require them to have proactive policies to prevent adverse impacts during biomass harvests.

*7.3. Threats of BHGs to Viability of Biomass Harvesting.* Woody biomass is currently a low-value product, and we found that respondents believed the market was too volatile to introduce further guidelines on the harvesting process. Forest managers and loggers identified the transportation costs of woody biomass as one of the most challenging factors of adding a chipping component to an operation. These findings are in line with those of Aguilar and Garrett [12], who reported that the costs of harvesting and transporting were identified by professional forest managers as the biggest challenge to harvesting woody biomass as a renewable energy feedstock.

Similarly, Dirkswager et al. [38] conducted a phone survey with logging business owners in Minnesota and found low product prices, high equipment and fuel costs, the lack of material, and environmental regulations were primary barriers in the harvesting of biomass. They also found that landowners believed increased forest regulations would threaten sustainability in timber harvesting. It is not clear which group will bear costs of BHGs. Loggers will spend more time implementing BHGs which will result in increased fuel costs and equipment usage. This suggests that the price landowners receive for woody biomass material may be reduced with BHG implementation.

*7.4. BMPs Reflect Public Distrust of Forest Industry.* Respondents believed that forest guidelines often were formulated by policy makers based on the public's distrust of logging rather than sound science. A review of the literature suggests the polarization between those in the forest industry and the public is not a new concept. Foresters have had decreasing public esteem for years. For example, the association of British Columbia Professional Foresters sponsors opinion polls of the public's perception of foresters each year and found that in 1997, 66 percent of the public believed it was "very important" to have forestry reserved for professional foresters. This percentage decreased each year, and in 2002, this number dropped to 40 percent [50].

Thomas [51] as cited in Luckert [52] wrote of American foresters and stated, "Twenty or so years ago, foresters were among the most respected and trusted professionals in the United States. Sadly that is no longer so." Keefer et al. surveyed loggers in Pennsylvania, and 70 percent felt the "negative public image of the logging industry" was the most significant pressure faced in the logging business [53, page 91]. These sentiments of forest managers and loggers illuminate the isolation and lack of effective communication between those working in the forestry industry and the public.

Forest managers frequently mentioned the public's fear of having a timber harvest site left as a denuded landscape of stumps. This perception was a reason forest managers cited as why policy makers believed that biomass harvesting needed to be regulated although, according to respondents, this bleak observation was only a temporary scene which is quickly replaced by new growth of the forest. The public's fear of a desolate site is explored throughout the literature and can be explained by the public's negative perception of clearcutting. Egan et al. [54] investigated tree farmer opinions in West Virginia and discovered that 55 percent believed clearcutting should be banned. Bliss [55] investigated the public's perceptions of clearcutting and found that several opinion polls have continually reported the widespread disapproval. Bliss also describes that Americans find clearcutting to be "aesthetically offensive" and the practice leads to conjured images of vast deforestation and degradation. Understanding the relationship between the public's negative view of forestry and forest regulations is necessary in the development of forest guidelines.

*7.5. Logging Economic Hardships.* The economic challenges of operating a logging business during a poor economy were discussed frequently by loggers throughout the interview process. Recent literature discusses the difficulties of the logging business. For example, Bolding and others [15] documented the challenges of logging in Virginia which include an aging workforce, recent mill closures, and volatility of the market. Egan and Taggart [56] found 69 percent of loggers surveyed in New England did not encourage their children to enter the logging business, and only 51 percent of respondents stated that they would still be logging in five years. Loggers are constrained by low prices, lack of markets, and high operation and fuel costs and expressed the desire to capitalize on available business opportunities. This suggests that the poor economic climate for loggers paired with the volatile woody biomass market will make it difficult for the forestry community to accept any regulations that further constrain profit margins for loggers. Additional research is needed to better understand the economics of harvesting operations using BHGs.

*7.6. Definition of Biomass.* Our results show that forest landowners are concerned with the conflicting definitions of woody biomass in political discussions. A number of landowners believed that definitions of woody biomass should not be restrictive based on the type of material or if the material is located on public or private land. A small number of landowners mentioned that they did not want the woody biomass definitions to include whole-tree chipping. There have been ongoing political debates regarding the formulation of a clear definition of woody biomass, particularly as it relates to renewable energy standards. Benjamin et al. [3] noted that the terms "forest biomass" and "biomass harvests" often create confusion among forest professionals and those in academia. Aguilar and Garrett [12] surveyed state foresters, state energy biomass contacts, and members of the National Council of Forestry Association Executives to understand perspectives of the definition of biomass and found that respondents believed the definition of biomass "should not differentiate between naturally regenerated forest stands and plantations or private and public forestlands." The current ambiguity in woody biomass definitions, particularly in North Carolina legislation, will likely impede the implementation of additional forest guidelines.

*7.7. Private Property Rights.* Forest landowners were adamant about their private property rights and believed they deserved the freedom to manage their forest as they wished, without an increase in environmental protections. These perceived landowner rights were the main reasons given for landowner respondents' opposition to BHGs. These results are comparable to Williams and others [57] who found a majority of Arkansas NIPF landowners believed they should have the ability to use their land as they chose without regulations although the landowners supported environmental protection.

In contrast, Bliss et al. [58] surveyed southeastern landowners and reported 76 percent felt their property rights should be limited if it was essential for the protection of the environment. In a similar study, Bliss et al. [18] reported rural residents, urban residents, and forest landowners believed that private property rights were necessary but secondary to environmental protection, and there was no significant difference between the groups. These two studies suggest landowners may support BHGs despite the opposition to additional forest protections, if guidelines ensure the protection of the forest.

## 8. Conclusions

Forest managers, forest landowners, and loggers in North Carolina were reluctant to accept any additional forestry guidelines, particularly related to the harvesting of woody biomass, due to the perceived economic and social impact of increased regulation. Reconciling these perspectives on woody biomass harvests with public concern over environmental impacts is necessary for effective expansion of biomass markets, and public policies that will facilitate the use of this potentially important natural resource.

Most forest manager, logger, and landowner respondents in this study believed that there was not sufficient scientific literature or evidence to support claims that biomass harvesting threatens wildlife habitat or ecosystem function. Available science related to the negative effects of woody biomass harvesting is either not well known by the forestry community or not believed. This suggests that the introduction of science has had little practical influence on the perceptions, or at least that literature has not been extended well to the forestry sector respondents in the field.

The opposition to BHGs among forest managers, loggers, and landowners parallels the resistance that the public typically has for additional environmental policies and regulations. However, this study provided unique results in that respondents expressed positive opinions and acceptance of BMPs yet antipathy towards additional forest guidelines. The differences among sectors were not always as anticipated. For example, consulting foresters could potentially profit from additional forest guidelines yet still opposed BHGs. Loggers would likely incur more costs from the implementation of BHGs, at least in the short run, than landowners, yet were more positive regarding adoption of BHGs.

Our interviews in North Carolina, and indeed the general literature, have not examined the role of market factors such as corporate social responsibility, environmental management systems (EMS) such as ISO 14000, external corporate sustainability rankings for investors and consumers, and forest certification in driving the demand for BHGs. Large electric and wood chip companies have extensive environmental management programs and must submit and demonstrate environmental compliance in all their activities. Woody biomass production will be required to meet these same strictures, and BHGs, forest certification, chain of custody, or other approaches surely will be necessary to obtain investor approval and bank financing for these projects. Thus the market drivers may lead the traditional forestry sector toward BHGs despite stakeholder concerns.

The findings from this qualitative study point to several recommendations in developing appropriate forest policies and incentives for practicing sustainable forestry on privately owned land. Recommendations for policy makers include the following.

(i) Bridge the divide between public and the forestry community's discourse regarding biomass policy to lessen conflict between environmental groups, forest managers, forest landowners, and loggers.

(ii) Consider the forestry community's reluctance to adopt additional guidelines.

(iii) Use market drivers of environmental management systems, corporate sustainability rankings, and certification to convey the need for BHGs.

(iv) Develop guidelines based on research and focus on harvesting impacts that are not already covered by BMPs, such as biodiversity, wildlife, and site productivity.

(v) Provide enhanced outreach and extension about the science regarding potential adverse impacts of biomass harvesting and the role of BHGs in ameliorating problems.

(vi) Develop standard protocols for measuring the amount of biomass initially and the amount to be left on site.

(vii) Specify local relevance of woody BHGs.

(viii) Define the term *woody biomass* clearly in appropriate policy.

(ix) Engage all stakeholders in discussions and development of BHGs, including environmental nongovernment organizations and bioenergy companies.

(x) Develop new, low-cost, innovative outreach methods for young and "tech-savvy" landowners, foresters, and loggers to participate in biomass markets and BHGs.

Future research needs to include assessing the perceptions of environmental groups and energy companies and users regarding BHGs. Other future studies could include a cost analysis to assess the financial impact of BHGs on logging operations. In conclusion, this study can aid in making informed policy decisions around BHGs that are not only sustainable, but also integrate the suggestions of the forestry community and scientific literature into guideline formulation. This information can also contribute to management decisions regarding sustainable BHGs and certification standards. The qualitative data and recommendations provided can be used for further education and outreach for the southeastern forest community and for the evaluation of specific solutions for the barriers of harvesting woody biomass.

## Appendix

See Tables 2, 3, and 4.

Table 2: Forest manager respondents.

| | Pseudonym | Age | Gender | Education | Job type |
|---|---|---|---|---|---|
| 1 | Ronald | 49 | M | Graduate school | Consultant |
| 2 | Jonathan | 53 | M | College | Consultant |
| 3 | Patrick | 53 | M | College | Consultant |
| 4 | Peter | 51 | M | Graduate school | Government agency |
| 5 | Beverly | 44 | F | College | Private industry |
| 6 | Joseph | 39 | M | College | Private industry |
| 7 | Andrew | 53 | M | College | Private industry |
| 8 | Wayne | 58 | M | Graduate school | Consultant |
| 9 | Patricia | 39 | F | College | Private industry |
| 10 | Jackson | 54 | M | College | Consultant |
| 11 | Wesley | 38 | M | College | Consultant |
| 12 | Michael | 49 | M | Graduate school | Government agency |
| 13 | Brad | 38 | M | College | Government agency |
| 14 | Matthew | 55 | M | College | Consultant |
| 15 | Adam | 62 | M | College | Consultant |
| 16 | Robert | 56 | M | College | Consultant |
| 17 | Lucas | 56 | M | Graduate school | Private industry |
| 18 | James | 48 | M | Graduate school | Private industry |
| 19 | Phillip | 60 | M | College | Private industry |
| 20 | Felix | 56 | M | Graduate school | Private industry |
| | Average | 50.5 | | | |
| | Standard deviation | 7.4 | | | |

Table 3: Logger respondents.

| | Pseudonym | Age | Gender | Education | Owns chipper |
|---|---|---|---|---|---|
| 1 | Spencer | 50 | M | College | No |
| 2 | Walter | 60 | M | 8th grade | No |
| 3 | Clayton | 45 | M | High school | No |
| 4 | Blake | 56 | M | Some college | Yes |
| 5 | Hugh | 52 | M | College | Yes |
| 6 | Drew | 50 | M | Some college | No |
| 7 | Harrell | 59 | M | Some college | Yes |
| 8 | Anthony | 52 | M | High school | Yes |
| 9 | Austin | 32 | M | High school | Yes |
| 10 | Max | 55 | M | Some college | No |
| 11 | Shaun | 75 | M | High school | Yes |
| 12 | Connor | 67 | M | College | No |
| 13 | Owen | 51 | M | High school | Yes |
| 14 | Alex | 31 | M | Some college | Yes |
| 15 | Kevin | 54 | M | High school | No |
| 16 | Charles | 47 | M | High school | Yes |
| 17 | Jason | 63 | M | Graduate school | Yes |
| 18 | Gary | 40 | M | High school | Yes |
| 19 | Jack | 51 | M | College | Yes |
| 20 | Edward | 48 | M | High school | Yes |
| | Average | 51.9 | | | |
| | Standard deviation | 10.5 | | | |

TABLE 4: Landowner respondents.

| | Pseudonym | Age | Gender | Education | Acreage of woodlands | Management plan |
|---|---|---|---|---|---|---|
| 1 | Duke | 76 | M | Graduate school | 164 | Yes |
| 2 | Cory | 59 | M | College | 107 | Yes |
| 3 | Mark | 44 | M | Graduate school | 180 | Yes |
| 4 | Stephen | 71 | M | College | 3000 | Yes |
| 5 | Ronald | 67 | M | College | 355 | Yes |
| 6 | Thomas | 68 | M | College | 250 | Yes |
| 7 | Ken | 55 | M | Graduate school | 70 | Yes |
| 8 | Brandon | 67 | M | College | 5000 | Yes |
| 9 | Elijah | 72 | M | High school | 750 | Yes |
| 10 | Evelyn | 59 | F | Graduate school | 650 | Yes |
| 11 | Blaine | 59 | M | Graduate school | 2500 | Yes |
| 12 | Verl | 60 | M | Graduate school | 125 | Yes |
| 13 | Henry | 48 | M | Graduate school | 2900 | Yes |
| 14 | William | 73 | M | Some college | 250 | Yes |
| 15 | Simon | 52 | M | High school | 25 | Yes |
| 16 | Margaret | 69 | F | High school | 110 | Yes |
| 17 | Todd | 80 | M | Graduate school | 94 | Yes |
| 18 | Chuck | 65 | M | High school | 400 | No |
| 19 | Roger | 58 | M | College | 7500 | Yes |
| | Average | 63.2 | | | 1285.7 | |
| | Standard deviation | 9.6 | | | 2041.3 | |

## Acknowledgments

The authors would like to thank the forest manager, logger, and landowner interview respondents who shared their opinions with them; the reviewers who provided helpful insight; the United States Department of Agriculture (USDA) and the Agriculture and Food Research Initiative (AFRI) for funding this research.

## References

[1] C. S. Galik, R. Abt, and Y. Wu, "Forest biomass supply in the southeastern United States—implications for industrial roundwood and bioenergy production," *Journal of Forestry*, vol. 107, no. 2, pp. 69–77, 2009.

[2] R. C. Abt, K. L. Abt, F. W. Cubbage, and J. D. Henderson, "Effect of policy-based bioenergy demand on southern timber markets: a case study of North Carolina," *Biomass and Bioenergy*, vol. 34, no. 12, pp. 1679–1686, 2010.

[3] J. G. Benjamin, R. J. Lilieholm, and C. E. Coup, "Forest biomass harvesting in the Northeast: a special-needs operation," *Northern Journal of Applied Forestry*, vol. 27, no. 2, pp. 45–49, 2010.

[4] D. Damery, M. Kelty, D. Benjamin, and R. J. Lilieholm, "Developing a sustainable forest biomass industry: Case of the US Northeast," *Ecology and Environment*, vol. 122, pp. 141–152, 2009.

[5] Pinchot Institute and The Heinz Center, *Forest Sustainability in the Development of Wood Bioenergy in the U.S.*, The Pinchot Institute for conservation and The H. John Heinz III Center for Science, Economics and the Environment, Washington, DC, USA, 2010.

[6] Southern Group of State Forester's, "Woody Biomass Harvesting Guidelines," Services, Utilization and Marketing Task force and Water Resources Committee Technical Paper, 2009.

[7] A. M. Evans, R. T. Pershel, and B. A. Kittler, *Revised Assessment of Biomass Harvesting and Retention Guidelines*, Forest Guild, Santa Fe, NM, USA, 2010.

[8] C. W. Woodall and V. J. Monleon, *Sampling Protocol, Estimation, and Analysis Procedures for the Down Woody Materials Indicator of the Fia Program*, GTR-NRS-22, USDA Forest Service, Northern Research Station, Newtown Square, Pa, USA, 2008.

[9] J. G. Benjamin, *Considerations and Recommendations for Retaining Woody Biomass on Timber Harvest Sites in Maine*, University of Maine, Maine Agricultural and Forest Experiment Station, Orono, Me, USA, 2010.

[10] L. J. Schulte, L. J. Tyndall, R. Hall, and K. Grubh, "Rapid assessment of woody biomass capabilities in three regions of the U.S. Midwest," 2008, http://www.nrem.iastate.edu/landscape/Publications/FinalWoodyBiomassReport.pdf.

[11] A. Enrich, D. Greene, and S. Baker, "Status of Harvesting & Transportation for Forest Biomass – Preliminary Results of a National Survey of Logging Contractors, Procurement Foresters, Wood Dealers and Forest Managers," 2009, http://frec.vt.edu/cofe/documents/2010/Enrich_COFE_BiomassSurvey.pdf.

[12] F. Aguilar and H. E. Gene Garrett, "Perspectives of woody biomass for energy: survey of state foresters, state energy biomass contacts, and national council of forestry association executives," *Journal of Forestry*, vol. 107, no. 6, pp. 297–306, 2009.

[13] D. Abbas, D. Current, K. Brooks, and D Arnosti, "Chapter 6: The Loggers' Voice in Harvesting Fuel: Cutting Costs and

Reducing Forest Fire Hazards Through Biomass Harvest," The Institute for Agriculture and Trade Policy, 2008, http://www.upwoodybiomass.org/downloads/harvesting_fuel_arnosti_et_al_2008_report.pdf.

[14] S. J. Milauskas and J. Wang, "West Virginia logger characteristics," *Forest Products Journal*, vol. 56, no. 2, pp. 19–24, 2006.

[15] M. Chad Bolding, S. M. Barrett, J. F. Munsell, and M. C. Groover, "Characteristics of Virginia's logging businesses in a changing timber market," *Forest Products Journal*, vol. 60, no. 1, pp. 86–93, 2010.

[16] D. R. Becker, S. M. McCaffrey, D. Abbas, K. E. Halvorsen, P. Jakes, and C. Moseley, "Conventional wisdoms of woody biomass utilization on federal public lands," *Journal of Forestry*, vol. 109, no. 4, pp. 208–211, 2011.

[17] T. R. Peterson and C. C. Horton, "Rooted in the soil: how understanding the perspectives of land-owners can enhance the management of environmental disputes," *Quarterly Journal of Speech*, vol. 81, no. 2, pp. 139–166, 1995.

[18] J. C. Bliss, S. K. Nepal, R. T. Brooks, and M. D. Larsen, "In the mainstream: environmental attitudes of mid-south forest owners," *Southern Journal of Applied Forestry*, vol. 21, no. 1, pp. 37–43, 1997.

[19] D. D. Dutcher, J. C. Finley, A. E. Luloff, and J. Johnson, "Landowner perceptions of protecting and establishing riparian forests: a qualitative analysis," *Society and Natural Resources*, vol. 17, no. 4, pp. 319–332, 2004.

[20] M. K. Measells, S. C. Grado, H. G. Hughes, M. A. Dunn, J. Idassi, and B. Zielinske, "Nonindustrial private forest landowner characteristics and use of forestry services in four southern states: results from a 2002-2003 mail survey," *Southern Journal of Applied Forestry*, vol. 29, no. 4, pp. 194–199, 2005.

[21] A. J. Londo, "An assessment of Mississippi's nonindustrial private forest landowners knowledge of forestry BMP's," *Water, Air, and Soil Pollution: Focus*, vol. 4, no. 1, pp. 235–243, 2004.

[22] R. A. Kluender, T. L. Walkingstick, and J. C. Pickett, "The use of forestry incentives by non-industrial forest landowner groups: is it time for a reassessment of where we spend our tax dollars?" *Natural Resources Journal*, vol. 39, no. 4, pp. 799–818, 1999.

[23] M. A. Kilgore and C. R. Blinn, "Policy tools to encourage the application of sustainable timber harvesting practices in the United States and Canada," *Forest Policy and Economics*, vol. 6, no. 2, pp. 111–127, 2004.

[24] M. A. Kilgore, J. L. Greene, M. G. Jacobson, T. J. Straka, and S. E. Daniels, "The influence of financial incentive programs in promoting sustainable forestry on the nation's family forests," *Journal of Forestry*, vol. 105, no. 4, pp. 184–191, 2007.

[25] A. L. Paula, C. Bailey, R. J. Barlow, and W. Morse, "Landowner willingness to supply timber for biofuel: results of an Alabama survey of family forest landowners," *Southern Journal of Applied Forestry*, vol. 35, no. 2, pp. 93–97, 2011.

[26] M. B. Miles and A. M. Huberman, Eds., *An Expanded Sourcebook, Qualitative Data Analysis*, Sage, Thousand Oaks, Calif, USA, 2nd edition, 1994.

[27] J. Corbin and A. Strauss, *Basics of Qualitative Research: Techniques and Procedures for Developing Grounded Theory*, Sage, Thousand Oaks, Calif, USA, 2008.

[28] D. E. Gray, *Doing Research in the Real World*, Sage, London, UK, 2004.

[29] A. L. Husak, S. C. Grado, and S. H. Bullard, "Perceived values of benefits from Mississippi's forestry Best Management Practices," *Water, Air, and Soil Pollution: Focus*, vol. 4, no. 1, pp. 171–185, 2004.

[30] P. V. Ellefson, M. A. Kilgore, and M. J. Phillips, "Monitoring compliance with BMPs: the experience of state forestry agencies," *Journal of Forestry*, vol. 99, no. 1, pp. 11–19, 2001.

[31] G. G. Ice, E. Schilling, and J. Vowell, "Trends for forestry best management practices implementation," *Journal of Forestry*, vol. 108, no. 6, pp. 267–273, 2010.

[32] M. A. Arthur, G. B. Coltharp, and D. L. Brown, "Effects of Best Management Practices on forest streamwater quality in eastern Kentucky," *Journal of the American Water Resources Association*, vol. 34, no. 3, pp. 481–495, 1998.

[33] J. L. Schuler and R. D. Briggs, "Assessing application and effectiveness of forestry best management practices in New York," *Northern Journal of Applied Forestry*, vol. 17, no. 4, pp. 125–134, 2000.

[34] W. M. Aust and C. R. Blinn, "Forestry best management practices for timber harvesting and site preparation in the eastern United States: an overview of water quality and productivity research during the past 20 years (1982-2002)," *Water, Air, and Soil Pollution: Focus*, vol. 4, no. 1, pp. 5–36, 2004.

[35] B. Holt, *Perception to inception: assessing contractor capacity to utilize woody biomass for energy production in the Southern Willamette Valley, Oregon*, M.S. thesis, 2008.

[36] S. K. Eliason, C. R. Blinn, and J. A. Perry, "Natural resource professional continuing education needs in Minnesota: focus on forest management guidelines," *Northern Journal of Applied Forestry*, vol. 20, no. 2, pp. 71–78, 2003.

[37] T. Dietz, P. C. Stern, and R. W. Rycroft, "Definitions of conflict and the legitimation of resources: the case of environmental risk," *Sociological Forum*, vol. 4, no. 1, pp. 47–70, 1989.

[38] A. L. Dirkswager, M. A. Kilgore, D. R. Becker, C. Blinn, and A. Ek, "Logging business practices and perspectives on harvesting forest residues for energy: a minnesota case study," *Northern Journal of Applied Forestry*, vol. 28, no. 1, pp. 41–46, 2011.

[39] M. E. Harmon, J. F. Frankin, F. J. Swanson et al., "Ecology of course woody debris in temperate ecosystems," *Advances in Ecological Research*, vol. 15, pp. 133–302, 1986.

[40] B. Freedman, V. Zelazny, D. Beaudette et al., "Biodiversity implications of changes in the quantity of dead organic matter in managed forests," *Environmental Reviews*, vol. 4, no. 3, pp. 238–265, 1996.

[41] M. L. Hunter, *Maintaining Biodiversity in Forest Ecosystems*, Cambridge University Press, Cambridge, UK, 1999.

[42] W. Jia-bing, G. De-xin, H. Shi-jie, Z. Mi, and J. Chang-jie, "Ecological functions of coarse woody debris in forest ecosystem," *Journal of Forestry Research*, vol. 16, no. 3, pp. 247–252, 2005.

[43] D. B. Botkin, *Discordant Harmonies: A New Ecology for the Twenty-First Century*, Oxford University Press, New York, NY, USA, 1990.

[44] T. R. Peterson, *Sharing the Earth: The Rhetoric of Sustainable Development*, University of South Carolina Press, Columbia, SC, USA, 1997.

[45] G. R. Hess and D. Zimmerman, "Woody debris volume on clearcuts with and without satellite chip mills," *Southern Journal of Applied Forestry*, vol. 25, no. 4, pp. 173–177, 2001.

[46] C. E. Moorman, K. R. Russell, G. R. Sabin, and D. C. Guynn, "Snag dynamics and cavity occurrence in the South Carolina Piedmont," *Forest Ecology and Management*, vol. 118, no. 1–3, pp. 37–48, 1999.

[47] D. A. Patrick, M. L. Hunter, and A. J. K. Calhoun, "Effects of experimental forestry treatments on a Maine amphibian community," *Forest Ecology and Management*, vol. 234, no. 1–3, pp. 323–332, 2006.

[48] Pennsylvania Department of Conservation and Natural Resources, "Guidance from Harvesting Woody Biomass for Energy in Pennsylvania," 2008, http://www.dcnr.state.pa.us/PA_Biomass_guidance_final.pdf.

[49] R. E. Matland, "Synthesizing the implementation literature: the ambiguity-conflict model of policy implementation," *Journal of Public Administration Research and Theory*, vol. 5, no. 2, pp. 145–174, 1995.

[50] "Market Facts. Market Trend. July 2002 British Columbia-wide public opinion tracking poll: Public perceptions and attitudes towards RPFs and forest management," 2002, http://www.abcfp.ca/publications_forms/publications/documents/2002_opinion_poll.pdf.

[51] J. W. Thomas, "Are there lessons for Canadian foresters lurking south of the border?" *Forestry Chronicle*, vol. 78, no. 3, pp. 382–387, 2002.

[52] M. K. Luckert, "Why are enrollments in Canadian forestry programs declining?" *Forestry Chronicle*, vol. 80, no. 2, pp. 209–214, 2004.

[53] M. J. Keefer, J. C. Finley, A. E. Luloff, and M. E. McDill, "Characterizing loggers' forest management decisions," *Journal of Forestry*, vol. 100, no. 6, pp. 8–15, 2002.

[54] A. F. Egan, C. C. Hassler, and S. T. Grushecky, "Logger certification and training: a view from West Virginia's logging community," *Forest Products Journal*, vol. 47, no. 7-8, pp. 46–50, 1997.

[55] J. C. Bliss, "Public perceptions of clearcutting," *Journal of Forestry*, vol. 98, no. 12, pp. 4–9, 2000.

[56] A. Egan and D. Taggart, "Who will log? Occupational choice and prestige in New England's North woods," *Journal of Forestry*, vol. 102, no. 1, pp. 20–25, 2004.

[57] R. A. Williams, D. E. Voth, and C. Hitt, "Arkansas' NIPF landowners' opinions and attitudes regarding management and use of forested property," in *Symposium on Non-Industrial Private Forests: Learning for the Past, Prospects for the Future*, pp. 230–237, 1996.

[58] J. C. Bliss, S. K. Nepal, R. T. Brooks Jr., and M. D. Larson, "Forestry community or grandfalloon?" *Journal of Forestry*, vol. 92, no. 9, pp. 6–10, 1994.

# Theorizing the Implications of Gender Order for Sustainable Forest Management

**Jeji Varghese[1] and Maureen G. Reed[2]**

[1] *Sociology and Anthropology, University of Guelph, Guelph, ON, Canada N1G 2W1*
[2] *School of Environment and Sustainability and Department of Geography and Planning, University of Saskatchewan, Saskatoon, SK, Canada S7N 5A6*

Correspondence should be addressed to Jeji Varghese, varghese@uoguelph.ca

Academic Editor: I. B. Vertinsky

Sustainable forest management is intended to draw attention to social, economic, and ecological dimensions. The social dimension, in particular, is intended to advance the effectiveness of institutions in accurately reflecting social values. Research demonstrates that while women bring distinctive interests and values to forest management issues, their nominal and effective participation is restricted by a gender order that marginalizes their interests and potential contributions. The purpose of this paper is to explain how gender order affects the attainment of sustainable forest management. We develop a theoretical discussion to explain how women's involvement in three different models for engagement—expert-based, stakeholder-based, and civic engagement—might be advanced or constrained. By conducting a meta-analysis of previous research conducted in Canada and internationally, we show how, in all three models, both nominal and effective participation of women is constrained by several factors including rules of entry, divisions of labour, social norms and perceptions and rules of practice, personal endowments and attributes, as well as organizational cultures. Regardless of the model for engagement, these factors are part of a masculine gender order that prevails in forestry and restricts opportunities for inclusive and sustainable forest management.

## 1. Introduction

In Canada, it can no longer be assumed that timber is the sole product of forestry. The broadening of interests in forests has been characterized by increased pressures from diverse interest groups to be involved in sustainable management of public resources. For example, forestry advisory committees across the country have been established to contribute directly into management decisions about forestry and thereby contribute local knowledge to the social, economic, and ecological dimensions of forest sustainability. Social sustainability includes consideration of society's responsibility, which is one of six criterion of sustainable forest management (SFM) in Canada. This criteria "*addresses the effectiveness of institutions in managing resources in ways that accurately reflect social values, the responsiveness of institutions to change as social values change, how we deal with the special and unique needs of particular cultural and/or socio-economic communities, and the extent to which the allocation of our scarce resources can be considered to be fair and balanced* (page 17 [1])". This social sustainability criterion includes fair and effective decision making (Criteria 6.4) and raises questions about inclusiveness, that is, how to ensure that a broad range of interests and values are included in these processes, given the heterogeneity of the population, particularly in terms of differences in terms of influence, dependency, and, cross-cutting variables such as gender and ethnicity.

In this conceptual article, we identify three models of engagement in SFM—expert-based, stakeholder-based, and civic engagement—and consider the ways in which each model includes issues and concerns of one social group historically excluded from forestry management—women. This focus does not presuppose that women are the only or most important consideration for sustainable forest management. Nor does it suggest that all forestry issues are segregated by gender. Nevertheless, empirical research conducted in Canada, India, Sweden, Kenya, Nepal, and Thailand has suggested that the interests of women and men in forestry are significantly different and that these differences are typically not reflected in forest policy, see [2]. We suggest that the gap between interests and outcomes may arise because the forestry sector is subject to a "gender order" that privileges men's contributions to forestry, constrains women's participation in forestry management, and ultimately reduces the capacity of the forestry sector to achieve inclusive forestry management as a key component of social sustainability.

We begin our discussion by drawing on previous research that demonstrates that women bring distinctive interests and values to forestry issues. Next, we provide a brief discussion of gender order and inclusive engagement. The consideration of inclusive engagement hinges on conceptualizations of participation and representation. We consider different definitions and how they apply. We then define the different models for involvement—expert-based, stakeholder-based, and civic engagement. We consider, in theory, how women's involvement in each model might be advanced or constrained. We then analyze experiences of women under each model according to five criteria. Finally, we draw some conclusions on how the gender order of forestry affects opportunities for inclusive forest management.

*1.1. Method/Approach.* We approach this paper by conducting a meta-analysis of a range of empirical studies to illustrate key factors that limit women's involvement across a range of engagement options. We draw on research we have participated in across the country from the mid 1990's to the present as well as research conducted by other academics at Canadian universities and researchers at the Canadian Forest Service to examine expert and stakeholder models. For example, to examine expert models we focus on studies of women foresters in British Columbia, Alberta and Saskatchewan see [3–5]. To examine stakeholder models, we focus on research of Burns Lake and Revelstoke Community Forests in British Columbia see [6, 7], the Prince Albert Model Forest in Saskatchewan see [8], in-depth case studies of three public advisory committees in Alberta see [9] and two committees in Manitoba and Nova Scotia see [10] and a national survey of both members and chairpersons of forest advisory committees that was conducted by a network of social scientists in 2004 see [11, 12]. To examine civic engagement models we draw mainly on published studies from outside of Canada, as this model is less prevalent within Canada.

## 2. Rationale: Why Women's Voices Matter in Forestry Debates

Despite the fact that most Canadian women do not use forest resources for subsistence purposes, they still hold interests and perspectives in forest management that are distinctive from men's and hence, their active involvement in giving advice about how forests should be managed could influence decisions about sustainability. Furthermore, forestry communities are experiencing rapid economic, social, and ecological changes that affect both women and men. At the community level, for example, climate change may have significant effects including reduced health status of residents during extreme events, altered paid and unpaid work patterns within communities, changes in livelihood and household relations, and long term health concerns for indigenous residents who have traditionally relied on country foods of the boreal forest. Because there remain marked differences in the roles, activities and expectations of men and women living in forestry-based communities, women and men will have different capacities to adapt to changing conditions [13].

Research has long documented that women express greater concern about the environment than men and express greater support for the protection of forest ecosystems [12–18]. They also typically express greater risk aversion to and concern for climate change—an environmental concern that will be significant for those living in boreal forests of Canada [19, 20]. Additionally, in Sweden, researchers have observed that women perceive social impacts of forest management differently than men [21], while women employed in the USDA Forest Service are more likely than men to hold positive expectations and higher levels of trustworthiness regarding environmentalists, range users, citizen activists, and tribal representatives [22].

Yet, women have not mobilized these perspectives, and concerns into collective social or political actions affecting forestry directly. With respect to the environment, women have been found working in grassroots organizations [23], engaging more in environmentally friendly behaviors that can be integrated into their everyday life [24], and working actively with forest certification programs [25]. With respect to forestry, typically women are not leaders of forestry companies or national or international environmental organizations [23, 26, 27] and they do not participate as actively in other decision-making positions to advance these interests [14].

Thus, while there is ample reason to believe that women bring different values, perspectives, expectations, and concerns to forestry issues, women have yet to advance these concerns directly. We suggest that this discrepancy can be explained by understanding how a particular gender order shapes ideas about what interests become represented in forest management forums. In the next section, we define gender order, describe how it operates within forestry communities, and illustrate how it shapes ideas and processes associated with inclusive forest management.

## 3. Defining Key Concepts

Our theoretical approach is taken from feminist scholarship on gender order, public involvement in environmental management, and inclusivity as a social indicator of sustainable forest management.

*3.1. Theory on Gender Order.* Theoretical work on gender order has focused on organizations and work sites. It is also applicable to "communities" in the sense of territorial and occupational communities related to forestry. A gender order refers to the "dichotomous order of gender whereby maleness and femaleness are perceived as opposites and attributed different forms of behavior, different roles, and different places (page 246 [28])". A gender order is not a natural order, but one that is created and recreated in everyday work interactions through organizational cultures that establish the normative rules, values, and meanings in the workplace. According to Gherardi and Poggio [28] these organizational cultures both establish and institutionalize gender positionings, frequently with the effect of "keeping women in "their place" (page 246 [28])".

That forestry work is highly gendered has long been documented. *"Canadian forestry continues to be dominated by a masculine gender order that separates men and women and favours male workers in general... [whereby] potential contributions women might make to management and planning for the sustainability of forestry and forestry communities are overlooked (page 78 [29])".* In North America and Scandinavia, forestry occupations have created and elevated the importance of a working *man*'s culture or community [30–34]. This bias has been generated by the mystique of logging that created an image of forestry as hard, outdoor, physical labor that is dirty, dangerous, and ultimately masculine. Despite an increase in the range of forestry occupations in the late 20th century that includes planning, regulation, as well as the introduction of labor saving devices, the idea of forestry as men's work still prevails. In forestry communities, this image is married to a traditional division of labor where men have "rightfully" enjoyed forestry jobs and high incomes and have served their families as primary breadwinners, while women have stayed at home to raise children or have only worked for "pin money" in forestry households. While this "traditional" (Traditional is set within quotations because of the debate over whether traditional refers to long-standing and prevalent gender roles or ones that are relatively recently instituted or locally naturalized [35].) division of labour has crumbled in urban settings, it has remained remarkably resilient in Canadian rural places [3]. For example, even where they are currently employed in "nontraditional" forestry occupations, women in forestry communities still carry the disproportionate share of juggling work-home schedules to meet childcare and other domestic duties [5, 36].

Furthermore, organizations, including advisory committees, model forests and other participatory venues, may be shaped by local practices, activities, norms, and attitudes that are taken for granted in everyday life. For example, in a study of Australian executive culture, Sinclair [37] described the organizational culture as a masculine domain "dominated by values, norms, symbols, and ways of operating that are oriented to men (page 6)" and that these characteristics were essential to the way men constructed their identity as leaders. The problem is "not that women's interests and men's interests are opposed—frequently they coincide—but rather it is to conduct their business as though they are always in synch (page 6)" [37]. Forest management operates within an organizational and occupational culture that is highly gendered. It has historically been biased toward male norms and thus can reproduce gender inequality even today—if unintentionally [22, 34, 38, 39]. So even though women serve on advisory tables, as minority representatives, women are faced with male occupational and organizational cultures that shape the extent to which women's participation is effective. One might suppose that this is a reflection of a deeper structural gender order that permeates society, but other previously male-dominated fields (e.g., business, medicine, engineering) have made positive strides to being more inclusive of women [40] in a manner that is not currently reflected to the same degree within forestry.

This ordering of gender in the structure of forestry and forestry communities has become institutionalized as data used to describe the industry and forestry communities are limited, researchers typically do not segregate their findings on work-labor studies by gender, and government policy makers assume that forestry workers are male [29]. Together, these practices give the impression that women are not numerically, economically, or socially important to forestry, and consequently, women's potential contributions to management and planning for the sustainability of forestry and forestry communities are frequently overlooked. This pattern also affects decisions about when, how, and who to involve in sustainable forest management.

*3.2. Inclusive Engagement.* An enormous literature across a range of topic areas has dealt with public involvement in political and social life. Environmental management in general and forestry in particular are no exception. "Natural resource scholars and practitioners who are interested in public engagement often refer to concepts such as representation, participation, or involvement, whereas democratic theorists often refer to the term inclusion (page 533)" [41]. Our conception of inclusion is that it involves both "nominal" and "effective" participation.

Nominal participation refers to the simple demographic representation of particular groups in society. Representatives are assumed to share the values, attitudes, and socioeconomic characteristics of those they represent; in a sense, they are mirrors of the larger population [42, 43]. Wellstead et al. [42] and Pitken [43] refer to this form of representation as descriptive or mirror representation. Participants may be seen as "standing for" the larger group. Applying this to forest management, if women are present in advisory committees, they are assumed to represent women's interests. This is a fairly limited and often inaccurate portrayal of how people actually operate. The limitation of nominal participation is that it tends to assume that individuals embody and represent a series of static characteristics rather than act on their beliefs. In the case of gender and forestry, a focus solely on nominal participation would assume that only women will bring forward concerns related to their gender and/or that women may restrict their contributions to such concerns. Despite their small numbers, we know that women come to decision making forums with knowledge and perspectives that are not necessarily relevant to gender (e.g., knowledge of the biophysical system), while they may have concerns about the topics of discussion or the procedures by which decisions are made that are different from those of their male counterparts.

Effective participation requires that participants are active and engaged in forest management. An assessment of effective participation would consider the rules of decision making, such as whether a process is viewed as accessible, transparent, fair, and consistent with existing laws, regulations, or policies [8, 44]. It might also consider the attributes of individuals such as their knowledge of particular issues, their ability to mobilize resources, and their comfort in speaking out on particular issues. Thus, effective participation involves assessing the ability of members to enhance the equity and efficiency of decisions and to advance their goals.

Effective participation also gives consideration to how power operates within SFM processes and the influence of power on the capacity of individuals and groups to advance their interests [3, 6]. Biases can be subtle, almost invisible, and yet, they can influence effective participation. They can emerge in procedures that determine selection criteria and processes, identify sources and relevance of data, value alternative experiences and knowledge, set the location and timing of meetings and payments for attendance, and provide level and type of agency support from government for the process [45, 46].

We suggest that nominal participation is linked to effective participation, particularly where participating groups are minorities. That is, the number of people from particular social groups will influence the "rules of entry" and "rules of practice"—both formal and informal—and thereby alter the conditions for effective participation. The absence of particular social groups may give rise to the assumption that the rules and procedures adopted by the more restricted group represents the "norm" against which any variation is considered deviant. Here, the notion of critical mass is relevant. For instance, Dahlerup [47] observed that among Scandinavian women politicians, once women became a significant minority (passing a threshold of some 30% seats in Parliament or local councils), there was less stereotyping and openly exclusionary practices by men, a less aggressive tone in discussions, a greater accommodation of family obligations in setting meeting times, and a greater weight given to women's concerns in policy formulation. The necessity of critical mass within forest advisory committees has also been documented more recently by Richardson et al. [10].

Effective participation may also be achieved through two forms of representation (subjective attached and objective unattached) noted by Pitken [43] and Wellstead et al. [42] whereby participants are "standing for" or viewed as acting on behalf of other groups. Subjective attached representation refers to *acting for those with a subjective interest* [43]. In the context of forestry, most advisory committees operate on this model; that is, participants are acting on behalf of others who share a stake in forestry matters. In the context of issues with a gender dimension, those with a subjective attached form of representation would be sensitive to gender issues and would acknowledge the differential stake that women may have in SFM. Hence, those perspectives would still be taken into account.

Objective unattached representation refers to *acting for those who are uninterested* [43]. This form of representation would involve individuals who might not be directly interested in forestry, but who might represent interests on behalf of all Canadians or even on behalf of the forest itself. Participants who speak for intrinsic values or future generations would fall into this category. With respect to gender, this form of representation would also imply that even if women were not present, there would be sensitivity to differences in perspectives by gender by those who participate. There is not much experience with this form of objective unattached representation in the Canadian forest sector. Value-based committees (with which we have little experience in Canada), citizen juries documented in the United States (again, little experience in Canada) or broader-based venues that are less engaged (e.g., open houses, questionnaire surveys, voting), and social science research (a few limited examples in Canada) may be seen as examples of attempts to achieve objective unattached representation.

Each of these forms of inclusion implies different ways in which gender-based issues are understood and brought to the table. Nominal participation/descriptive representation would require that women be present to ensure their interests are met, while subjective attached and objective unattached representation suggest that women need not necessarily be present. They do, however, require that on the part of participants there is an awareness of and an ability to understand and speak for the different perspectives that women and men may have.

## 4. Inclusiveness of Expert-Based, Stakeholder-Based, and Civic Engagement Models

Within this subsection, we examine three models of engagement for their extent of inclusiveness in SFM. To do this, we compare how the role of women across expert-based, stakeholder-based, and civic engagement models differ and examine factors that constrain women's access and effectiveness in each model.

The pyramid in Figure 1 illustrates fewer numbers of people involved in more elite models at the top and greater number of people involved as we move toward the bottom. Hence, by definition, the models at the top of the pyramid offer smaller degrees of inclusion of the range of views than do the models towards the bottom of the pyramid. Yet, the level of input is less detailed and less sustained as the scope of involvement widens. Nevertheless, in all three models, both nominal and effective participation of women may be constrained (or enhanced) by several factors including rules of entry, divisions of labour, social norms and perceptions and rules of practice, personal endowments and attributes, as well as organizational cultures.

These factors provide points of analysis or assessment of each of the three models described in more detail below. It is important to note that these factors act on different scales and can interact with each other. Rules of entry refer to how access is gained or constrained for nominal participation. Effective participation is enabled or constrained by the other factors. Division of labour considers how paid and unpaid work as well as specialization within paid work may be divided along gender lines. Social norms and perceptions and rules of practice include broader societal expectations about acceptable individual behavior within social settings as

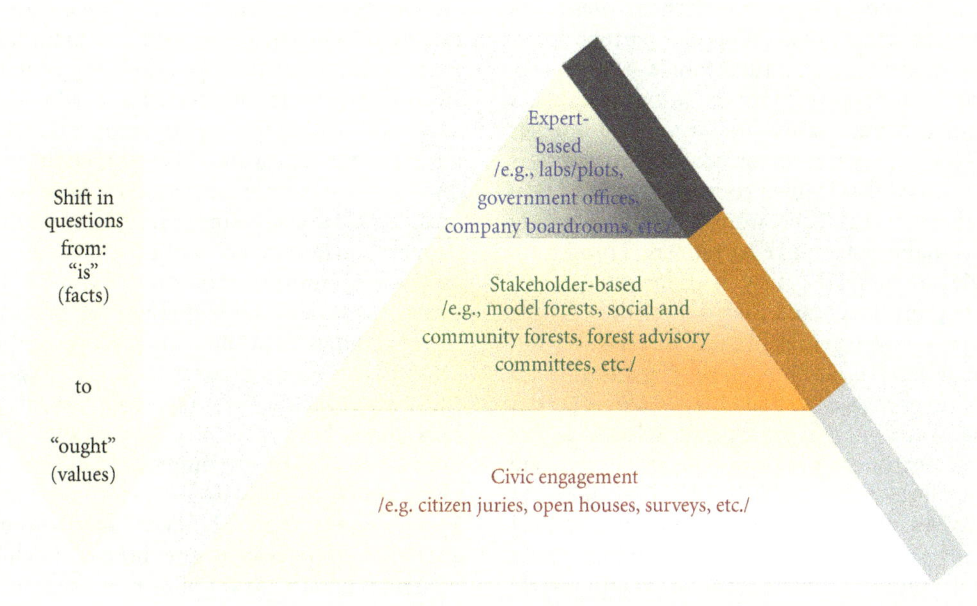


FIGURE 1: Three Models of SFM Engagement.

well as informal guidelines about how to behave collectively. Personal endowments and attributes include education level, property status, marital status, age, and so forth. Organizational culture refers to the shared values and norms that guide behavior within an organization and between the organization and others. We argue that regardless of the model for engagement, these factors are part of a masculine gender order that prevails in forestry.

*4.1. Expert-Based Models of Engagement.* Expert-based models provide representation and voice to experts (whether foresters, scientists, policy makers, etc.); thus privileging their knowledge in SFM. The underlying assumption within this model is that expert knowledge produces "facts," and that decision making needs to rely on these "objective facts" or "truths." In order to be involved in expert-based models of forest management, expertise is typically restricted to knowledge about forestry science, regulation or policy, and business. While decisions made about forestry affect forestry communities more broadly, it is rare that expertise is sought about the social dimensions of forestry communities, including public health, social welfare, culture, and so on that may be affected by changes in forestry or land use policy.

*Rules of entry* into expert panels require that participants have experience in positions of specific forestry expertise and/or authority. Very few experts in forestry science and business are women. For example, between 4 and 16 percent of registered professional foresters are female in each province [48]. Furthermore, Teske and Beedle [48] noted that in that year, there were no women Chief Executive Officers in any of the forest companies, although there were several women vice presidents. Despite increases in the number of women with formal education in Canada, women continue to be significantly underrepresented in management positions in the public and private sectors [5]. Thus, within expert-based models, the *gendered division of labor* and *rules of entry* typically restrict women's participation and reduce the range of issues deemed relevant to community sustainability. The lack of women in management positions within the industry and regulation of forestry translates directly into their absence in expert panels.

*Gendered social norms and perceptions* and associated *rules of practice* also influence expert-based models. There remains a social construction of appropriate female behavior (e.g., soft speech, deference to males, making the coffee) that operates even in the absence of overt structures. Even when entry rules are favourable, women may not participate as effectively due to a range of preexisting social norms that reinforce stereotypes about acceptable female behavior, appropriate ways of communicating, and social interaction between men and women. For example, female registered professional foresters (RPFs) working in BC in the 1990s reported being allowed to attend policy meetings to serve the coffee rather than to provide expertise [3].

*4.2. Stakeholder-Based Models of Engagement.* Stakeholder-based models give representation and voice to stakeholders, organizations/groups, and so forth that have a stake in forest management. In theory, stakeholder-based models are viewed as being more inclusive than expert-based models, in that they include local people and their knowledge. But of course this raises questions about who determines who has a stake and how groups may be represented? In terms of stakeholder-based models of engagement in Canada, such as within-community forests, model forests, and forest advisory committees, women have had a limited nominal and effective participation. In terms of nominal participation, for example, while 5 of 15 managers (staff) of model forests are female, only 2 of 15 model forest board chairs/presidents (decision makers) are female. In addition, few partner representatives within model forests and stakeholders are female. A case study of a community forest in British Columbia revealed no woman representative [6]. At a much broader scale, an analysis of 102 forest sector advisory committees across Canada revealed that women comprised only 17 percent of the membership [12]. These women held significantly different values about forestry and environmental protection than their male counterparts, they had lower levels of trust for industry, and they rated their experiences on the committee less favorably than men [12].

Formal *rules of entry* affect nominal participation within stakeholder models of engagement and may reinforce gender bias. Stakeholder representatives are typically selected as part of their employment or they may be selected if they are representatives of formal domains such as unions, management representatives, and chambers of commerce. Women are more likely to be part of organizations that address public health, education, children's sports, and community events that are not as hierarchically organized and/or are less likely to be considered part of "forestry." Men are more likely to be part of unions or business associations that are organized toward forestry employment are thus more likely to be considered stakeholders to represent local interests [5, 10, 29]. Women that do attend are more likely to attend because of their work as ENGO representatives [12]. In addition, Richardson et al. [10] quoted one female member of a forest management advisory committee as saying "*because we do not have a lot of women in the industry we do not naturally gravitate towards thinking about or suggesting women to be on the committee (page 527).*" This quotation demonstrates that it becomes normalized to think that women are not interested or knowledgeable, and therefore, to overlook opportunities to engage women in these processes. Thus, while nominal participation may be constrained by a lack of women in professional networks of the forestry/resource sector, this gap gains positive reinforcement by prevailing assumptions about women's desire or ability to participate in stakeholder advisory groups.

Nominal and effective participation within stakeholder models of engagement are also affected by a gendered *division of labour*. Forestry as a profession is subdivided into a number of different area, some involves hard physical labour, in rough inaccessible terrain and remote areas and others in increasingly technical and office-related functions. This diversity of function, in combination with the socialization of men and women, has led to stereotypes about what type of work men and women in forestry prefer. As a result, certain aspects of forestry are seen as more suitable to males or females. Fullerton [49] and Martz et al. [5] found that even though on average women have higher levels of education than men, within the forest industry, women are overrepresented in clerical and administrative occupations and the ranks of unemployed, and that women are underrepresented in operations, scientific, and management categories. In a gender analysis of Alberta foresters, Varghese et al. [4] found that there was a perception of male domination in operations, whereas silviculture (which is considered to have less arduous fieldwork since it takes place in harvested areas) was perceived to be equally accessible to both men and women. Even within government positions related to forestry, women are overrepresented in clerical and administrative occupations and underrepresented in operations, scientific, and management categories [49]. In addition, the gendered division of labour in society continues to shape women's involvement in community and related service work. Despite stronger environmental sentiments, women are less able to be involved in direct activism or extra-curricular voluntary efforts because of the division of labour. Reproductive responsibilities limit the nature, scope, and efficacy of their activism or community work [24, 50]. These responsibilities "*not only structure one's disposable time, they also shape the costs and risks of movement participation (page 913)*" [24]. Even where women are employed full-time, women, particularly young married women, tend to be more constrained by their family responsibilities. Thus, the timing of meetings and the need to balance other household/childcare work can be serious barriers to women's participation

*Personal endowments and attributes* can also shape expert women's contributions to forestry management. These factors include education levels, role of the family and in the community, and socioeconomic status. These personal attributes combine with social norms to provide entry points or barriers for some groups over others. For example, women living in rural areas typically have higher rates of formal education than men, but they have less formal employment experience in forestry. In a local culture where education-by-apprenticeship is more highly valued, women's educational status may not grant them greater access to or effectiveness in forestry management processes. Richardson et al. [10] demonstrated how female members of forest management advisory committees were conscientious of their minority positions and felt uncomfortable offering their views on issues unless they were confident in their views.

*Social norms and perceptions and rules of practice* reinforce the division of labour and rules of entry. Related to the division of labour is a perception that women do not have a direct stake in forestry management or are not qualified to represent community interests. For example, women on Vancouver Island were shut out of some forestry activities and associated discussions because they were told that they "could not handle the language" [3]. If women were to participate, they may try to raise issues that are deemed irrelevant. For example, when women on Vancouver Island tried to engage men in discussions about local economic development, they were told that men in the community were not interested in participating in women's knitting circles. Richardson et al. [10] found masculine norms were taken for granted within forest management advisory committees. Many participants within her study felt that women on these advisory committees displayed behaviors typically attributed to men (e.g., being outspoken and confident) rather than behaviors typically attributed to women (e.g., shyness or drawing on emotional arguments).

These perceptions and practices reinforce and reproduce a masculine *organizational culture* that limits women's nominal and effective participation and reproduces gender inequality—if unintentionally. Gherardi [51] notes that even where masculine organizational cultures appear to include women, participants typically express their distrust of equality and implicitly threaten those who seek to change the rules. All of these challenges reduce the number and efficacy of women participants in stakeholder-based models that operate within highly masculinized cultures.

*4.3. Civic Engagement Models of Engagement.* Civic engagement models reflect more of an increased acknowledgement of the public nature of the bulk of forest resources and hence are based more on representing the diverse set of values that are present among the general public, not only the local communities. Civic engagement models however are not well developed in the Canadian forest sector, so we turn to examples in other jurisdictions. Citizen juries, which were originally developed in Europe [52, 53], have been used in the United States to incorporate the concerns of "ordinary citizens" in contested public debates [54].

*Rules of entry.* Generally, citizen juries are randomly selected with some consideration of balancing across selected characteristics such as age, education, and gender [53], however there are also examples where marginalised populations, such as women, are targeted [55]. However, in the latter case, this "process of selecting jurors may be contested—for example, those who do not feel it is important for women to participate directly (page 79 [56])" reflecting gendered *social norms/perceptions* and *rules of practice* impacting nominal participation.

*Social norms/perceptions* and *rules of practice* and *organizational cultures* can also diminish effective participation. As citizens' juries are modelled on an idealisation of the legal process [57] with an emphasis on expert testimony and rationalism [58], Ward et al. argue that it is "easy to be patronising, tolerating "non-rational" forms of argumentation from "others" (e.g., from women or indigenous groups), or as a second-best when rational argumentation fails (page 286 [59])" Pickard's [60] criticisms of citizen juries related to their lack of accountability, authority, legitimacy, and representativeness raises broader questions about the role of civic engagement in forestry management.

Other mechanisms that would involve a wide range of participants who speak for others include open houses, surveys, and workshops. Although people may speak on their own behalf, the commitment by each individual is less sustained and the input is less detailed than that of the other mechanisms. In addition, although women may secure nominal participation, effective participation is marred by the lack of influence of most civic engagement models [60].

Table 1 provides an overview of some types of measures that would enable us to assess women's nominal and effective participation for each form of representation within each model of engagement. Citizen engagement models also seem to share many of the drawbacks of stakeholder models. As we move from expert to civic models, we improve the chances for nominal participation of women because the criteria for selection become less specialized, however, none of the models addresses effective participation because the barriers are systemic, insomuch that the barriers are often invisible.

Table 2 highlights some examples of how the gender order pervades each of the three models of engagement. Women historically have been absent (lack of nominal participation) or present but effectively silenced (limited effective participation) in forestry management. In comparing engagement models, we see that there has been limited access and limited effective participation for women in all three models. We have shown that a masculine gender pervades regardless of the engagement model.

## 5. Discussion and Conclusion: How the Gender Order of Forestry Affects the Opportunities for Inclusive Sustainable Forest Management

Figure 2 is a concept map that provides an overview of the links among the concepts we have discussed in this paper. By starting at the bottom of the concept map and working our way to the top, we examined how gendered rules of entry, divisions of labour, social norms and perceptions, personal endowments and attributes, and organizational cultures result in and reinforce a particular gender order that constrains both nominal and effective participation within three engagement models for forestry management. We did this by examining the links among the different types of participation and representation and the implications of these differences for the engagement models. Despite different models or strategies, these factors remain as persistent and often invisible barriers. We argued that engagement models are embedded within a masculine gender order that severely restricts women's nominal and effective participation.

Like in forestry itself, women are distinctly underrepresented in SFM. One reviewer raised an interesting question about possible causes for this underrepresentation: *"is it possible that there exists a deliberate resistance to women's involvement on the part of some actors on the basis of the likelihood that women's perceptions and concerns will challenge dominant narratives in the industry? In other words, perhaps it is not women per se who are being excluded but rather the set of values and concerns they have a propensity to hold?"* Inclusiveness across a range of social groups (representing diverse values and concerns) is necessary to work toward achieving social sustainability of forests. In order to enhance inclusiveness in SFM, we need to pay attention to how women and other marginalized groups may be excluded from SFM and/or effectively participating in SFM. This requires a mix of strategies that would increase the nominal participation of women and also improve the opportunities for women to be effective. Moving beyond a simple add women and stir approach, we need to address those constraints that limit effective participation by identifying and addressing the underlying masculine gender order in forestry. Because of the link between nominal and effective participation, we need strategies that address both.

Simple practical strategies to address social constraints to SFM recognize the competing time interests of women. For example, providing an honorarium, childcare/eldercare (or costs), considering the timing of meetings recognizes time constraints. In addition, ensuring a "critical mass,"

Table 1: Measures to assess female participation and representation in SFM engagement models.

| Participation | Forms of Representation | SFM engagement models: Expert | Stakeholder | Civic |
|---|---|---|---|---|
| Nominal | Descriptive | For example, numbers of female "experts" | For example, numbers of female stakeholder representatives and/or gender issue focused stakeholders | For example, numbers of females in forest advisory committees |
| Effective | Descriptive | For example, female experts are able to voice their concerns in a meaningful way and be heard | For example, female stakeholders reps are able to voice their concerns in a meaningful way and be heard | For example, females are able to voice their concerns in citizen juries in a meaningful way and be heard |
| | Subjective attached | For example, experts act on behalf of women forestry experts whether present or not | For example, stakeholder representatives act on behalf of women/women's groups who have a stake in forestry whether present or not | For example, citizen juries act on behalf of women's/women's groups who have an interest in forestry whether present or not |
| | Objective unattached | For example, experts acting on behalf of women who do not have a direct stake in forestry even when they are not present | For example, stakeholder representatives act on behalf of women's groups who do not have a direct stake in forestry even when they are not present | For example, citizen juries act on behalf of women who do not have a direct stake in forestry even when they are not present |

Table 2: Examples of influence of gender order on inclusivity of engagement models.

| Processes of gender order | Engagement Models: Expert *(e.g., expert panels)* | Stakeholder *(e.g., advisory committees)* | Civic *(e.g., citizen juries)* |
|---|---|---|---|
| Rules of entry | The few female experts in forestry science and business limit nominal participation | Stakeholders often represent formal domains limiting nominal participation as women often involved in informal domains considered external to forestry. | Process of selecting citizen juries contested based on social norms. |
| Division of labour | Few females in senior management positions limit nominal participation. | Female foresters are often assigned to office-related forestry work. Women often involved in community and related service. | |
| Social norms & perceptions and rules of practice | Pre-existing social norms that reinforce gender stereotypes limits effective participation | Perception that women do not have a direct stake in forestry decisions or are not qualified to represent community interests. | Citizen juries can be patronizing of nonrational argument. |
| Personal endowments and attributes | Stereotype that effective managers require masculine skills. | Lack of critical mass may influence confidence of women. | |
| Organizational cultures | Gender order within forestry as a profession | Gender order within industry and society. | Emphasis on rationalism in citizen juries excludes other forms of arguments |

providing gender sensitivity training may address some of the constraints women face in actively engaging in these processes. Capturing gender differences in statistics provides essential information to acknowledge and address possible gender issues. But we recognize that these solutions focus on simplistic ideas about representation and inclusiveness. Consequently, we also advocate for more difficult changes, such as raising the profile of social sustainability alongside ecological and economic sustainability; addressing how systemic biases within forestry communities and the forest industry consider production, consumption, and appreciation of forest resources as gender-neutral; giving greater attention to how power relations affect the definition and use of local community knowledge, with particular attention to how that knowledge is produced and how participation is constrained; greater attention to strategies for incorporating the resulting diversity into SFM (e.g., gender differences across different models, such as consensus, conflict resolution, and majority wins). Greater participation in this case by women may not yield more democratic processes if we do not also have

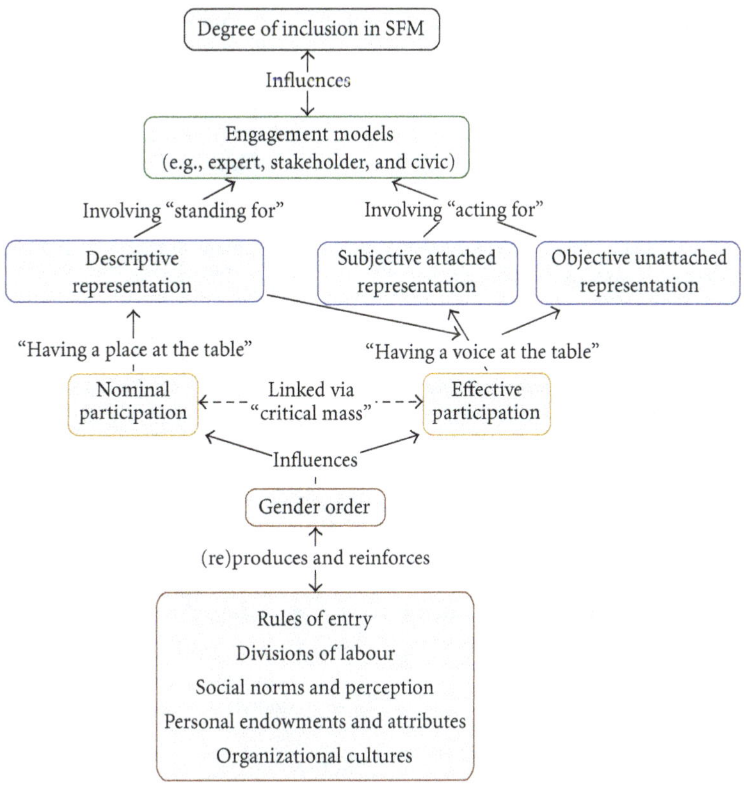


Figure 2: Concept map highlighting framework for considering gender order in SFM.

the right mix of participants, if we have not addressed the ground rules for debate, or if we ignore the way in which power operates at the negotiating table including how power enters speech itself. For those who are motivated to increase women's participation in forestry governance, increased awareness of the gender order and a realization that forestry does not mirror society but rather appears to be an outlier generally would also enable consideration of how other sectors have addressed some of the structural issues. One of the benefits of our meta-analysis is that we can develop "new processes" with our eyes open in advance and try to address some of the "structural" challenges rather than simply "tinker" with the engagement models hoping it will fix things.

Before closing, we acknowledge that not all women want to participate in forestry governance, just as not all men do. But those women who do want to participate may be constrained by a number of factors, including the masculine nature of direct venues as well as the organization preferences of women. As we strive to improve processes for sustainable forest management, it is important to continually increase our understanding of how our policies and daily practices construct gender. Particularly as gender positioning affects the formation and implementation of policy and has enormous influence in the differential capacity of women and men to contribute to ecological, economic, and social sustainability. A gender focus will help to sensitize researchers and policy makers to multiple inequities and help create opportunities for more inclusive concepts, analyses, and ultimately more inclusive policies and practices that place equal value on the contributions of women and men, and particular social groups without privileging any particular one. In this way, we can gain a better understanding of what constitutes inclusive forest management for all residents of forestry communities and try to identify strategies to deliver it.

While we discuss gender as a point of exclusion, we realize that the category of women is a heterogeneous one. Some groups of women are not marginalized, some groups may face forms of marginalization that are similar to other groups, while others may face a double jeopardy. For example, indigenous people and those associated with nontimber forest products have also been marginalized and may share similar forms of exclusion. Women within these social groups may experience multiple forms of exclusion from forestry management due to their gender and to the social status of their social or cultural group [61]. The exclusions can perpetuate overlapping and/or cumulative invisible losses, whereas inclusionary forest management can increase transparency and make these losses visible [62]. It is possible that the framework we have developed may apply to other marginalized or excluded groups, however, only empirical research would reveal its suitability.

## Acknowledgments

The authors would like to acknowledge and thank their research participants over the years. Funds for research reported here were gratefully received from the Sustainable Forest Management Network. Students and colleagues in

Canada and Sweden have helped hone their ideas. They also thank the anonymous reviewer for their insightful comments.

## References


[1] Canadian Council of Forest Ministers (CCFM), *Defining Sustainable Forest Management in Canada: criteria and Indicators 2003*, Natural Resources Canada, Ottawa, Canada, 2003.

[2] G. Lidestav and M. Reed, "Preface: gender and forestry," *Scandinavian Journal of Forest Research*, vol. 25, no. 9, pp. 1–5, 2010.

[3] M. G. Reed, *Taking Stands: gender and the Sustainability of Rural Communities*, UBC Press, Vancouver, Canada, 2003.

[4] J. Varghese, P. Nelson, and R. Boehm, *Towards Equal Participation: Women Foresters in Alberta*, Gender and Policy Action Committee, Alberta, Canada, 1999.

[5] D. Martz, M. G. Reed, I. Brueckner, and S. Mills, *Hidden Actors, Muted Voices: The Employment of Rural Women in Canadian Forestry and Agri-food Industries, Final Report to the Status of Women*, Policy Research Fund, Ottawa, Ontario, 2006.

[6] M. G. Reed and K. McIlveen, "Toward a pluralistic civic science?: assessing community forestry," *Society and Natural Resources*, vol. 19, no. 7, pp. 591–607, 2006.

[7] J. Varghese, *Impacts of, and challenges for, local ownership in the forest sector*, Ph.D. Dissertation, University of Alberta, 2005.

[8] J. Varghese, *Prince Albert Model Forest Public Participation Review. A report prepared for the PAMF Local Level Indicators Working Group*, Saskatchewan, Canada, 2000.

[9] J. R. Parkins and D. J. Davidson, "Constructing the public sphere in compromised settings: environmental Governance in the Alberta Forest Sector," *Canadian Review of Sociology*, vol. 45, no. 2, pp. 177–196, 2008.

[10] K. Richardson, A. J. Sinclair, M. G. Reed, and J. R. Parkins, "Constraints to participation in Canadian forestry advisory committees: a gendered perspective," *Canadian Journal of Forest Research*, vol. 41, no. 3, pp. 524–532, 2011.

[11] J. R. Parkins, L. Hunt, S. Nadeau, M. Reed, J. Sinclair, and S. Wallac, "Public participation in forest management: results from a national survey of advisory committees," Information Report NOR-X-409, Northern Forestry Centre Information Report Series, 2006.

[12] M. G. Reed and J. Varghese, "Gender representation on Canadian forest sector advisory committees," *Forestry Chronicle*, vol. 83, no. 4, pp. 515–525, 2007.

[13] M. G. Reed, "Environmental governance and gender in canadian resource industries and communities," in *Resource and Environmental Management in Canada: Addressing Conflict and Uncertainty*, B. Mitchell, Ed., pp. 526–553, Oxford University Press, New York, NY, USA, 4th edition, 2010.

[14] K. D. V. Liere and R. E. Dunlap, "The social bases of environmental concern: a review of hypotheses, explanations and empirical evidence," *Public Opinion Quarterly*, vol. 44, no. 2, pp. 181–197, 1980.

[15] P. Mohai, "Men, women, and the environment: an examination of the gender gap in environmental concern and activism," *Society & Natural Resources*, vol. 5, no. 1, pp. 1–19, 1992.

[16] D. J. Davidson and W. R. Freudenburg, "Gender and environmental risk concerns: a review and analysis of available research," *Environment and Behavior*, vol. 28, no. 3, pp. 302–339, 1996.

[17] D. B. Tindall, "Social values and the contingent nature of public opinion and attitudes about forests," *Forestry Chronicle*, vol. 79, no. 3, pp. 692–705, 2003.

[18] H. Uliczka, P. Angelstam, G. Jansson, and A. Bro, "Non-industrial private forest owners' knowledqe of and attitudes towards nature conservation," *Scandinavian Journal of Forest Research*, vol. 19, no. 3, pp. 274–288, 2004.

[19] D. J. Davidson, T. Williamson, and J. R. Parkins, "Understanding climate change risk and vulnerability in northern forest-based communities," *Canadian Journal of Forest Research*, vol. 33, no. 11, pp. 2252–2261, 2003.

[20] G. Johnsson-Latham, *A Study on Gender Equality as a Prerequisite for Sustainable Development*, Environment Advisory Council, Stockholm, Sweden, 2007.

[21] S. Arora-Jonsson, "Relational dynamics and strategies: men and women in a forest community in Sweden," *Agriculture and Human Values*, vol. 21, no. 4, pp. 355–365, 2004.

[22] K. E. Halvorsen, "Relationships between national forest system employee diversity and beliefs regarding external interest groups," *Forest Science*, vol. 47, no. 2, pp. 258–269, 2001.

[23] J. Seager, *Earth Follies: Coming to Terms with the Global Environmental Crisis*, Routledge, New York, NY, USA, 1993.

[24] D. B. Tindall, S. Davies, and C. Mauboules, "Activism and conservation behavior in an environmental movement: the contradictory effects of gender," *Society and Natural Resources*, vol. 16, no. 10, pp. 909–932, 2003.

[25] L. K. Ozanne, C. R. Humphrey, and P. M. Smith, "Gender, environmentalism, and interest in forest certification: Mohai's paradox revisited," *Society and Natural Resources*, vol. 12, no. 6, pp. 613–622, 1999.

[26] B. Livesey, "The politics of Greenpeace," *Canadian Dimensions*, vol. 28, pp. 7–12, 1994.

[27] S. Müller, "Report on the professional and family situation of women working in the environmental field," in *Proceedings of the 2nd European Feminist Research Conference, Feminist Perspectives on Technology, Work and Ecology*, T. Eberhart and C. Wachter, Eds., IFF IFZ Interdisciplinary Research Center for Technology, Work and Culture, Graz, Austria, July 1994.

[28] S. Gherardi and B. Poggio, "Creating and recreating gender order in organizations," *Journal of World Business*, vol. 36, no. 3, pp. 245–259, 2001.

[29] M. G. Reed, "Reproducing the gender order in Canadian forestry: the role of statistical representation," *Scandinavian Journal of Forest Research*, vol. 23, no. 1, pp. 78–91, 2008.

[30] T. Dunk, *It's a Working Man's Town: Male Working-class Culture in Northwestern Ontario*, McGill-Queen's University Press, Quebec, Canada, 1991.

[31] M. S. Carroll, *Community and the Northwestern Logger: Continuities and Changes in the Era of the Spotted Owl*, University of Colorado Press, Boulder, Colo, USA, 1995.

[32] B. Brandth and M. S. Haugen, "Breaking into a masculine discourse women and farm forestry," *Sociologia Ruralis*, vol. 38, no. 3, pp. 427–442, 1998.

[33] B. Brandth and M. S. Haugen, "From lumberjack to business manager: masculinity in the Norwegian forestry press," *Journal of Rural Studies*, vol. 16, no. 3, pp. 343–355, 2000.

[34] UNECE/FAO Team of Specialists on Gender and Forestry, Ed., *Time for Action: Changing the Gender Situation in Forestry Report*, Food and Agriculture Organization of the United Nations, Rome, Italy, 2006.

[35] J. K. Gibson-Graham, "Stuffed if I know!: reflections on post-modern feminist social research, Gender," *Place and Culture*, vol. 1, no. 2, pp. 205–224, 1994.

[36] V. Preston, D. Rose, G. Norcliffe, and J. Holmes, "Shift work, childcare and domestic work: divisions of labour in Canadian paper mill communities," *Gender, Place and Culture*, vol. 7, no. 1, pp. 5–29, 2000.
[37] A. Sinclair, *Trials at the Top*, The Australian Centre, University of Melbourne, Melbourne, Australia, 1994.
[38] P. Davidson and R. Black, "Women in natural resources management: finding a more balanced perspective," *Society and Natural Resources*, vol. 14, no. 8, pp. 645–656, 2001.
[39] B. Brandth, G. Follo, and M. S. Haugen, "Women in forestry: dilemmas of a separate women's organization," *Scandinavian Journal of Forest Research*, vol. 19, no. 5, pp. 466–472, 2004.
[40] WinSETT Centre, "QuickStats # 11: Women in Paid Work," 2011, http://www.ccwestt.org/Portals/0/publications/QuickStatsNo11%20Jul%202011.pdf.
[41] J. R. Parkins and R. E. Mitchell, "Public participation as public debate: a deliberative turn in natural resource management," *Society and Natural Resources*, vol. 18, no. 6, pp. 529–540, 2005.
[42] A. M. Wellstead, R. C. Stedman, and J. R. Parkins, "Understanding the concept of representation within the context of local forest management decision making," *Forest Policy and Economics*, vol. 5, no. 1, pp. 1–11, 2003.
[43] H. Pitken, *The Concept of Representation*, University of California Press, Berkeley, Calif, USA, 1967.
[44] A. Conley and M. A. Moote, "Evaluating collaborative natural resource management," *Society and Natural Resources*, vol. 16, no. 5, pp. 371–386, 2003.
[45] M. Alston and J. Wilkinson, "Australian farm women—shut out or fenced in? The lack of women in agricultural leadership," *Sociologia Ruralis*, vol. 38, no. 3, pp. 391–408, 1998.
[46] B. Agarwal, "Participatory exclusions, community forestry, and gender: an analysis for South Asia and a conceptual framework," *World Development*, vol. 29, no. 10, pp. 1623–1648, 2001.
[47] D. Dahlerup, "From a small to a large majority: women in Scandinavian Politics," *Scandinavian Political Studies*, vol. 11, pp. 275–298, 1988.
[48] E. Teske and B. Beedle, "Journey to the top—breaking through the canopy: Canadian experiences," *Forestry Chronicle*, vol. 77, no. 5, pp. 846–853, 2001.
[49] M. Fullerton, "Gender structures in forestry organizations: Canada," in *Time for Action: Changing the Gender Situation in Forestry*, UNECE/FAO Team of Specialists on Gender and Forestry, Ed., p. 2026, Food and Agriculture Organization of the United Nations, Rome, Italy, 2006.
[50] S. McGregor, *Beyond Mothering Earth: Ecological Citizenship and the Politics of Care*, UBC Press, Vancouver, Canada, 2006.
[51] S. Gherardi, *Gender, Symbolism and Organizational Cultures*, Sage, London, Uk, 1995.
[52] University of Newcastle upon Tyne. "Citizens" Jury calls for moratorium on commercialisation of GM to continue University of Newcastle upon Tyne, 2005, http://www.ncl.ac.uk/press.office/press.release/content.phtml?ref=1063031063.
[53] G. Smith and C. Wales, "The theory and practice of citizens' juries," *Policy and Politics*, vol. 27, no. 3, pp. 295–308, 1999.
[54] Jefferson Center, "Citizen Jury Handbook," 2005, http://www.jefferson-center.org/.
[55] M. P. Pimbert and W. T. Prajateerpu, *A Citizens Jury/Scenario Workshop on Food and Farming Futures for Andhra Pradesh, India*, International Institute for Environment and Development, London, UK, 2002.
[56] N. Kanji and S. F. Tan, "Understanding local difference: gender (plus) matters for NGOs," *Participatory Learning and Action*, vol. 58, no. 1, pp. 74–81, 2008.
[57] J. Gobert, *Justice Democracy and the Jury*, Ashgate Publishing Company, Aldershot, UK, 1997.
[58] A. Coote and J. Lenaghan, *Citizens' Juries: Theory into Practice*, IPPR, London, UK, 1997.
[59] H. Ward, A. Norval, T. Landman, and J. Pretty, "Open citizens' juries and the politics of sustainability," *Political Studies*, vol. 51, no. 2, pp. 282–299, 2003.
[60] S. Pickard, "Citizenship and consumerism in health care: a critique of citizens' juries," *Social Policy and Administration*, vol. 32, no. 3, pp. 226–244, 1998.
[61] M. G. Reed and D. Davidson, "Terms of Engagement: the involvement of Canadian rural communities in sustainable forest management," in *Reshaping Gender and Class in Rural Spaces*, B. Pini and B. Leach, Eds., pp. 199–220, Ashgate Publishing Company, Aldershot, UK, 2011.
[62] N. J. Turner, R. Gregory, C. Brooks, L. Failing, and T. Satterfield, "From invisibility to transparency: identifying the implications," *Ecology and Society*, vol. 13, no. 2, article 7, 2008.

# Afforestation for the Provision of Multiple Ecosystem Services: A Ukrainian Case Study

**Maria Nijnik,[1] Arie Oskam,[2] and Anatoliy Nijnik[3]**

[1] *Social, Economic and Geographical Sciences, The James Hutton Institute, Craigiebuckler, Aberdeen AB15 8QH, Scotland, UK*
[2] *Agricultural Economics and Rural Policy, Wageningen University, Hollandseweg 1, 6706 KN Wageningen, The Netherlands*
[3] *Environmental Network Ltd., The Hillocks, Tarland, Aboyne AB34 4TJ, Scotland, UK*

Correspondence should be addressed to Maria Nijnik, mnijnik@hotmail.co.uk

Academic Editor: Frits Mohren

This paper presents an economic analysis of the planting of trees on marginal lands in Ukraine for timber production, erosion prevention, and climate mitigation. A methodology combining econometric analysis, simulation modelling, and linear programming to analyse the costs and benefits of such afforestation has been adopted. The research reveals that, at discount rates lower than 2%, establishment of new forests is economically justified in the majority of forestry zones. Incorporating the effects of afforestation on mitigating climate change increases social benefits. However, the results indicate that whilst soil protection benefits to agriculture from afforestation in the Steppe are expected to be high, carbon sequestration and timber production activities in the Steppe are cost inefficient due to low rates of tree growth and relatively high opportunity costs of land. The opportunity costs of land are also high in the Polissja where afforestation is cost inefficient at 2% and higher discount rates.

## 1. Introduction

Forestry is multifunctional by nature, but has traditionally been a sector of the economy whose primary objective is to maximize profits from timber production. Today, the focus of forestry is much wider, with evaluation of a broad range of ecosystem services being the mainstream in forestry economics. It is being increasingly recognised that, in addition to timber production, woodlands contribute to global carbon budgeting, provide biodiversity and aesthetic values, and serve as the basis for developing local entrepreneurial opportunities, tourism, recreation, and rural livelihoods [1]. The manifold ecosystem services do not represent a minor issue but rather are an important precondition for diverse human activities which predetermine multifunctionality of forests [2]. The relative societal weight of forestry and its manifold contribution to livelihoods are most clearly reflected in policies focusing on the development of new forests [1, 3]. Af–forestation for multiple purposes is reflected in policies of many countries [4] where multifunctional development of forestry is evidenced through institutional analyses and public opinion surveys [5] and is expressed in real terms on the ground [6].

The origin of discussions on "what forest's multifunctionality is" and "how to manage forests for multiple purposes" goes back to the 1940s. An important question is whether multifunctionality should be considered in the vertical sense, with each lot of land or forest stand fulfilling two or more functions, or whether the term should be used to describe a pattern of diversity in the horizontal sense, in which different areas are dedicated to different functions. Dana implied [7] a vertical interpretation of multifunctionality, while Pearson's view [8] was horizontal: "effective multiple use is merely organized and coordinated specialization." The vertical vision of multifunctionality seemed to be dominant in the last part of the 20th century, although the horizontal interpretation also had its advocates [9]. Bowes and Krutilla [10], for example, argue that forestry is capable of producing successfully a long list of outputs, many of which are complementary in production, whereas Sedjo questions [11] whether there is a role for specialization in multifunctional forest management.

In the current paper, forest multifunctionality is considered in the vertical sense [7]. Characterised in this way, multifunctionality is seen as an economic concept, capturing the multiple processes taking place in forests, and shifting the emphasis of forestry away from maximising production of material (commodity) goods towards its broader objectives, that is, the provision of environmental services (erosion alleviation and climate change mitigation). Climate change, changing socioeconomic conditions, and fragile ecosystem stability have resulted in a need for a more enlarged and continuous forest cover in countries like Ukraine [12] and in new, adaptive approaches to forest management [13, 14]. A suite of forest management alternatives can be ranging from intensive even-aged management to close-to-nature forestry, and nonintervention systems [15]. The selection of an appropriate management system is considered to be fundamental to implementing multifunctional forestry [16]. Multifunctionality implies that the objectives of numerous forestry stakeholders are incorporated into an adaptive and participatory planning process so that ecosystem services are enhanced across the landscapes [14, 17].

Decision Support Systems (DSSs) are suitable platforms for the integration of information, models, and methods required to support complex forest management solutions including those of afforestation [18, 19]. With its focus on North-Western Europe, the AFFOREST DSS seeks to answer on where, how, and how much to afforest in order to maximize carbon sequestration and groundwater recharge and to minimize nitrate leaching [20]. ENCOFOR gives support on where and how to afforest land in developing countries in order to combine successful greenhouse gas mitigation with poverty alleviation and environmental restoration [21]. The SimForTree DSS is designed to compare different forest management and policy scenarios at different spatial scales under changing environmental conditions, particularly climatic [22]. Large-scale scenario models incorporating economic considerations of forest creation and management are becoming available [23].

Afforestation has been widely analysed in the literature, and it is commonly considered by contemporary forestry science among the most straightforward policy measures, including those addressing climate change [24]. It is relatively cheap (cost efficient), clean (it may concurrently provide other ecosystem services), proven (many countries have the legacy of tree growing), effective in the short-term (providing almost immediate effect after the tree planting) and is a rather low labour and energy consuming strategy [25]. It can be incorporated in multifunctional forest use to simultaneously enlarge timber production and bring a variety of other benefits [26].

The purpose of afforestation entails the choice of species to be planted, as does the location, tree-growing conditions and other factors. Carbon sequestration positively correlates with the growth rates of trees; it is therefore advocated to plant the most fast-growing tree species where appropriate or to establish short rotation plantations for the purpose of climate change mitigation alone [27, 28]. However, there is sufficient evidence that forests significantly reduce soil erosion and prevent soil run-off after heavy rains [12, 29]. Tree plantations are still the main providers of materials for construction and industries, and of fuel for economies and households [30]. Forests provide multiple ecosystem services, and in consideration of a more holistic approach to afforestation, a number of recent studies have a wider focus by addressing several forest functions in one instance.

Michetti and Nunes [31] analyse the role of afforestation and timber management activities in stabilizing climate. Lindner et al. [32] cover the capacity of forests to provide economic, social, and ecological services (wood production; nonwood forest products; carbon sequestration; biodiversity; recreation; protective functions) under the impacts of climatic change. The European Forestry Institute used the European Forest Information (EFISCEN) modelling approach [33] to gain insights into the effects of changing management practices in forests across 30 countries under different scenarios [34]. The need to maintain the national diversities that constitute European forestry within harmonised strategies is being recognised.

Ukrainian scientists contributed to the scenario analysis through the development of long-term predictions for their forests. Their scenarios are consistent with the projections developed at the International Institute for Applied Systems Analysis (IIASA), the Ukrainian National Agricultural University, and the Ukrainian National University of Forestry [35–37]. Numerous forestry studies support the idea that a larger forest area would be attractive in Ukraine [36, 38], but limited knowledge is available about the economics of creating multifunctional forests to address wider sustainability objectives.

Thus, based upon both international and national studies on multipurpose afforestation, the current paper considers the economics of tree planting in Ukraine for the provision of timber production, soil protection and carbon sequestration services. Firstly, "user" benefits of afforestation which accrue to two primary sectors of rural economy, that is, forestry and agriculture, are considered, with evaluation of the role of forests in soil protection complementing an assessment of the timber supply benefits accruing from establishment of forest plantations. The paper then defines key objectives of woodlands expansion and analyses the potential for tree planting, estimating its costs and benefits. This is followed by a pilot cost-benefit analysis (CBA) of the establishment of forest plantations: if afforestation adds to the welfare of society, it is economically sound. This is observed in some areas in Ukraine, and we discuss where and why this is the case. Finally, by considering carbon sequestration in trees under the storage policy scenario, the paper goes beyond the "user" benefits of afforestation that accrue to the forestry and agriculture sectors, linking the planting of trees in Ukraine to global climate change mitigation policy objectives.

## 2. Background

The history of establishment of forest plantations in Ukraine dates back to the 17th century; however, until the 1920s, tree-planting activities were isolated and episodic [39]. The protection of land through woodland creation was considerably hampered by economic and social conditions and then by the

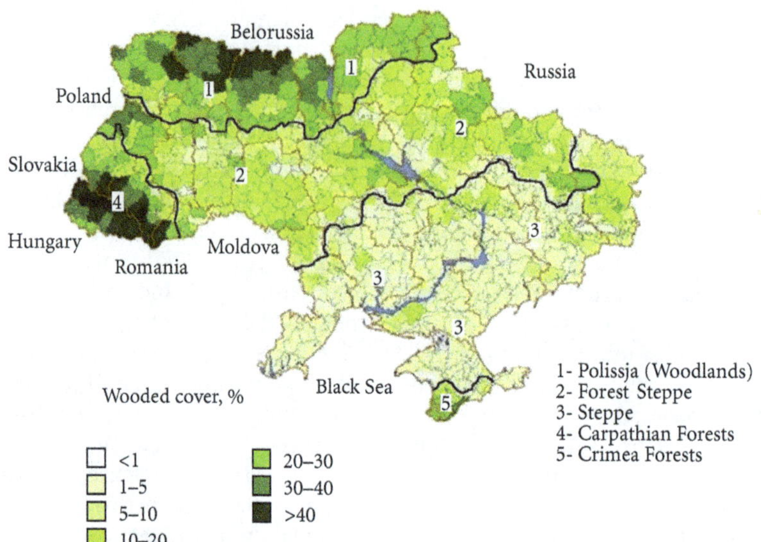


FIGURE 1: Forestry zoning of Ukraine Source: adapted from [39] and the records of the National Academy of Sciences [47].

shift in forest policy from afforestation towards natural forest regeneration. In the 1990s, the decreasing scale of afforestation was also caused by difficulties of the transition process from a command-and-control towards a market economy, such as the lack of well-defined and ensured property rights; shortage of investment and economic incentives; increasing attention paid by forest management to short-term financial objectives [40].

Nevertheless, the reforestation coefficient remains at 94%; planting on forestland amounts to about 28.5 kha; protective plantings on eroded and unproductive agricultural lands total 7 kha; shelterbelt plantings total 1.1 kha per year [36]. The necessity of afforestation is stressed in the President's Decree [41], which aimed to reform forestry in Ukraine, and in the State Programme "Forests of Ukraine, 2002–2015" [42]. Forestry legislation rests on the principle of sustainable forest management, and multiple forestry objectives are recognised by law [43].

Ukraine has good forest growing conditions and productive forests [44]. The territory was extensively covered with forests a millennium ago. Today, forest cover comprises 16.5%, and this is among the lowest estimates in Europe [45]. Given that 15% of Ukraine's territory is under extreme anthropogenic pressures, and considering the role of forests for environmental quality, the development of woodlands is considered important [46]. Ukraine has a level of cultivation of 54.8% by area and faces partial erosion on 35% of its arable land. Annually, 4 Mt of fertile soil are washed away. The damage to agriculture from erosion exceeds 8 M€ [39]. The intensity of soil erosion varies across the territory for which the annual increment of eroded land is 90 kha. According to the National Academy [47] wooded cover should increase to 20%, since this would alleviate spatial spreading of erosion and its intensity.

In recent years, afforestation in Ukraine has played an important role also as a carbon dioxide reduction measure [48]. Since 2001, afforestation has become an eligible component of climate policy, meaning that over and above emissions reduction, an enhancement of GHG "sinks" and "reservoirs" via tree planting is becoming considered as increasingly important. Also, as a consequence of its afforestation and sustainable forestry strategy, Ukraine could become self-sufficient in wood and become a price setter in the European wood production. This is because the demand for wood in Europe is rising by 0.8–2.6% per year, with Europe likely to remain a high wood cost region. The potential niche for Ukraine's forestry in an international perspective is therefore its low cost of delivered wood [49].

## 3. Methodology and the Results

A methodology combining econometric analysis, simulation modeling, and linear programming for a cost-benefit analysis of afforestation has been adopted, as described in the following sections, which also present the results of the research. The modelling approach is straightforward, practical, and policy relevant. It could be easily modified, adjusted, and applied to other cases and in other countries to review evidence and the opportunities available with respect to the delivery of multiple services from land use policy decisions by assessing existing forest management options and identifying new ones.

*3.1. Afforestation Potential.* Afforestation potential was assessed across Ukraine's forestry zones. We developed comprehensive forestry zoning in an earlier, separate study (Figure 1) to provide a background for consideration of policy measures to enhance sustainability of forest use [39].

This spatial classification divides the territory of Ukraine into the following main forestry areas: the Polissja (Woodland), Forest Steppe, Steppe, Crimea, and the Carpathians, and further divides these main forestry areas into spatial units of lower levels of hierarchy using various factors of

TABLE 1: Potential for afforestation by land use across zones, kha.

| Zones | State forest fund | | | Agricultural land | | | Total |
|---|---|---|---|---|---|---|---|
| | Ravines | Sand | Rocky | Eroded | Deflated | Rocky | |
| Polissja (Woodland) | 65.0 | 82.0 | 0.5 | 73.7 | 0.7 | 26.1 | 248.0 |
| Forest Steppe | 95.0 | 84.0 | 0.6 | 451.6 | 18.3 | 61.0 | 710.5 |
| Steppe | 24.0 | 64.0 | n.a. | 669.4 | 40.6 | 137.5 | 935.5 |
| Carpathians | 1.6 | n.a. | 1.4 | 24.6 | n.a. | 143.4 | 171.0 |
| Crimea | 0.8 | n.a. | 1.8 | 13.1 | 1.8 | 206.8 | 224.3 |
| Ukraine (totally) | 186.4 | 230.0 | 4.3 | 1232.4 | 61.4 | 574.8 | 2289.3 |

*Source.* Computed on the basis of data from the National Academy of Sciences of Ukraine [47].

TABLE 2: Net annual returns to current agricultural activities, €/ha.

| Forestry zone | | Forage and pasture | Wheat |
|---|---|---|---|
| Polissja (Woodland) | Eroded and deflated land | 8.0 | 37.8 |
| Forest Steppe | Rocky land | 7.8 | n.a. |
| | Eroded land | 10.0 | 52.1 |
| | Deflated lands | 9.2 | 14.7 |
| Steppe | Rocky land | 8.0 | n.a. |
| | Eroded land | 20.0 | 61.5 |
| | Deflated land | 6.0 | 27.2 |
| Carpathians | Rocky land | n.a. | n.a. |
| | Eroded land | 7.8 | 0 |
| Crimea | Rocky land | 7.0 | 0 |
| | Eroded and rocky | 7.0 | 0 |

*Source.* Based on the data from the National Academy of Sciences of Ukraine [28].

land use and forestry development. The land suitable for tree planting was deemed to include bare land of the "State Forest Fund" (SFF, that is, public forest estate) that is under the management of the State Committee of Forests. The land deemed suitable for afforestation also included bare and marginal agricultural land currently used for forage and pasture and some land used for wheat production for which net returns are low (Table 1).

The table shows that a total area of 2.29 Mha was initially identified as suitable for tree planting. After estimating the net present value (NPV) of afforestation, this area was reduced, to account for land in which the opportunity costs appear to be comparatively high (see below for further details). Given tree growing conditions and assumptions based on interviews with forest specialists [44], pine was considered appropriate for planting in the Steppe, Crimea and Polissja; pine and oak in the Forest Steppe; beech, fir, and/or spruce in the Carpathians.

*3.2. Costs of Afforestation.* The costs of afforestation of marginal and bare land in the SFF include tree-planting costs (€100–200/ha) and silvicultural expenses (€12.5–30/ha annum). These costs vary; but given that within each zone conditions are relatively stable, costs are assumed to be the same within each zone. Marginal agricultural land has alternative options to afforestation, therefore, in addition, net returns associated with current use of land were considered. Net annual returns for current wheat production were computed on the basis of land productivity data, costs of wheat production, and output prices. Estimation of costs for land used for forage and pasture was based on land productivity, and prices which agricultural enterprises pay for the equivalent cattle feed (Table 2).

Computation was in Ukraine's currency Hryvnya (which in 2007 corresponded to 0.14 €). The present value (PV) costs occurring over 100 years of stipulated ages of timber harvesting were estimated at several discount rates based on [36], as seen in Table 3.

The results show that at a 4% discount rate, the PV of afforestation costs is €484 per ha on average for the country. The highest PV of costs is in the Steppe at €609.5 per ha, with the lowest, at €288 per ha, in the Carpathians. The divergence in cost estimates across zones is explained by the diversity of tree-growing and socioeconomic conditions.

*3.3. Timber Supply Benefits*. A sum of monetary value from additional timber yield and monetary estimates of soil protection pertaining to arable land comprise the total benefits of afforestation (when only "user benefits" to forestry and agriculture are considered). For the monetary value of timber yield changes, the model multiplies estimates of a physical crop change based on acreage in production by the price of timber [50]. This implies that timber use and prices are constant. Allowing in the long run for a stable average annual timber cut of 2 $m^3$/ha (ca. 50% of mean annual increment, MAI), about 4.6 $Mm^3$ of additional timber could

Table 3: Afforestation costs, M€.

| Forestry zone | Annual costs by zone | | | PV costs | | | |
|---|---|---|---|---|---|---|---|
| | Opportunity | Planting | Care and protection | $r = 0\%$ | $r = 2\%$ | $r = 4\%$ | $r = 6\%$ |
| Polissja (Wooland) | 1.4 | 16.1 | 2.0 | 356.3 | 162.7 | 99.5 | 72.7 |
| Forest Steppe | 6.4 | 32.8 | 4.1 | 1084.3 | 486.0 | 290.5 | 207.5 |
| Steppe | 14.1 | 49.8 | 7.1 | 2173.3 | 965.0 | 570.2 | 402.7 |
| Carpathians | 0.8 | 7.5 | 0.9 | 177.9 | 80.9 | 49.2 | 43.8 |
| Crimea | 0.8 | 19.6 | 2.5 | 345.0 | 159.9 | 99.4 | 73.7 |
| Ukraine | 23.5 | 125.8 | 16.6 | 4136.8 | 1854.5 | 1108.8 | 792.4 |

*Source.* Computed on the basis of data from the National Academy of Sciences of Ukraine [28].

Table 4: Estimates of the returns from timber harvesting.

| Zone | Species | Stock of stands in 100 years, $m^3$/ha | Returns in the year of harvesting €/ha | PV returns €/ha, 4% | PV returns by zone M€ 0% | 2% | 4% | 6% |
|---|---|---|---|---|---|---|---|---|
| Polissja | Pine | 250 | 1250 | 24.75 | 310.0 | 42.8 | 6.1 | 0.9 |
| Forest Steppe | pine | 350 | 1750 | 34.65 | 612.9 | 84.6 | 12.1 | 1.8 |
| | Oak | 350 | 1750 | 34.65 | 612.9 | 84.6 | 12.1 | 1.8 |
| Steppe | pine | 250 | 1250 | 24.75 | 584.7 | 27.7 | 11.6 | 1.7 |
| Carpa-thians | beech | 350 | 1575 | 31.18 | 134.7 | 18.6 | 2.7 | 0.4 |
| | fir | 400 | 2000 | 39.6 | 171.0 | 23.6 | 3.4 | 0.5 |
| Ukraine | | | | | 2304.0 | 318.0 | 45.6 | 6.8 |

be produced, bringing annual returns of 23 M€, if the stumpage value of timber is ca. €5/$m^3$ [49].

Benefits were also computed across zones over a 100-year period. Table 4 shows the results when only commercial timber cut is taken into account. The following assumptions were made: stand composition in the Forest Steppe comprises 50% pine and 50% oak trees; half of the Steppe is planted with trees which would be harvested; beech stands in the Carpathians are planted on 50% of the area, as are fir stands. The Crimea, where forests play primarily a role in enhancing environmental quality, timber harvesting was not considered.

The table suggests that, at the discount rate of 0%, PV returns from timber harvesting are 23.04 M€ (comparable with the annual returns of 23 M€ estimated above). The highest returns per acreage are in the Forest Steppe and the Carpathians. However, timber benefits alone do not justify tree planting in any of the zones.

*3.4. Soil Protection Benefits.* The notion that the scale of erosion depends on forest cover [47] was put to an empirical test by using a semilogarithmic regression (Figure 2). The robustness of the functional form adopted was justified, and economic attractiveness of planting trees to mitigate erosion was then assessed.

The results of the estimation show a statistically significant (at 1% significance level) negative relationship between

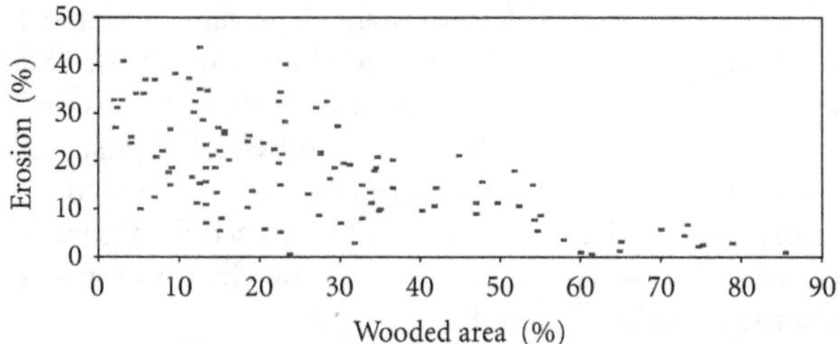


Figure 2: Relationship: wooded cover—erosion.

the share of eroded land ($E$, %) and the share of wooded land ($W$, %):

$$\log(E) = \underset{(29.13)}{3.4653} - \underset{(-9.38)}{0.0329} * W; \quad \text{or} \quad E = 31.986e^{-0.0329W}, \quad R^2 = 0.45 \tag{1}$$

The $t$-statistic of −9.38 suggests that the negative coefficient on $W$ is significantly different from 0, and with the increase of forest cover, the erosion rates decrease. The value of $R^2$ indicates that more factors influence erosion rates and there is room for improvement of the model.

Table 5: Simulated rates of soil erosion.

| Wooded area (W), % | Erosion (E), Ukraine, % | Erosion (E), Carpathians, % | Elasticity, Ukraine, % | Elasticity, Carpathians, % |
|---|---|---|---|---|
| 0 | 32.0 | 79.1 | −1.05 | −4.13 |
| 5 | 27.1 | 60.9 | −0.89 | −3.18 |
| 10 | 23.0 | 46.9 | −0.76 | −2.45 |
| 15 | 19.5 | 36.1 | −0.64 | −1.89 |
| 20 | 16.6 | 27.8 | −0.54 | −1.45 |
| 25 | 14.1 | 21.4 | −0.46 | −1.12 |
| 30 | 11.9 | 16.5 | −0.39 | −0.86 |
| 35 | 10.1 | 12.7 | −0.33 | −0.66 |
| 40 | 8.6 | 9.8 | −0.28 | −0.51 |
| 45 | 7.3 | 7.5 | −0.23 | −0.39 |
| 50 | 6.2 | 5.8 | −0.20 | −0.30 |
| 55 | 5.2 | 4.4 | −0.17 | −0.23 |
| 60 | 4.4 | 3.4 | −0.15 | −0.18 |
| 65 | 3.8 | 2.6 | −0.12 | −0.14 |
| 70 | 3.2 | 2.0 | −0.11 | −0.11 |
| 75 | 2.7 | 1.6 | −0.09 | −0.08 |
| 80 | 2.3 | 1.2 | −0.08 | −0.06 |
| 85 | 2.0 | 0.9 | −0.06 | −0.05 |
| 90 | 1.7 | 0.7 | −0.05 | −0.04 |
| 95 | 1.4 | 0.5 | −0.05 | −0.03 |
| 100 | 1.2 | 0.4 | −0.04 | −0.02 |

For the Carpathian Mountains, where the conditions differ substantially from elsewhere in the country (lowland), the estimated equation is as follows:

$$\log(E) = 4.3702 - 0.0523 * W; \quad \text{or} \quad E = 79.059e^{-0.0523W},$$
$$(5.46) \quad (-3.99) \qquad R^2 = 0.5 \tag{2}$$

Simulated rates of erosion at different levels of wooded cover are shown in Table 5.

From the estimated equations, marginal changes in erosion rates relative to marginal changes in wooded cover rates are for Ukraine: $dE/dW = -0.0329E$, and for the Carpathians: $dE/dW = -0.0523E$. These estimations show the "elasticity" of erosion with respect to wooded cover. These results indicate that until wooded cover reaches 27% in the Carpathians and ca. 2% in Ukraine, the erosion is elastic, that is, when wooded cover is increasing marginally, the erosion is reduced proportionally as much. This is observed up to the point when the share of eroded land is around 30% in Ukraine, and as far as it falls below 19% in the Carpathians. The results suggest that if there were no woods in rural areas the share of eroded lands would comprise 79% in the Carpathians and 32% on average in Ukraine.

By using the results of the regression analysis, indicative estimates of the soil protection role of forests were computed. In the Polissja where wooded cover comprises ca. 26%, the "elasticity" of erosion is −0.43%. This means that a 1% increase in wooded cover leads to a 0.43% decrease in erosion rates. A 1% increase of forest cover, that is, an increase of 0.029 Mha, will mitigate erosion on 0.2 Mha of land. The net annual returns from that land were then calculated on the basis of data from Table 2 and were considered as measures of soil protection benefits to agriculture from marginal expansion of forests in the Polissja. The corresponding estimations were made for other zones, and the equation is:

$$X = \varepsilon \cdot \frac{E}{W}, \tag{3}$$

where $X$ is the indicative measure of soil protection benefits to agriculture from marginal expansion of forest cover; $\varepsilon$ "elasticity" of erosion with respect to forest cover, that is, 1% increase in $W$ leads to $\varepsilon$% decrease in $E$, % (Table 5); $W$ share of wooded land, %; $E$ share of eroded agricultural land, %.

Forests start providing protection benefits after the age of 5 years, and with their gradual regeneration, forests keep providing these benefits indefinitely [12]. These considerations were taken into account when computing economic estimates of the soil protection function of forestry for agriculture. The potential for forest expansion was taken from Table 1, and the assumption was made that, in nonmountainous areas, 30% of agricultural land is used for wheat production. Given the assumptions presented above, the soil protection benefits of afforestation for agriculture are shown in Table 6 and are the highest in the Steppe.

*3.5. Economic Analysis of Afforestation for Timber Production and Erosion Alleviation.* The economic analysis of afforestation was carried out at various hierarchical levels. First,

Table 6: Estimates of soil protection benefits to agriculture.

| Forestry zone | Annual average benefits, €/ha | | Annual benefits |
|---|---|---|---|
| | Wheat | Forage/pasture | M€/zone |
| Polissja (Woodland) | 7.6 | 1.6 | 0.8 |
| Forest Steppe | 33.0 | 9.0 | 11.5 |
| Steppe | 58.2 | 17.0 | 27.5 |
| Carpathians | 0 | 9.7 | 1.7 |
| Crimea | 0 | 12.2 | 2.7 |
| Ukraine | | | 44.2 |

"user" net present value (NPV) benefits of afforestation to forestry and agriculture in Ukraine were considered. The general idea of the LP model discussed by us in detail in [44] was to consider the establishment of future forests on bare and marginal agricultural lands in Ukraine, in such a way that allows the attainment of maximum cumulative NPV of benefits from new plantations over the period of 100 years, subject to constraints. The land available for conversion into forest and tree species most suitable for growing in the particular conditions, identified earlier in the current paper, were considered, together with the following forest management regimes. The first forest management regime was a basic silviculture that is based on quick replanting of trees, after timber harvesting. The second regime was that of planting trees and the attendant silvicultural operations recommended by Ukraine forest legislation. The third regime was that of a basic silviculture, but with the timber rotation period corresponding to maximum sustainable yield.

Major constraints of this model are acreage. The model implies, for instance, that in the Carpathian Mountains, beech forests do not grow at high altitudes or that there is no land suitable for wheat production in the mountains.

The results of LP modelling provide evidence that under the assumptions and at a discount rate of 4%, it is economically sound to establish monoculture plantations on the perceived bare land in the Forest Steppe (0.28 Mha), Steppe (0.13 Mha), and the Carpathians (0.01 Mha). The shadow price of bare land (245.2 €/ha) is highest in the Steppe. Establishment of monoculture plantations appeared to be a more economically sensible solution, with basic silviculture proven to perform better economically in all forestry zones. These results can be explained by the fact that the scope of this phase of the research was limited to user values that accrue to Ukraine's forestry and agriculture. The purpose was to form a basis for economic appraisal of practical aspects of land use management decisions. The analysis has been further extended, and in the following section, the research is presented for a higher hierarchical scale, so that in addition to "user" benefits of afforestation, carbon sequestration benefits in trees under a storage policy scenario are incorporated in the analysis.

*3.6. Comparing Costs and Benefits of Afforestation.* In addition to domestic gains to forestry and agriculture, afforestation provides climate change mitigation benefits. The economics of carbon sequestration forestry scenarios for Ukraine as a stand-alone analysis is considered in Nijnik [51]. In the current paper, economic evaluation of tree planting for multiple purposes, including for carbon sequestration under the storage policy scenario, is presented. The storage policy scenario presumes a one-time planting of trees for a period of 100 years, without accounting for future use of wood and land [52]. The analysis was carried out across forestry zones with maximising of NPV of afforestation as the criterion. It determines the present value of net benefits by discounting the stream of benefits ($B$) and costs ($C$) back to the beginning of the base year $t = 0$:

$$\mathrm{NPV} = \sum_{t=0}^{n} B_t/(1+r)^t - \sum_{t=0}^{n} C_t/(1+r)^t. \quad (4)$$

Benefits of afforestation are expected to accrue over a long period of time, and a period of 100 years was chosen to capture most of these benefits and costs. In addition to timber production and soil protection benefits from the potential forests to alleviate soil erosion discussed earlier, the carbon (C) uptake benefits from afforestation have been approximated using the following procedure.

The functional forms for stand growth of tree species were estimated, using the equations provided by the authors in Nijnik [53]. The coefficients of Lakida et al. [54] were used to translate the stem biomass into total above-ground biomass. The volume of stem wood was multiplied by 0.2 $tC/m^3$ for its translation into carbon [55]. Carbon sequestered by the root was estimated, depending on tree species, either on the basis of the relationships presented in Van Kooten and Bulte [52] or in Lakida et al. [54]. Then, based on Nijnik [44, 51], the sequestered carbon was computed across zones. The price of 15 € per tonne of carbon was assumed to be stable and was used to calculate carbon uptake benefits based on Sandor and Skees [56]. The fact that an increase in the amount of carbon credits available to buy, especially pertaining to "hot air," is pushing the price of carbon credits down was not taken into account. The discount settings of 0%, 2%, and 4% were applied when calculating the carbon storage option. The PV benefits from afforestation are shown in Table 7.

The results presented in Table 7 suggest that, for example, in the Polissja the highest benefits would accrue from the increased timber production and carbon uptake, whilst in the Steppe the highest benefits would accrue from the forest function of soil protection.

TABLE 7: Afforestation benefits, PV M€.

| Forestry zone | r% | Production | Soil protection | Carbon uptake |
|---|---|---|---|---|
| Polissja (Woodland) | 0 | 310 | 84 | 362.6 |
| | 2 | 42.8 | 36.2 | 49.1 |
| | 4 | 6.1 | 20.6 | 6.6 |
| Forest Steppe | 0 | 1125.8 | 1150 | 1255.9 |
| | 2 | 169.2 | 495.6 | 170.1 |
| | 4 | 24.2 | 281.8 | 23.0 |
| Steppe | 0 | 584.7 | 2750 | 1237.4 |
| | 2 | 80.7 | 1185.2 | 167.5 |
| | 4 | 11.6 | 673.9 | 22.7 |
| Carpathians | 0 | 305.7 | 170 | 660.7 |
| | 2 | 42.2 | 73.3 | 89.4 |
| | 4 | 6.1 | 41.7 | 12.2 |
| Crimea | 0 | 0 | 270 | 437.3 |
| | 2 | 0 | 116.4 | 59.2 |
| | 4 | 0 | 66.2 | 8.0 |

TABLE 8: Economic evaluation of afforestation across zones, PV M€.

| Forestry zone | r% | Total benefits | Costs | NPV |
|---|---|---|---|---|
| Polissja (Woodland) | 0 | 756.6 | 356.3 | 400.3 |
| | 2 | 128.1 | 162.7 | −34.6 |
| | 4 | 33.3 | 99.5 | −66.2 |
| Forest Steppe | 0 | 3531.7 | 1084.3 | 2447.4 |
| | 2 | 834.7 | 486 | 348.7 |
| | 4 | 329.0 | 290.5 | 38.5 |
| Steppe | 0 | 4572.1 | 2173.3 | 2398.8 |
| | 2 | 1433.4 | 965 | 468.4 |
| | 4 | 696.6 | 570.2 | 126.4 |
| Carpathians | 0 | 1136.4 | 177.9 | 958.5 |
| | 2 | 204.9 | 80.9 | 124 |
| | 4 | 59.9 | 49.2 | 10.7 |
| Crimea | 0 | 707.3 | 345 | 362.3 |
| | 2 | 175.6 | 159.9 | 15.7 |
| | 4 | 74.2 | 99.4 | −25.2 |

The results vary substantially across the territory of Ukraine and depend upon the applied discount rates. The NPV of afforestation is positive in the majority of zones for discount rates of up to 2%. At these discount rates, creation of multifunctional forest plantations would enlarge social benefits, including those of climate change mitigation, and would add to the welfare of society. At a discount rate of 4%, the area of forest plantations would be 1.82 Mha (excluding the Crimea and the Polissja). In the Carpathian and Crimean Mountains, commercial timber harvesting is restricted, and economic benefits from timber are therefore modest. Agricultural production is also limited in the mountainous regions. Consequently, the benefits that accrue to agriculture from the forest function of soil protection are moderate (Table 8).

## 4. Conclusions

Afforestation in Ukraine, where vast areas are currently suitable for tree planting, is seen as a means to contribute to sustainable land management and climate change mitigation. The afforestation costs are fairly low, apparently due to good forest growing conditions and relatively low labour costs. An expansion of forest cover is important for soil protection. Annually, 1 ha of forest in Ukraine provides soil protection benefits to agriculture of €1.6 to €58.2. Such a broad range can be explained by the variety of conditions. A low share of forest cover might be among the causes of erosion, and planting trees is a possible measure to alleviate this, particularly in the Steppe. When only timber production gains and gains from the protection of agricultural land against erosion are taken into account, at 2% through 4% discount rates, the benefits from afforestation are high in the Steppe, Forest Steppe, and the Carpathians, where the tree planting is economically justified on ca. 1.82 Mha of land. When a discount rate of 4% is used, economic justification of tree planting would be limited to bare land in these zones, with a total area of 0.42 Mha.

The results are more positive when afforestation considerations include the rewards for climate change mitigation. But with or without consideration of carbon uptake, at discount rates lower than 2%, afforestation costs are covered by the returns in the majority of regions. The results indicate that whilst soil protection benefits from afforestation in the Steppe are expected to be high, carbon sequestration and timber production activities are not cost efficient due to low rates of tree growth and relatively high opportunity costs of land. The opportunity costs of land are also high in the Polissja where afforestation is cost-inefficient at 2% and higher discount rates.

## 5. Discussion

In Ukraine, and many other former command-and-control economies (to which this study would convey), large-scale agriculture under the previous regime supported the conversion of forest or grassland to agricultural land. Currently, the decreasing agriculture [45] will likely cause the increase of abandoned land. An expansion of woodlands and, consequently, a rising role of forestry could therefore be predicted. International regulations and national programmes supporting the conversion of agricultural land back to forest focus largely on remote areas. Therefore, afforestation projects need to be incorporated in regional schemes of sustainable rural development, where socioeconomic, environmental, and climate change related components of land use changes are to be considered jointly.

In order to be viable afforestation projects need to be coherent, effective, cost efficient, widely acceptable by the public and consistent with other aspects of sustainable development policy [28]. Numerous examples indicate that climate policy measures will likely be accepted by the public and will consequently be implemented, if they are consistent with the programmes that focus on issues, other than climatic stresses [57]. Many scholars support this view, emphasising the "win-win" opportunities of forest carbon projects, which may all at once provide biodiversity conservation and rural development benefits [58]. Policy measures then should aim the "win-win" situations, which would benefit rural development, the environment, people, and the economy all together.

Afforestation for multiple purposes could enlarge total benefits and prevent potential conflicts related to the trade-offs between biodiversity and carbon sequestration benefits or between landscape amenity values and those of carbon uptake. Although a multipurpose afforestation may result in lower rates of carbon sequestration, it is expected to be more attractive to people, because in the majority of cases it will provide additional benefits and promote sustainable rural development. A multi-purpose afforestation is often seen as a sustainable way of land reclamation and of increasing of the productivity of abandoned land [59, 60], whilst utilization of biomass from new plantations can provide employment opportunities and create new options for land development, being also a sustainable alternative for nuclear energy.

The potential for afforestation in Ukraine appeared to be significant. This allows us to argue that forestry activities can contribute to climate change mitigation and can provide additional social and environmental benefits by contributing to economic development in remote regions, which are most strongly affected by the transition processes. However, new insights are needed into the connection between climate policies and strategies to promote sustainable forestry and sustainable land use and sustainable rural development. Efficient and feasible forestry-based initiatives and inter-sectoral cooperation need to be well embedded into existing policy areas, and if flexible mechanisms are implemented, a considerable scope exists for multifunctional land use systems and "win-win" solutions.

In transition countries affected by regional socioeconomic disparities, the economic and social issues of forestry development are particularly important in view of accumulating financial assets and providing social opportunities for sustainable policy and promarket reforms. Afforestation seeks to enlarge economic gains to forestry through an enlarged wood production and to the agricultural sector, because of soil protection and hydrological forest functions. It also deemed to be beneficial in a broader sense, for example, concerning the mitigation of climate change.

The Kyoto Protocol provides opportunities for countries to cope with the changing climate from an economic perspective. Its flexible mechanisms were designed to help Annex B countries, including Ukraine, to meet their emissions reduction targets at least cost [61]. It allows (non-EU) countries like Ukraine to sell carbon offsets to industrialised countries and, therefore, raise funds needed for its forestry sector and a wider economy. Carbon sequestration through afforestation in Ukraine could be beneficial also for Annex B countries in view of stabilising their collective emissions in the cheapest possible way via the trading of carbon-offset services. European investors are showing their interest to invest in afforestation projects in Ukraine. However, the potential gains from international projects are not seen as priorities for climate policies in either of these countries [62, 63]. Therefore, unless the necessary institutional infrastructure is developed and the barriers for investment are identified and addressed, Ukraine cannot expect to benefit widely from carbon crediting.

Thus, there is a difference between the benefits of afforestation identified in this paper and the benefits for those planting the trees. Although tree planting would enlarge social gains, welfare maximisation conditions would hardly be guaranteed because of market failures. The problem: "who pays for tree planting and who receives the benefits" can scarcely be solved in Ukraine through the market and would therefore need to be regulated by government [37]. It seems that government would have to justify public policy concerning tree planting and balance costs and benefits to provide incentives for those planting the trees. Also, afforestation would have to be elaborated in close dialogue with stakeholders and with public involvement including in-depth consideration of various scenarios for woodlands expansion. Such deliberative processes will increase the capabilities of policy actors, assisting them in the delivery of sustainability objectives to forest management practices on the ground.

## Endnotes

1. The Millennium Ecosystem Assessment [1] grouped multiple ecosystem services into supporting services, such as nutrient cycling, oxygen production, and soil formation; provisioning services of food, fibre, fuel, and water; regulating services, including climate regulation, water purification, and flood protection; cultural/social services, such as education, recreation, and aesthetic value.

2. Afforestation is an expansion of forest area on lands which more than 50 years previously contained forest but which were later converted to another use. Reforestation is a restoration of degraded or recently (within 20–50 years previously) deforested lands [64]. In this paper, these terms are considered synonymously.

3. The economics of water erosion and flooding in the Carpathians are beyond the limits of this paper. This is ongoing research, and findings on these issues will be discussed separately.

4. The economics of carbon sequestration through afforestation in Ukraine, with an assessment of the carbon storage scenario, is considered in a stand-alone paper by Nijnik [51].

5. In this paper, we seek to assess (also, in monitory terms) the biophysical potential (upper benchmark) for afforestation. In practice, however, but this is beyond the scope of this paper, tree-planting activities will be constrained by numerous economic, social, and environmental factors (e.g., land use planning; land use change and climatic change; economic and institutional development; financial considerations). Further, the potential of regulatory carbon offset trading, for example, will be limited to carbon balances, resulting from the eligible mitigation forestry projects subject to cap, as well as by the costs of GHG inventory preparation and usually high transaction costs. Also, the use of NPV criteria alone may lead to overestimation [65]. Further, in traditional forestry (where timber is the primary objective) maximising the NPV implies a comparison of the net benefits from postponing harvesting with the net benefits from harvesting timber and investing the profits. However, as shown by Nijnik [53], the maximising of NPV could promote volumes of harvesting higher than the net growth of forest stands. It follows that in search of sustainable forestry which is also to be economically efficient the establishment of fast-growing monoculture plantations would be encouraged. But this would endanger biological diversity, health, and vitality of forest ecosystems. Often also, this would enlarge costs related to care and protection of monoculture forest stands that are less stable biologically. There is a threat therefore that in the real world situation (in conditions of noninternalised externalities and market imperfections) as it would apply to monopurpose forestry, the social and environmental components of sustainability would remain undermined.

6. "Reforestation coefficient" is the share (%) of reforested land in the total area of land previously covered with forest but that has been cut down or died (see also http://www.ukrref.org.ua/, http://http://www.ukrndnc.org.ua http://www.bestreferat.ru/referat-160465.html for more discussion). According to the State Committee of Forestry of Ukraine annual planting in Ukraine was as follows: in 1949–1965, 100000–200000 ha, in 1966–1990, 55000–100000 ha and in 1995–2005, 35000–40000 ha per year [36].

7. These projections are based on ecological/environmental criteria, with the focus primarily on hydrological and soil protection forest functions. These forest functions are seen as the priorities in Ukraine, and the 20% share of wooded cover is deemed to deliver a sustained ecological balance.

8. However, prospects and requirements with regard to land use changes and forest development would be case specific and varying for different end users, depending on their key objectives, and on natural and economic conditions of the location (e.g., forest and soil characteristics and level of erosion, climatic drivers, and climate mitigation strategies) and in terms of the content, scope, scale, risks, and uncertainties, dynamics and sequencing of afforestation and on the range of multiple forest ecosystem services anticipated.

9. According to the MENR [46], 1.14 Mha were to be planted with trees until the year 2010. This includes the creation of forest shelterbelts and tree planting on steep banks, sands, river, and reservoir banks, as well as on eroded and contaminated land.

10. The discount settings of 2% and 4% were kept as central in this research.

11. For example, together with the share of wooded cover, spatial distribution and the distance between fields and woods in rural landscapes play an important role in erosion mitigation, particularly in low-forested areas, but this is not taken into account in the model as it stands.

12. The figures on $W$ and $E$ are already given in percentages, thus it is not a precise computation of elasticity.

13. In this way soil protection benefits to agriculture from marginal expansion of forests were converted into monetary values. It is assumed that the protection benefits remain constant for all forest age classes over 5 years.

14. Conversion of forest to row crops increases erosion by a factor of 20–1000 [66]. Lampietti and Dixon [67] found that the monetary value of forest function to alleviate erosion is around $30/ha.

15. A substantial potential for tree planting and terrestrial carbon offsetting/trading has attracted attention of foreign investors, for example, from The Netherlands, concerning a pilot afforestation project of 5 kha in Central, Eastern, and Western Ukraine, and a project designed to regenerate forests on lands contaminated after the Chernobyl nuclear accident (Northern part of the country, in Polissja). The creation of forests on these lands will help prevent the distribution of radioactive contamination and spreading of soil erosion [68].

## References

[1] Millennium Ecosystem Assessment (MEA), *Millennium Ecosystem Assessment Synthesis Report*, 2005, http://maweb.org/documents/document.356.aspx.pdf.

[2] C. Folke, J. Colding, and F. Berkes, "Synthesis: building resilience and adaptive capacity in social-ecological systems," in *Navigating Social-Ecological Systems: Building Resilience for Complexity and Change*, F. Berkes, J. Colding, and C. Folke, Eds., pp. 352–387, Cambridge University Press, 2003.

[3] European Commission (EC), *Environmental Policy Review: Communication from the Commission to the European Parliament* COM, vol. 195. Brussels, Belgium, 2007, http://www.nhlbi.nih.gov/meetings/workshops/cardiorenal-hf-hd.htm.

[4] M. Kaljonen, E. Primmer, G. De Blust, M. Nijnik, and M. Kulvik, "Multifunctionality and biodiversity conservation—intitutional challenges," in *Nature Conservation Management: From Idea to Practical Issues*, T. Chmelievski, Ed., pp. 53–69, Helsinki-Aarhus, Lublin, Poland, 2007.

[5] M. Nijnik, L. Zahvoyska, A. Nijnik, and A. Ode, "Public evaluation of landscape content and change," *Land Use Policy*, vol. 26, no. 1, pp. 77–86, 2008.

[6] A. Mather, G. Hill, and M. Nijnik, "Post-productivism and rural land use: cul de sac or challenge for theorization?" *Journal of Rural Studies*, vol. 22, no. 4, pp. 441–455, 2006.

[7] S. T. Dana, "Multiple use, biology and economics," *Journal of Forestry*, vol. 41, pp. 625–627, 1943.

[8] G. A. Pearson, "Multiple use in forestry," *Journal of Forestry*, vol. 42, pp. 243–249, 1944.

[9] J. R. Vincent and C. S. Binkley, "Efficient multiple-use forestry may require land-use specialization," *Land Economics*, vol. 69, no. 4, pp. 370–376, 1993.

[10] M. D. Bowes and J. V. Krutilla, *The Multiple-Use Economics of Public Forest lands: Resources for the Future*, Washington, DC, USA, 1989.

[11] R. Sedjo, "Transgenic trees and trade problems on the horizon," *Resources*, vol. 155, pp. 9–13, 2004.

[12] S. Gensiruk and S. Ivanytsky, *Forest Management and Setting up a Proper Share of Forest Cover*, Lviv Publisher, L'viv, Ukraine, 1999.

[13] European Environmental Agency (EEA), "Europe's ecological backbone: recognising the true value of our mountains," EEA Report 6/2010, p. 248, Copenhagen, Denmark, 2010.

[14] L. Fazey, J. Gamarra, J. Fischer, M. Reed, L. Stringer, and M. Christie, "Adaptation strategies for reducing vulnerability to future environmental change," *Frontiers in Ecology and the Environment*, vol. 8, no. 8, pp. 414–422, 2010.

[15] J. Carnus, J. Parrotta, E. Brockerhoff et al., "Planted forests and biodiversity," *Journal of Forestry*, vol. 104, no. 2, pp. 65–77, 2006.

[16] G. Buttoud, "Multipurpose management of mountain forests: which approaches?" *Forest Policy and Economics*, vol. 4, no. 2, pp. 83–87, 2002.

[17] G. Weiss, "The political practice of mountain forest restoration—comparing restoration concepts in four European countries," *Forest Ecology and Management*, vol. 195, no. 1-2, pp. 1–13, 2004.

[18] J. G. Borges, A. Falcão, C. Miragaia, P. Marques, and M. Marques, "A decision support system for forest resources management in Portugal," in *System Analysis in Forest Resources*, G. J. Arthaud and T. M. Barrett, Eds., vol. 7 of *Managing Forest Ecosystems*, pp. 155–164, Kluwer Academic, 2003.

[19] K. M. Reynolds, M. Twery, M. J. Lexer et al., "Decision support systems in natural resource management," in *Handbook on Decision Support Systems*, F. Burstein and C. Holsapple, Eds., vol. 2 of *International Handbooks on Information Systems Series. Handbook on Decision Support System*, pp. 499–534, Springer, 2008.

[20] S. Gilliams, J. Van Orshoven, B. Muys, H. Kros, G. W. Heil, and W. Van Deursen, "AFFOREST sDSS: a metamodel based spatial decision support system for afforestation of agricultural land," *New Forests*, vol. 30, no. 1, pp. 33–53, 2005.

[21] L. Verchot, R. Zomer, O. Van Straaten, and B. Muys, "Implications of country-level decisions on the specification of crown cover in the definition of forests for land area eligible for afforestation and reforestation activities in the CDM," *Climatic Change*, vol. 81, no. 3-4, pp. 415–430, 2007.

[22] B. Muys, J. Hynynen, M. Palahi et al., "Simulation tools for decision support to adaptive forest management in Europe," *Forest Systems*, vol. 19, no. 1, pp. 86–99, 2011.

[23] T. Eid and K. Hobbelstad, "AVVIRK-2000: a large-scale forestry scenario, model for long-term investment, income and harvest analyses," *Scandinavian Journal of Forest Research*, vol. 15, no. 4, pp. 472–482, 2000.

[24] The Stern Review, "The Economic of Climate Change," 2006, http://www.icaew.com/en/library/subject-gateways/environment-and-sustainability/stern-review/.

[25] M. Nijnik, "Carbon capture and storage in forests," in *Carbon Capture: Sequestration and Storage*, R. E. Hester and R. M. Harrison, Eds., vol. 29 of *Issues in Environmental Science and Technology*, pp. 203–238, The Royal Society of Chemistry, Cambridge, UK, 2010.

[26] F. Schmithüsen, "Multifunctional forestry practices as a land use strategy to meet increasing private and public demands in modern societies," *Journal of Forest Science*, vol. 53, no. 6, pp. 290–298, 2007.

[27] G. C. van Kooten, *Climate Change Economics*, Edward Elgar, Cheltenham, UK, 2004.

[28] M. Nijnik and L. Bizikova, "Responding to the kyoto protocol through forestry: a comparison of opportunities for several countries in Europe," *Forest Policy and Economics*, vol. 10, no. 4, pp. 257–269, 2008.

[29] D. Anderson, "Economic aspects of afforestation and soil conservation projects," *The Annals of Regional Science*, vol. 21, no. 3, pp. 100–110, 1987.

[30] H. Phillips, *Secure Resource Supply for the European Wood Industry*, Inventory scenarios for timber resources in Ireland. COST E44 workshop, Dublin, Ireland, 2006.

[31] M. Michetti and R. Nunes, *Afforestation and Timber Management Compliance Strategies in Climate Policy. A Computable*

*General Equilibrium Analysis*, vol. 2011.004 of *FEEM Note di lavoro*, 2011.

[32] M. Lindner, M. Maroschek, S. Netherer et al., "Climate change impacts, adaptive capacity, and vulnerability of European forest ecosystems," *Forest Ecology and Management*, vol. 259, no. 4, pp. 698–709, 2010.

[33] G. Nabuurs, R. Paivinen, and H. Schanz, "Sustainable management regimes for Europe's forests—a projection with EFISCEN until 2050," *Forest Policy and Economics*, vol. 3, no. 3-4, pp. 155–173, 2001.

[34] G. Nabuurs, R. Paivinen, M. J. Schelhaas et al., "Nature-oriented forest management in Europe: modeling the long-term effects," *Journal of Forestry*, vol. 99, no. 7, pp. 28–34, 2002.

[35] A. Shvidenko and Andrusishin, *Ukraine: The Conditions and Prospects of the Forest Sector*, IIASA, Laxenburg, Austria, 1998, Unpublished manuscript.

[36] State Committee of Forestry of Ukraine (SCF), *State Records*. Kyiv, 2007.

[37] I. Soloviy, "New forests for Europe: afforestation at the turn of century," in *Proceedings of the EFI*, vol. 35 of *Afforestation in Ukraine—potential and restrictions*, pp. 195–211, February 2000.

[38] A. Strochinskii, Y. Pozyvailo, and S. E. Jungst, "Forests and forestry in Ukraine: standing on the brink of a market economy," *Journal of Forestry*, vol. 99, no. 8, pp. 34–38, 2001.

[39] S. Gensiruk and M. Nijnik, *Geography of Forests in Ukraine*, Lviv Publisher, L'viv, Ukraine, 1995.

[40] M. Nijnik and G. C. Van Kooten, "Forestry in the Ukraine: the road ahead?" *Forest Policy and Economics*, vol. 1, no. 2, pp. 139–151, 2000.

[41] The President's Decree aimed to reform the forestry sector of Ukraine, 2004.

[42] The State Programme, "Forests in Ukraine, 2002–2015," Kyiv, Ukraine, 2002, http://www.rada.gov.ua.

[43] The Forest Code of Ukraine, 2006, http://zakon.rada.gov.ua/cgi-bin/laws/anot.cgi?nreg=3852-12.

[44] M. Nijnik, *To sustainability in forestry: the Ukraine's case*, Ph.D. thesis, Wageningen University, 2002.

[45] FAO, Forstat, 2010, http://www.fao.org/forestry/country/en/ukr/.

[46] Ministry of Environment and Natural Resources (MENR), "The National Report. Kyiv," 2003, http://myland.org.ua/index.php?id=&lang=en.

[47] National Academy of Sciences (NAS), *Ukraine: the Prognosis for the Development of Production Forces*, Kyiv, 2002.

[48] R. Bun, M. Gusti, L. Kujii et al., "Spatial GHG inventory: analysis of uncertainty sources. A case study for Ukraine," in *Accounting for Climate Change: Uncertainty in Greenhouse Gas Inventories—Verification Compliance and Trading*, D. Lieberman, M. Jonas, Z. Nahorski, and S. Nilsson, Eds., pp. 63–74, Springer, Amsterdam, The Netherlands, 2007.

[49] S. Nilsson and A. Shvidenko, "The Ukrainian Forest Sector in a Global Perspective," IR-99-011/March. IIASA, Laxenburg, Austria, 1999.

[50] N. Hanley and C. Spash, *Cost-Benefit Analysis and the Environment*, Edward Elgar, 1993.

[51] M. Nijnik, "Economics of climate change mitigation forest policy scenarios for Ukraine," *Climate Policy*, vol. 4, no. 3, pp. 319–336, 2004.

[52] G. Van Kooten and E. Bulte, *The Economics of Nature: Managing Biological Assets*, Blackwell Publications, Oxford, UK, 2000.

[53] M. Nijnik, "To an economist's perception on sustainability in forestry-in-transition," *Forest Policy and Economics*, vol. 6, no. 3-4, pp. 403–413, 2004.

[54] P. Lakida, S. Nilsson, and A. Shvidenko, "Estimation of Forest Phytomass for Selected Countries of the Former European USSR," WP-95-79, IIASA, Laxenburg, Austria, 1995.

[55] A. Jessome, "Strength and related properties of woods grown in Canada. Forestry," Tech. Rep. 21, Ottawa, Canada, 1977.

[56] R. Sandor and J. Skees, "Creating a market for carbon emissions opportunities for US farmers," *Choices Quarter*, vol. 1, pp. 13–18, 1999.

[57] S. Huq and H. Reid, "Mainstreaming adaptation in development. IDS bulletin," *Climate Change and Development*, vol. 35, pp. 15–21, 2004.

[58] D. Klooster and O. Masera, "Community forest management in Mexico: carbon mitigation and biodiversity conservation through rural development," *Global Environmental Change*, vol. 10, no. 4, pp. 259–272, 2000.

[59] K. Naka, A. L. Hammett, and W. B. Stuart, "Forest certification: stakeholders, constraints and effects," *Local Environment*, vol. 5, no. 4, pp. 475–481, 2000.

[60] E. Fernandez, L. Rojo, M. N. Jimenez et al., "Afforestation improves soil fetiliyu in south-eastern Spain," *European Journal of Forest Research*, vol. 194, no. 4, pp. 707–717, 2010.

[61] M. Nijnik, B. Slee, and G. Pajot, "Opportunities and challenges for terrestrial carbon offsetting and marketing, with some implications for forestry in the UK," *South-East European Forestry Journal*, vol. 1, no. 2, pp. 69–79, 2011.

[62] S. Fankhauser and L. Lavric, "The investment climate for climate investment: joint Implementation in transition countries," *Climate Policy*, vol. 3, no. 4, pp. 417–434, 2003.

[63] E. Petkova and G. Faraday, "Good Practices in Policies and Measures for Climate Change Mitigation," REC and WRI, Szentendre, Hungary, 2001.

[64] IBN-DLO, *Resolving Issues on Terrestrial Biospheric Sinks in the Kyoto Protocol*. Dutch National Research Programme on Global Air Pollution and Climate Change, Wageningen, The Netherlands,1999.

[65] R. N. Lubowski, A. J. Plantinga, and R. N. Stavins, "Land-use change and carbon sinks: econometric estimation of the carbon sequestration supply function," *Journal of Environmental Economics and Management*, vol. 51, no. 2, pp. 135–152, 2006.

[66] G. Van Kooten, *Land Resource Economics and Sustainable Development. Economic Policies and the Common Good*, UBC Press, Toronto, Canada, 1993.

[67] J. Lampietti and J. Dixon, *To See the Forest for the Trees: A Guide to non-Timber Forest Benefits*, World Bank, Washington, DC, USA, 1995.

[68] P. Lakyda, I. Buksha, and V. Pasternak, "Opportunities forfulfilling Joint Implementation projects in forestry in Ukraine," in *Forest, Climate and Kyoto*, vol. 222 of *Unasylva*, pp. 1–6, 2005.

# Permissions

The contributors of this book come from diverse backgrounds, making this book a truly international effort. This book will bring forth new frontiers with its revolutionizing research information and detailed analysis of the nascent developments around the world.

We would like to thank all the contributing authors for lending their expertise to make the book truly unique. They have played a crucial role in the development of this book. Without their invaluable contributions this book wouldn't have been possible. They have made vital efforts to compile up to date information on the varied aspects of this subject to make this book a valuable addition to the collection of many professionals and students.

This book was conceptualized with the vision of imparting up-to-date information and advanced data in this field. To ensure the same, a matchless editorial board was set up. Every individual on the board went through rigorous rounds of assessment to prove their worth. After which they invested a large part of their time researching and compiling the most relevant data for our readers. Conferences and sessions were held from time to time between the editorial board and the contributing authors to present the data in the most comprehensible form. The editorial team has worked tirelessly to provide valuable and valid information to help people across the globe.

Every chapter published in this book has been scrutinized by our experts. Their significance has been extensively debated. The topics covered herein carry significant findings which will fuel the growth of the discipline. They may even be implemented as practical applications or may be referred to as a beginning point for another development. Chapters in this book were first published by Hindawi Publishing Corporation; hereby published with permission under the Creative Commons Attribution License or equivalent.

The editorial board has been involved in producing this book since its inception. They have spent rigorous hours researching and exploring the diverse topics which have resulted in the successful publishing of this book. They have passed on their knowledge of decades through this book. To expedite this challenging task, the publisher supported the team at every step. A small team of assistant editors was also appointed to further simplify the editing procedure and attain best results for the readers.

Our editorial team has been hand-picked from every corner of the world. Their multi-ethnicity adds dynamic inputs to the discussions which result in innovative outcomes. These outcomes are then further discussed with the researchers and contributors who give their valuable feedback and opinion regarding the same. The feedback is then collaborated with the researches and they are edited in a comprehensive manner to aid the understanding of the subject.

Apart from the editorial board, the designing team has also invested a significant amount of their time in understanding the subject and creating the most relevant covers. They scrutinized every image to scout for the most suitable representation of the subject and create an appropriate cover for the book.

The publishing team has been involved in this book since its early stages. They were actively engaged in every process, be it collecting the data, connecting with the contributors or procuring relevant information. The team has been an ardent support to the editorial, designing and production team. Their endless efforts to recruit the best for this project, has resulted in the accomplishment of this book. They are a veteran in the field of academics and their pool of knowledge is as vast as their experience in printing. Their expertise and guidance has proved useful at every step. Their uncompromising quality standards have made this book an exceptional effort. Their encouragement from time to time has been an inspiration for everyone.

The publisher and the editorial board hope that this book will prove to be a valuable piece of knowledge for researchers, students, practitioners and scholars across the globe.

# List of Contributors

**Daniel Gagnon**
Fiducie de Recherche sur la Foret des Cantons-de-lEst, Eastern Townships Forest Research Trust, 1 Rue Principale, St-Benoıt-du-Lac, QC, Canada
Centre dEtude de la Foret (CEF), Universite du Quebeca Montreal, C.P. 8888, Succursale Centre-Ville, Montreal, QC, Canada
Department of Biology, University of Regina, 3737 Wascana Parkway, Regina, SK, Canada

**Normand Chevrier**
Departement des Sciences Biologiques, Universite du Quebeca Montreal, C.P. 8888, Succursale Centre-Ville, Montreal, QC, Canada H3C 3P8

**Julien Fortier**
Fiducie de Recherche sur la Foret des Cantons-de-lEst, Eastern Townships Forest Research Trust, 1 Rue Principale, St-Benoıt-du-Lac, QC, Canada
Centre dEtude de la Foret (CEF), Universite du Quebeca Montreal, C.P. 8888, Succursale Centre-Ville, Montreal, QC, Canada H3C 3P8

**Benoit Truax and France Lambert**
Fiducie de Recherche sur la Foret des Cantons-de-lEst, Eastern Townships Forest Research Trust, 1 Rue Principale, St-Benoıt-du-Lac, QC, Canada

**Xia Huang and Timothy M. Young**
Center for Renewable Carbon, University of Tennessee, Knoxville, TN 37996-4570, USA

**James H. Perdue**
USDA Forest Service, Southern Research Station, 2506 Jacob Drive, Knoxville, TN 37996-4570, USA

**Kusaga Mukama**
District Natural Resources Office, Liwale District Council, P.O. Box 23, Liwale, Tanzania

**Irmeli Mustalahti**
Department of Political and Economic Studies, University of Helsinki, P.O. Box 59, 00014 Helsinki, Finland

**Eliakimu Zahabu**
Department of Forest Mensuration and Management, Sokoine University of Agriculture, P.O. Box 3011, Morogoro, Tanzania

**W. Mark Ford**
Department of Fish and Wildlife Conservation, Virginia Polytechnic Institute and State University, Blacksburg, VA 24061, USA
US Geological Survey, Virginia Cooperative Fish and Wildlife Research Unit, Blacksburg, VA 24061, USA

**Nathan R. Beane and Eric R. Britzke**
Environmental Laboratory, US Army Engineer Research and Development Center, 3909 Halls Ferry Road, Vicksburg, MS 39180, USA

**Joshua B. Johnson**
Pennsylvania Game Commission, 2001 Elmerton Avenue, Harrisburg, PA 17110, USA

**Alexander Silvis**
Department of Fish and Wildlife Conservation, Virginia Polytechnic Institute and State University, Blacksburg, VA 24061, USA

**Peter F. Newton**
Canadian Wood Fibre Centre, Canadian Forest Service, Natural Resources Canada, Sault Ste. Marie, ON, Canada

**Graeme S. Cumming**
Percy Fitz Patrick Institute, DST-NRF Centre of Excellence, University of Cape Town, Rondebosch, Cape Town 7701, South Africa

**Xanic J. Rondon**
Percy Fitz Patrick Institute, DST-NRF Centre of Excellence, University of Cape Town, Rondebosch, Cape Town 7701, South Africa
Center for Latin American Studies, University of Florida, Gainesville, FL 32611, USA

**Rosa E. Cossıo**
School of Natural Resources and Environment, University of Florida, Gainesville, FL 32611, USA

**Jane Southworth**
Department of Geography, University of Florida, Gainesville, FL 32611, USA

**R. Justin De Rose**
Forest Inventory and Analysis, Rocky Mountain Research Station, 507 25th Street, Ogden, UT 84401, USA

**Robert S. Seymour**
School of Forest Resources, University of Maine 5755 Nutting Hall, Orono, ME 04469, USA

**N. Mbahin**
Environmental Health Division, International Centre of Insect Physiology and Ecology (icipe), P.O. Box 30772-00100, Nairobi, Kenya
Department of Biological Sciences, Kenyatta University, P.O. Box 43844-00100, Nairobi, Kenya

**E. N. Kioko**
Biodiversity Conservation Division, National Museums of Kenya, P.O. Box 40658-00100, Nairobi, Kenya

**S. K. Raina**
Environmental Health Division, International Centre of Insect Physiology and Ecology (icipe), P.O. Box 30772-00100, Nairobi, Kenya

**J. M. Mueke**
Department of Biological Sciences, Kenyatta University, P.O. Box 43844-00100, Nairobi, Kenya

**Oliver Chikumbo**
Scion, Sustainable Design, Private Bag 3020, Rotorua 3046, New Zealand

**Russ Lea**
North Carolina State University, Raleigh, NC, USA
CEO NEON Inc., Boulder, CO, USA

**Paul Marsh**
1433 Lutz Avenue, Raleigh, NC 27607, USA

**R. C. Kellison**
North Carolina State University, Raleigh, NC, USA

**Bradley D. Pinno**
Alberta School of Forest Science and Management, Faculty of Agricultural Life and Environmental Sciences, University of Alberta, Edmonton, AB, Canada
Natural Resources Canada, Canadian Forest Service, Northern Forestry Centre, 5320 122 Street, Edmonton, AB, Canada

**Victor J. Lieffers and Simon M. Landhausser**
Alberta School of Forest Science and Management, Faculty of Agricultural Life and Environmental Sciences, University of Alberta, Edmonton, AB, Canada

**Doria R. Gordon**
The Nature Conservancy and Department of Biology, University of Florida, P.O. Box 118526, Gainesville, FL 32611, USA

**S. Luke Flory, Aimee L. Cooper and Sarah K. Morris**
Agronomy Department, University of Florida, P.O. Box 110500, Gainesville, FL 32611, USA

**Virginia H. Dale, Matthew H. Langholtz, Beau M. Wesh and Laurence M. Eaton**
Oak Ridge National Laboratory, Environmental Sciences Division, Center for Bio Energy Sustainability, Oak Ridge, TN 37831, USA

**Hans A. Persson**
Department of Ecology, Swedish University of Agricultural Sciences, P.O. Box 7044, 750 07 Uppsala, Sweden

**Emil Klimo, Jilí Kulhavý, Alois Prax and Ladislav Menšík**
Department of Forest Ecology, Mendel University in Brno, Zemedelska 3, 613 00 Brno, Czech Republic

**Pavel Hadaš**
Data Processing, Orechovka 1727, 696 62 Straznice, Czech Republic

**Oldlich Mauer**
Department of Silviculture, Mendel University in Brno, Zemedelska 3, 613 00 Brno, Czech Republic

**Diane Fielding, Frederick Cubbage, M. Nils Peterson, Dennis Hazel, Brunell Gugelmann and Christopher Moorman**
Department of Forestry and Environmental Resources, North Carolina State University, Raleigh, NC 27695-8008, USA

**Jeji Varghese**
Sociology and Anthropology, University of Guelph, Guelph, ON, Canada

**Maureen G. Reed**
School of Environment and Sustainability and Department of Geography and Planning, University of Saskatchewan, Saskatoon, SK, Canada

**Maria Nijnik**
Social, Economic and Geographical Sciences, The James Hutton Institute, Craigiebuckler, Aberdeen AB15 8QH, Scotland, UK

**Arie Oskam**
Agricultural Economics and Rural Policy, Wageningen University, Hollandseweg 1, 6706 KN Wageningen, The Netherlands

**Anatoliy Nijnik**
Environmental Network Ltd., The Hillocks, Tarland, Aboyne AB34 4TJ, Scotland, UK

www.ingramcontent.com/pod-product-compliance
Lightning Source LLC
Chambersburg PA
CBHW050443200326
41458CB00014B/5044

* 9 7 8 1 6 3 2 3 9 0 4 2 4 *